GLOBALIZING INFORMATION

The Economics of International Technology Trade

World Scientific Studies in International Economics

(ISSN: 1793-3641)

Vol. 22 International Trade Policy Formation: Theory and Politics
by Wolfgang Mayer (University of Cincinnati, USA)

Vol. 23 Priorities and Pathways in Services Reform:
Part I — Quantitative Studies
edited by Philippa Dee (Australian National University, Australia)

Vol. 24 Globalizing Information: The Economics of International Technology Trade
by Keith E Maskus (University of Colorado at Boulder, USA)

Vol. 25 Priorities and Pathways in Services Reform: Part II — Political Economy Studies
edited by Christopher Findlay (University of Adelaide, Australia)

Vol. 26 World Trade Organization and International Trade Law: Antidumping, Subsidies and Trade Agreements
by Gary N Horlick (Law Offices of Gary N Horlick, USA & University of Bern, Switzerland)

Vol. 27 European Economic Integration, WTO Membership, Immigration and Offshoring
by Wilhelm Kohler (University of Tübingen, Germany)

Vol. 28 Services Trade Reform: Making Sense of It
by Philippa Dee (Australian National University, Australia)

Vol. 29 The Social Value of the Financial Sector: Too Big to Fail or Just Too Big?
edited by Viral V Acharya (New York University, USA & Centre for Economic Policy (CEPR), UK), Thorsten Beck (Tilburg University, The Netherlands & Centre for Economic Policy (CEPR), UK), Douglas D Evanoff (Federal Reserve Bank of Chicago, USA), George G Kaufman (Loyola University Chicago, USA), & Richard Portes (London Business School, UK & Centre for Economic Policy (CEPR), UK)

Vol. 30 The Role of Central Banks in Financial Stability: How Has It Changed?
edited by Douglas D Evanoff (Federal Reserve Bank of Chicago, USA), Cornelia Holthausen (European Central Bank, Germany), George G Kaufman (Loyola University Chicago, USA) & Manfred Kremer (European Central Bank, Germany)

The complete list of the published volumes in the series can be found at
http://www.worldscientific.com/series/wssie

24 World Scientific Studies in International Economics

GLOBALIZING INFORMATION

The Economics of International Technology Trade

Keith E. Maskus

University of Colorado at Boulder, USA

World Scientific

NEW JERSEY • LONDON • SINGAPORE • BEIJING • SHANGHAI • HONG KONG • TAIPEI • CHENNAI

Published by

World Scientific Publishing Co. Pte. Ltd.
5 Toh Tuck Link, Singapore 596224
USA office: 27 Warren Street, Suite 401-402, Hackensack, NJ 07601
UK office: 57 Shelton Street, Covent Garden, London WC2H 9HE

British Library Cataloguing-in-Publication Data
A catalogue record for this book is available from the British Library.

World Scientific Studies in International Economics — Vol. 24
GLOBALIZING INFORMATION
The Economics of International Technology Trade

ISBN 978-981-4401-75-3

In-house Editor: Yvonne Tan

Typeset by Stallion Press
Email: enquiries@stallionpress.com

Printed in Singapore by World Scientific Printers.

Contents

List of Contributors

John H. Barton was the George E. Osborne Professor of Law, Emeritus, at Stanford University School of Law.

David L. Carr is Associate Professor of Economics, University of South Dakota.

Mattias Ganslandt is Director of the Center for European Law and Economics, Brussels.

Yin He is Associate Professor of Economics, University of International Business and Economics, Beijing.

Bernard M. Hoekman is Research Director in Global Economics, European University Institute, Florence.

Denise Konan is Professor of Economics, University of Hawaii.

Changying Li is Professor of Economics, Shandong University, Jinan, China.

James R. Markusen is Professor of Economics, University of Colorado at Boulder.

Keith E. Maskus is Professor of Economics, University of Colorado at Boulder.

Rebecca Neumann is Associate Professor of Economics, University of Wisconsin at Milwaukee.

Shuichiro Nishioka is Associate Professor of Economics, West Virginia University.

Mohan Penubarti (now Mohan Rao) is a Managing Partner of Navigant Consulting in Chicago, Illinois.

Kamal Saggi is Professor of Economics, Vanderbilt University.

Tobias Seidel is Professor of Economics, University of Duisburg-Essen.

Eina Wong is a Business Analyst at Sun Microsystems, Wellington, New Zealand.

Guifang Yang is a Managing Partner of Deloitte Consulting in Atlanta, Georgia.

Lei Yang is Assistant Professor of Economics, Hong Kong Polytechnic University.

Acknowledgments

This book represents the culmination of many years of researching, thinking and writing about global trade, investment, technology transfer and intellectual property rights. I could not have made nearly as much progress on these issues without the essential contributions of many co-authors over the years, both those represented in this volume and others. Thus, I owe the greatest debt of gratitude to all of these colleagues, including especially Mattias Ganslandt, Yin He, Denise Eby Konan, James Markusen, Mohan Rao, Guifang Yang and Lei Yang. I am especially grateful to the late John Barton, one of the intellectual giants among legal scholars regarding intellectual property rights, trade and technology diffusion. I also wish to thank numerous analysts and friends from international organizations and non-governmental organizations, who truly are the front-line analysts of important policy issues in global intellectual property and economic development.

In preparing this volume I had the great assistance of Meredith Kroenlein, without whom it would not be possible. I also thank Robert Stern and Zvi Ruder, who encouraged me to compile these papers into what I hope is a readable and coherent book. Readers interested in the complexities of innovation, licensing, trade, investment and intellectual property should find much to absorb and contemplate in this journey of discovery.

Keith E. Maskus
Boulder, Colorado, USA
September 2013

Chapter 1

Introduction and Overview

Keith E. Maskus

1. Introduction

In this volume I bring together a selection of significant papers I have published over the years, many with co-authors, covering aspects of empirical trade, foreign direct investment (FDI), technology transfer and intellectual property rights (IPRs). While these chapters encompass a wide range of theoretical, empirical and policy approaches to a multitude of economic questions, they share a fundamental theme, which is to understand how various aspects of international commercial activity affect the dissemination of knowledge across borders. This has been a strong interest of mine since the beginning of my career in the early 1980s. Over that period these issues have become increasingly important, with IPRs in particular rising to the pinnacle of global policy concerns.

Interest in this area has mushroomed for a variety of reasons. First, the relentless pace of globalization, driven by technological progress, declining costs of investment and trade, and reduced barriers to international economic activity, makes the operations of multinational enterprises (MNEs) ever more central to societies and economies. Second, how technological information is traded internationally, whether implicitly in goods or explicitly through FDI, joint ventures and licensing contracts, is critically affected by policy choices and a key determinant of globalized learning. Third, the expansion of globalization raises opportunities for both formal technology transfer through markets and informal learning and diffusion through imitation and copying. The efficiency and equity of these processes depend on a multitude of factors that deserve close analysis.

In particular, because IPRs — patents, trademarks, copyrights and trade secrets — deeply affect both the gains from formal information contracts and the costs of unauthorized imitation and trade, they have taken center stage in global policy arenas. These policies matter not only for global innovation and international technology transfer among commercial rivals and business networks but also for the costs of innovating and providing access to such key public goods as essential medicines and environmentally sound technologies. IPRs are a central component of the World Trade Organization (WTO), modern free trade agreements, investment treaties, and various international accords on the environment, agriculture and public health. They occupy the minds of scholars, activists, policymakers and nongovernmental organizations.

Hence, the title of this volume: *Globalizing Information: The Economics of International Technology Trade*. A deep and complex set of problems and tradeoffs surrounds the processes by which firms buy and sell technologies through trade, investment and contracts. Equally complex are the dynamics of growth and innovation in an environment where the costs of R&D and imitation depend crucially on policy parameters, including IPRs. Critical global policy questions must be addressed regarding how easily — and at what costs — technologies that can treat social objectives are transferred from rich to poor nations.

These general concerns are the backbone of the chapters presented in this volume. The articles are presented in four parts. The first part groups together five contributions covering theoretical and empirical studies of the impacts of IPRs in the global economy, ranging from trade, FDI and licensing to innovation and strategic pricing in an oligopoly setting. Many of the findings are novel and counterintuitive, pointing to the intricate workings of intellectual property rights in international markets. The second part analyzes the deep relationships between innovation and international technology transfer (ITT). Again, these chapters cover both formal theory, based on the dynamics of innovation and learning in the open economy, and empirical study of the determinants of innovation.

The third set of papers turns to more standard trade issues, analyzed within the backdrop of ITT. These articles cover the factor-endowment basis of international trade, the role of technological differences in how MNEs choose to locate production facilities, and the net gains from liberalizing trade in goods and services where those additional flows afford greater access to frontier global technologies. The final grouping turns to critical policy issues in the world economy with respect to IPRs and technology diffusion. The first article covers the basic question of whether IPRs should even be in the WTO agreements. Subsequent pieces analyze aspects of access to knowledge and provision of essential medicines.

Following is a brief overview of the individual chapters. Because these contributions are complex, these summaries should help readers understand the nature of each paper and how it may address issues of particular interest and concern. The book should be of interest to graduate students working in the areas of trade, trade policy, FDI and IPRs. It should also provide a useful overview and roadmap to policymakers interested in what economic analysis has to say about trade liberalization, IPRs reforms and innovation policies in the global economy.

2. Chapter Overview

Part I: Intellectual Property Rights

Much of my research focuses on specific questions regarding the impacts of IPRs, and how such rights are regulated, on international economic activity. The chapters in the opening part offer a selection of such studies. Is it true that differences in patent laws across countries have an effect on global trade? How do IPRs reforms in poor countries affect global innovation incentives and resulting growth rates? Does the existence of IPRs generate market power that can sustain differences across countries in prices of identical goods? How are such processes altered if policymakers allow competition from firms that attempt to arbitrage away these international price differences? Such questions are covered in this initial part, which serves as a useful introduction to these complex topics.

Chapter 2 — "How Trade-Related Are Intellectual Property Rights?" with Mohan Penubarti, Journal of International Economics, 1995

This chapter was the first article to pose the basic question of whether international variations in patent rights actually affect international trade patterns in a significant way. The issue is critical, for advocates of stronger global patent regimes, led by the U.S. and EU governments, argue forcefully that differential IPRs policies distort international trade. However, as the Uruguay Round negotiations proceeded, culminating in the Agreement on Trade-Related Aspects of Intellectual Property Rights (TRIPS), there was no systematic evidence that this assertion was true or false. As the chapter points out, there certainly could be important impacts on trade because IPRs generate market power on behalf of their owners. However, whether the effect of patent reforms in developing countries would increase or reduce trade was unclear. The reason is that MNEs encountering tighter protection of their technologies abroad could decide to ship more volumes as local rivals found it more difficult to compete — a *market-expansion effect.* However, they might also react to their stronger positions by cutting exports and raising prices — a *market-power effect.* Which of these would dominate is an empirical question.

Thus, the authors assemble data on bilateral exports of manufactured goods in detailed industries from the higher-income economies, who were members of the Organization for Economic Cooperation and Development, to similar countries and many developing economies. The data are a cross-section of trade flows in 1984, a time when there were considerable differences across countries in patent laws. Estimating a basic model of bilateral trade with differentiated goods, augmented by measures of patent rights, the authors find that stronger patents indeed expand bilateral imports significantly. This remains true even after accounting for the problem that IPRs could be endogenous, with more trade causing expanded patent rights as much as the other way around. Interestingly, the evidence suggests that this impact of patent reforms is stronger in medium-technology and low-technology goods than in evidently patent-intensive goods. Nonetheless, the effects exist throughout industrial sectors and the coefficients imply that the economic impacts are quite significant.

The essential finding, therefore, is that IPRs indeed are "trade-related" and importantly so. This paper became the foundation of a large and still-evolving empirical literature on how legal rights to protect technology, copyrighted goods and trademarks influence global trade and investment.

Chapter 3 — "Intellectual Property Rights, Licensing and Innovation in an Endogenous Product-Cycle Model," with Guifang Yang, Journal of International Economics, 2001

Many international economists conceive of IPRs as protected monopoly positions that slow the international diffusion of modern technologies and limit innovation and growth. The essential logic is that as countries adopt stronger patents they make it more difficult for local enterprises to reverse engineer and imitate the products and technologies of foreign firms. In turn, innovative international firms enjoy stronger and longer monopoly positions, reducing the pressure to develop new goods and thereby suppressing innovation. This was the message of a series of theoretical papers in the 1990s analyzing the role of IPRs in global product cycles.

The problem with this idea is that it entertains a gross simplification of what patents are and how they operate. While one effect of patent reforms is to raise imitation costs, another is to reduce the costs incurred by MNEs in formally transferring technologies to production facilities abroad. One reason is that such firms feel more confident that they can keep their technological information proprietary within their subsidiaries and another is that achieving and enforcing licensing contracts with local partners becomes easier. In turn, we would expect greater flows of both FDI and licensing, at least toward countries with the ability to absorb and use new technologies.

Thus, Chapter 3 offers a dynamic model of the technology product cycle in which stronger IPRs in the recipient country (the "South") both raise the costs of imitation and reduce the difficulties of successfully licensing a new technology. Because of the latter factor, innovative firms in the origin country (the "North") find it more profitable to transfer their technologies abroad quicker, leaving them freer to develop new goods at home. The theory develops a quality-ladders model in which firms engaging in R&D develop a higher-quality version of existing technologies and face the risks of competition from other innovative firms in the North and imitation in the South. However, because wages and production costs are lower in the South they have an incentive to license there, which is globally efficient. This incentive expands with stronger IPRs in the South, raising licensing and innovation flows in equilibrium. This article was the first to consider licensing in a general-equilibrium product-cycle theory and spurred a reconsideration of the full economic effects of patent reforms in the emerging economies.

Chapter 4 — "Vertical Distribution, Parallel Trade and Price Divergence in Integrated Markets," with Mattias Ganslandt, European Economic Review, 2007

The phrase "parallel trade" refers to legitimately produced goods that are shipped from one country to another without the authorization of the company that owns a patent, copyright or trademark on them. Parallel imports (PIs) are common in the world. For example, discount chain stores in the United States often feature cosmetics and consumer electronics that have come into the country through such channels. Whether such trade is legal depends on the legal principle called the exhaustion doctrine, which determines the point at which the original IPRs owner can no longer control further distribution. Within the European Union, for example, the policy is complete regional exhaustion: once a good is sold in one location within the Single Market, parallel traders may ship it freely to another. As a result, there are large intra-EU trade flows in medicines, automobile and machinery parts, consumer goods and many other products, as firms buy them where the wholesale prices are low and resell them where prices are high.

The curious aspect of this arbitrage is that PIs seem to have little ability to drive retail prices together across countries within the EU, which is not what we would expect where parallel trade is unimpeded. Thus, Chapter 4 sets out a novel theoretical model that helps explain this fact. The analysis starts from the observation that the vast bulk of PIs occur at the distributor, or wholesale, level, rather than the commonly imagined arbitrage by individual consumers. In this context, the original manufacturers (that is, the IPRs owners) of medicines and cosmetics are not powerless to deal with the problem. The authors put together a model in which an original manufacturer sells an IPRs-protected good to independent distributors in a high-priced and low-priced market. Parallel traders ship goods from the latter to the former. However, the original firm can reduce or deter this activity through its own price setting. In particular, it charges its distributor a higher wholesale price where markets are integrated by PIs than it would if markets were fully segmented. The degree to which it does so depends on market demands and the costs of shipping goods between markets. A counterintuitive outcome is that as these trade costs fall the wholesale and retail prices in the two markets can move apart from each other, rather than converge as expected. The reason is that the original firm can raise its price in the PIs-source nation so high as to shut down the local distributor, resulting in a monopoly that can sustain divergent prices. These results point to the critical importance of trade costs and market structure in understanding the actual effects of parallel trade.

Chapter 5 — "Parallel Imports and the Pricing of Pharmaceutical Products: Evidence from the European Union," with Mattias Ganslandt, Journal of Health Economics, 2004

Despite the fact that parallel imports are a fairly common phenomenon in Europe, there are very few studies of how such trade actually affects prices. The reason is that independent data on the amount of PIs are rarely collected by trade authorities and surveys rarely induce trading companies to reveal their activities. An important exception, however, is parallel trade in pharmaceuticals because the firms engaging in it must acquire approval from public-health authorities in the importing nation to sell the goods they procure elsewhere. These data also are rarely made public, but the Swedish Medical Products Agency was willing to share an extensive database from the late 1990s with the authors of Chapter 5. That article was the first in the literature to analyze empirically the impacts of PIs volumes on prices of patent-protected goods in a high-priced market.

Using this detailed product-level information on sales and wholesale prices of medicines sold by original manufacturers in Sweden, the authors estimate an equation relating prices to competition from

PIs. They show that the amounts of entry and sales by parallel-trading companies expand rapidly when parallel imports are legalized and grow most significantly in "blockbuster" drugs with high sales volumes. The econometric model finds that competition from any entry by PIs firms diminishes prices somewhat but this competitive effect increases with the number of such companies selling a particular product. Even controlling for the endogeneity of entry, the authors find that competition from parallel trade reduces wholesale prices earned by original manufacturers by between 12 and 18 percent, which is economically significant. They also note, however, that most of the gains from these price declines go to hospitals and pharmacists, rather than to ultimate consumers because retail prices are little affected. Furthermore, parallel trading companies make a significant economic margin on their activities, which is not necessarily beneficial from Sweden's standpoint.

Chapter 6 — "The Impact of Parallel Imports on Investments in Cost-Reducing Research and Development," with Changying Li, Journal of International Economics, 2006

Because parallel trade exists outside the channels of commerce authorized by IPRs owners, they compete with the activities of the original manufacturers and generally reduce profits of the latter. This fact raises the risk that PIs can diminish innovation incentives because originator firms develop new goods and technologies in the anticipation of making profits. If parallel trade significantly diminishes the returns on investments in R&D, they can slow down the development of new products, generating a dynamic loss in consumer welfare. This is the primary criticism of PIs in medicines, for example, as patent holders in the pharmaceutical industry argue they would invest less in R&D if the United States opened its market to such imports.

Thus, Chapter 6 offers a simple theoretical model analyzing the conditions under which this claim is true. In the model, an originator firm chooses in an early stage of competition how much to invest in R&D to generate a lower-cost process technology for the good it sells. In later stages this firm can sell the good to independent distributors in two markets, which may be segmented or integrated by parallel trade. The model is akin to that in Chapter 4, where the firm must choose a set of wholesale prices to manage the competition from PIs the best it can. However, the additional complication is that the firm wishes to maximize its profits over time through its R&D decisions.

The theory turns up interesting tradeoffs in the firm's decisions regarding investments and distributor prices. Compared to a situation where PIs are banned, permitting competition from parallel trade unambiguously reduces the returns to R&D and innovation falls. However, the extent to which this happens depends on market circumstances, including the costs of trading goods across borders through PIs. High trade costs imply that PIs do not reduce R&D very much, which is intuitive. However, even with very low trade costs, in a highly integrated market, rising PIs can be associated with higher volumes of R&D, since the need to compete becomes extreme for the originator firm. The analysis therefore describes an important complementarity between reducing trade costs through liberalization and deregulation, on the one hand, and the gains to R&D on the other.

Part II: Innovation and International Technology Transfer

The chapters in Part II take a far broader view than the basic focus on intellectual property rights. Rather, the questions posed here have to do with the fundamental determinants and effects of innovation and international technology transfer. What are the policy levers that encourage global dissemination of knowledge? What determines the extent to which a firm may be willing to transfer its proprietary know-how to a foreign competitor? How does financial development interact with decisions to invest in R&D? What are the implications for global innovation and growth if the South learns technology so well that it becomes a source of innovation itself?

Chapter 7 — "Transfer of Technology to Developing Countries: Unilateral and Multilateral Policy Options," with Bernard M. Hoekman and Kamal Saggi, World Development, 2005

Chapter 7 sets the stage for Part II by taking a broad global view regarding policies that can encourage or discourage ITT from the rich world to the poor world. The authors review data on the channels through which technology flows abroad, including trade, investment and licensing. These flows are rising rapidly as a result of globalization, though they are concentrated among rich nations and the larger emerging economies. Without question they are the source of considerable learning by firms in the recipient countries, whether through demonstration effects, reverse engineering, training of skilled managers or formal technology contracts.

The chapter discusses carefully how and why there may be failures in the global markets for technology, which implies there is considerable scope for national and international policy support. Thus, the authors sketch important linkages between trade and investment policies, R&D subsidies and other innovation and learning inducements for countries seeking to attract more inward technology flows. FDI and licensing, for example, can generate beneficial technology spillovers, which are more effective in countries where local firms also engage in their own R&D programs. Here there are significant interrelationships with IPRs and competition policies as well. The most important message is that each country needs to tailor its approach to its own needs and aspirations, rather than adopt a global "one size fits all" model. The chapter concludes with extensive comments on how the global policy arena can be optimized to encourage effective and welfare-enhancing ITT.

Chapter 8 — "Intellectual Property Rights, Technology Transfer and Exports in Developing Countries," with Lei Yang, Journal of Development Economics, 2009

Chapter 8 addresses the question of whether technology transfer can build export capacity in developing countries. The authors develop a simple oligopoly model in which a North firm with an advanced technology decides whether to license it to a South rival, knowing that doing so will reduce the latter's costs and make it a more competitive firm. The South firm decides whether to accept the licensing contract, knowing that it can compete without paying a license fee if it does not. Within the contract the North firm must choose how much of its secret know-how it is willing to share with its potential partner, knowing that the more it does so the greater the reduction in that firm's costs. A complex tradeoff is studied in which the extent of licensing, and the resulting cost reduction in the South enterprise, depends on trade costs, the strength of IPRs, the ability of the Southern firm to learn the know-how and other factors.

The key insight of this model is that it may be profitable for the company with the advanced technology to license it to its potential partner, even if that will generate a more competitive rival enterprise. The South firm may end up exporting a large volume of the good back to the Northern market as a result of the licensed patents and know-how, a novel finding for this literature. The authors carefully work out the welfare implications of various kinds of policies regarding the South's patent policy and North import restrictions. There are important interrelationships between these two kinds of policies that should interest policymakers in both rich countries and developing countries, including significant scope for agreements between them on trade policy and patents. The authors go on to relate these insights to the evolution of the TRIPS Agreement at the WTO.

Chapter 9 — "Southern Innovation and Reverse Knowledge Spillovers: A Dynamic FDI Model," with Yin He, International Economic Review, 2012

This very recent paper provides a deep theoretical treatment of the dynamics of innovation and FDI in a general-equilibrium product-cycle model between North and South. In the prior literature such models took Southern firms to be passive recipients or imitators of incoming technology that simply became locations of production, while Northern enterprises continued to innovate. However, the rapid emergence of MNEs headquartered in such emerging countries as China, India and Brazil suggests that a fuller dynamic operates: firms in the South actively absorb incoming technologies through imitation and learning by doing, which makes them more capable of innovation themselves. A full product cycle, as modeled here, should permit ITT from North to South through trade and FDI, a period of learning in the emerging economy, and ultimately innovation in both North and South, with both regions learning from the other.

Thus, in Chapter 9 the authors consider how to model and complete this entire product cycle, with both North and South innovation and learning occurring in steady-state equilibrium. Several cases are considered: FDI with no Southern learning; FDI with Southern learning and innovation; and FDI with Southern learning, innovation and reverse spillovers to the North. The first case is standard in the literature, with an expansion of FDI from North to South generating global efficiency gains. The second and third cases yield novel and interesting results, however. Where there is Southern innovation but no reverse spillover, the world sees slower innovation and growth as a result of stronger local patent rights. In effect, this possibility generates an inefficient specialization outcome, with the high-wage North engaged in too much production and the low-wage South performing too much innovation, both being the areas in which they have a comparative disadvantage. However, when newly innovated goods in the South are capable of being learned in the North, the reverse spillover restores an efficient equilibrium and expands innovation above the other cases. This is the first general-equilibrium model in the literature of dynamic competition in which both regions of the world innovate and learn simultaneously. It has powerful implications for such policies as restrictions on inward FDI and Southern IPRs. Liberalized investment regimes, combined with patent reforms, can expand global knowledge growth, but this outcome is most likely when there are mutual spillovers occurring among countries.

Chapter 10 — "How National and International Financial Development Affect Industrial R&D," with Rebecca Neumann and Tobias Seidel, European Economic Review, 2012

This chapter offers a deep empirical approach to another important question: how does access to foreign capital and technology affect strategic decisions by firms in various manufacturing sectors to invest in R&D? This is a vital question in economic development because capital-market liberalization is often claimed to encourage local R&D, either directly through incoming FDI or indirectly through spillover effects into improved profitability for investments. The broader issue analyzed is whether and how a range of different forms of domestic and international capital-market deepening influences R&D. These channels could include, *inter alia*, development of bond and equity markets, credit markets, international bond financing, and gaining capital from foreign FDI partners.

The paper sheds light on whether and how various forms of financial development raise or lower the R&D intensities of enterprises across manufacturing sectors. A particularly interesting question relates to the form of investment, whether in physical capital or intangible assets. The latter are generally reputation, processes and various forms of IPRs. In countries with strong constraints on financing resources, R&D may be restricted because it is difficult to borrow resources to invest in human capital, which cannot be collateralized. In turn, effective financial development can be particularly helpful to investments in intangibles as markets deepen. At the international level, the traditional conception of FDI is that it offers additional access to physical capital. However, the modern view of MNEs is that they are more likely to be sources of new technology. Indeed, there has been remarkable growth in the internationalization of R&D as multinational firms establish research affiliates abroad, suggesting that openness to FDI may be a particularly important means of encouraging domestic research capabilities.

The empirical results in the chapter are striking. The authors use an extensive database of R&D expenditures by industry across all OECD economies from 1990 to 2003 to estimate the relationship between research intensities and various measures of domestic and international financial development. The findings point to a strong association between all forms of domestic financial-market development and R&D intensity, indicating that improving and deepening capital markets is critical for supporting investments in knowledge assets. However, among sources from the international capital markets, only FDI seems to be a major factor in financing research and development. This finding supports the view that FDI, by establishing strong and durable investment partners, is more likely to encourage expansion of domestic R&D efforts. Greater access to other forms of arm's-length international capital markets is likely to favor expansion of physical capital assets, which are more easily collateralized and monitored on a global scale. A final observation is that these results again support the basic proposition that investment liberalization, particularly in FDI, is complementary to IPRs reforms in building domestic research capacity.

Part III: Trade in Factors and Technology

The four chapters in this part depart from formal work on technology transfer and IPRs, turning instead to more traditional questions in international trade and investment analysis. The topics covered include the factor-endowment basis for international trade, the knowledge-capital model of the multinational enterprise, and computational analysis of the gains from liberalizing restrictions on trade in services in a developing country. Despite the focus on standard trade questions, an important implicit theme is that trade in factor services, international FDI in knowledge capital and access to high-productivity services from abroad all incorporate significant amounts of technological information. Thus, the essential point of the papers in Part III is to consider trade and investment as important forms of technology transfer, which can improve the productivity of domestic firms.

Chapter 11 — "A Test of the Heckscher–Ohlin–Vanek Theorem: The Leontief Commonplace," Journal of International Economics, 1985

Perhaps the most central prediction in trade theory is that countries tend to specialize in, and export, goods that intensively use their most abundant factors. Thus, richer economies tend to export capital-intensive and skill-intensive goods, while labor-abundant countries tend to export products that use relatively more unskilled labor, and economies with large tracts of agricultural land export foodstuffs and primary commodities. While this prediction is not literally true for individual products, given the complexity of their production, it is true in theory for the amount of factor services implicitly traded in goods. Thus, a capital-abundant country should export more capital than it imports and a labor-abundant country should export more labor than it imports, a proposition known as the Heckscher–Ohlin–Vanek (HOV) theorem.

Despite the intuitive appeal of this prediction, it has proven hard to demonstrate that it holds in actual data, a problem dating back to the famous Leontief paradox. One reason is that traditional empirical approaches to the question did not actually compute the theoretically correct measures of factor-services trade. The article reprinted in Chapter 11 was the first in the literature to offer a fully specified and correct test of the HOV theorem, introducing the concepts of sign tests, rank tests and strong prediction tests. Using a U.S. database involving factor usages in an input–output table, international trade by commodity, and variables measuring American and international endowments of capital, human capital and unskilled labor, the computations find that the HOV predictions fail badly, even when using simple ranking exercises of revealed factor trade.

The article concludes by pointing out potential reasons for these failures, which may have been due to inadequate empirical measures or the inability of the HOV theory to capture other fundamental determinants of trade. The article is important because it helped stimulate a long and deep literature on methods to improve tests of both HOV and competing trade theories.

Chapter 12 — "Development-Related Biases in Factor Productivities and the HOV Model of Trade," with Shuichiro Nishioka, Canadian Journal of Economics, 2009

This recent article exemplifies current scholarship on the factor content of trade, extending the tradition of the analysis in Chapter 11. Here the authors study whether the HOV predictions fit actual data patterns closely if the figures are adjusted for the simple possibility that production factors vary in underlying productivity across nations. A number of theoretical extensions of the theory are modeled, with appropriate predictions about trade in factor services as a function of these adjustments. One basic specification permits Hicks-neutral productivity differences at the country level applied to all factors, while another is more general in permitting productivity to vary by both factors and countries. Such specifications are nested in a statistical testing framework that permits determination of which technology assumptions best fit the trade data.

An extensive database involving input–output tables, direct factor inputs and trade flows for the year 2000 is assembled for 29 countries, including both developed and developing economies. Estimation and testing finds that a simple model permitting technologies to vary by both country and factor fits the data quite well, without the need to resort to complex theories on the demand side. More fundamentally, the estimated productivities are strongly correlated with aggregate factor abundance. For example, workers in Japan, which is capital-abundant, are productive (relative to Japanese capital) because they have access to machines and computers that make them efficient. This systematic correlation between labor productivity and capital abundance is consistent with general principles of the HOV model of trade without factor-price equalization.

Moreover, the fact that countries have factor-specific productivity ratios is further evidence of skill-biased technological change. For example, where there is rapid growth of physical capital the impacts vary for different types of labor, while capital productivity and labor productivity evolve differently with the stages of economic development. In particular, capital-skill complementarity plays a key role because the efficient operation of highly productive capital in developed countries requires an extensive supply of skilled labor. In poor countries, where unskilled labor is the abundant factor, limited access to capital makes labor productivities systematically lower. Thus, a nuanced extension of HOV holds: both differences in factor productivities and factor endowments matter for determining the factor contents of trade.

Chapter 13 — "Estimating the Knowledge-Capital Model of the Multinational Enterprise," with David L. Carr and James R. Markusen, American Economic Review, 2001

The next chapter departs from trade analysis to focus on an important and novel theory of the multinational enterprise and FDI: the knowledge-capital model. Rather than conceiving of FDI as a form of capital flow, this model recognizes that the basis for most multinational activity is *knowledge capital*, or the intangible assets owned by the parent firm. Such assets could be a superior technology, production process, unique know-how or specialized and highly skilled labor, with many of these protected through patents, trademarks, copyrights and trade secrets. There are two essential foundations of this model. First, FDI is not really trade in physical or financial capital, as in older conceptions, but rather the international deployment of advanced knowledge and better technical information. Second, and crucially, such knowledge-based intangible assets are *non-rival*, which means they can be used simultaneously at low cost in many locations. A technology can be shared among production facilities, in the form of blueprints or chemical formulas, but a machine cannot.

These ideas support a rich and complex model that goes a long way in explaining the decisions MNEs make about where to locate their production facilities around the world. If the costs of making investments in subsidiaries are relatively small, then firms will rationally invest in multiple plants of similar factor intensities, sharing their knowledge assets among them, in order to avoid the costs of transporting goods. The resulting sales are made largely in local or regional markets. This phenomenon is referred to in the literature as "horizontal FDI" or "market-seeking FDI." In contrast, if trade costs are rather small, then headquarter firms are likely to invest in production facilities where labor costs are low and ship the output to high-wage locations. This case is called "vertical FDI" or "labor-seeking investment," which is typified by the location of assembly activities in electronics and clothing in such labor-abundant locations as China and Mexico. The knowledge-capital model offers theoretical predictions about market and national conditions under which each of these outcomes is most likely to transpire, as explained thoroughly in the article.

The analysis in Chapter 13 is an extensive empirical estimation of how well the actual data on international sales of U.S.-based multinational enterprises, and also on sales of foreign MNEs in the United States, are explained by this model. The authors assemble data on international affiliate sales in manufacturing in both directions, using figures from 36 countries over the period 1986–1994. These figures are related econometrically to fundamental economic determinants indicated by the knowledge-capital theory. These determinants include joint market size, differences in relative skill endowments, various measures of investment and trade costs, and important interaction terms among these variables to capture the model's non-linearities.

The results are quite consistent with the model's predictions, though horizontal FDI seems most prevalent in the data. Among the more interesting findings are, first, that increases in the costs of shipping goods to host countries sharply increase FDI activity as local affiliate sales substitute for exports, and, second, that as two countries become more equal in size they invest more in each other's market. This article was the seminal contribution to a now-substantial literature on how knowledge capital affects global FDI activity and the model is a workhorse in modern trade analysis.

Chapter 14 — "Quantifying the Impact of Services Liberalization in a Developing Country," with Denise Eby Konan, Journal of Development Economics, 2006

This article offers a considerable departure from the econometric studies of earlier chapters. It is a computational general equilibrium model of Tunisia, which is a small open economy that remains poor. Tunisia sustains significant restrictions against entry of foreign providers of transportation, tourism, finance, insurance, engineering and other key services. Because such services are critical inputs into manufacturing and other sectors, it is plausible that the absence of international competition, and the more competitive technologies it generally entails, is a roadblock to Tunisian productivity growth.

To assess this idea the authors use a Tunisian input–output table, with additional information on factor inputs, trade, taxes and other variables to construct a general-equilibrium representation of the economy. The model is used to simulate the output, trade, consumption and welfare impacts of liberalizing both trade in goods and restrictions on services. The key insight is that domestic service providers, being heavily protected, may be highly inefficient relative to best global practices. In this case, the import restrictions force domestic enterprises and consumers to pay high costs for unproductive services, a pure welfare loss that would be ameliorated by opening up the border. An alternative hypothesis is that domestic service firms do use efficient global technologies but simply make rents (profits) from their protected monopoly positions. In this case, liberalization would shift rents from providers to users but would not make the economy much more efficient. The authors develop a simple model based on these alternative assumptions to assess the impacts of removing trade restrictions and entry barriers.

Again, the results are striking. There are modest welfare gains overall from removing services barriers if these restrictions only protect the rents of otherwise productive Tunisian enterprises. At the other extreme, assuming that Tunisian firms do not employ efficient production techniques, entry by foreign firms introduces cost-saving innovations and services prices fall. In this case, domestic welfare increases dramatically, by nearly eight percent of initial consumption. This policy change also raises the real incomes of both capital and labor, an outcome that is quite different from what standard trade theory would predict, because liberalization of service barriers permits firms in all other sectors to take advantage of a more efficient and critical input, generating higher productivity. These findings point to the importance of securing access to superior international production technologies, where in this case the technology transfer comes through entry of foreign service producers.

Part IV: Policy Analysis of Global Public Goods and Technology Transfer

Ultimately, interest in innovation, technology transfer and IPRs arises because they affect creativity, access to information, and economic and social well-being. In the global arena it has become increasingly clear, for example, that a stronger international regime of patent protection could have the beneficial effect of expanding commercial technology markets but could also raise significant risks of diminishing access in poor countries to medicines, agricultural inputs, environmentally sound technologies and other key public goods. Economists have therefore placed growing emphasis on the practical policy implications of IPRs as regards global public goods. Are patents really an impediment to access and TT? How can IPRs be leveraged to improve innovation and access at the same time? The chapters in this part offer a series of policy analyses at the heart of such questions.

Chapter 15 — "Regulatory Standards in the WTO: Comparing Intellectual Property Rights with Competition Policy, Environmental Protection and Core Labor Standards," World Trade Review, 2002

One of the much-debated issues at the time the WTO was founded, and still controversial, is whether there is a sound justification for including IPRs — in the form of the TRIPS Agreement — in the global trading system. Multilateral trade agreements are built on the fact that mutual cuts in tariffs, quotas and other restrictions on trade in goods generate efficiency and welfare gains for all participants. However, patents, copyrights and other IPRs are very different from direct trade distortions. Making such regulations stronger through greater scope of protection and enforcement of exclusive rights is not necessarily welfare-increasing in all countries. Indeed, reforms to harmonize IPRs on a high level of protection are likely to reduce well-being in the medium run in countries that rely on imports of technology and new goods. Put differently, trade liberalization is mutually beneficial, while globalized IPRs may not be and could tilt the welfare imbalance against poor countries. As a result, many observers argue vociferously that TRIPS never should have entered the WTO agreements.

Another argument is that, unlike trade barriers, IPRs are "behind the border" commercial regulations that affect far more than trade flows. Patents matter for access to medicines and green technologies, copyrights affect the terms under which universities provide educational materials, and plant variety rights may raise the costs of agricultural inputs. How can an agreement focused on trade rules effectively deal with such complex questions, and should it even try? Would the attempt overwhelm the ability of the WTO to function? If IPRs fit, then why would not other forms of regulation that indirectly affect trade and investment?

The article in Chapter 15 was an early attempt to put some economic logic into such debates, which are highly polemical.

The analysis compares four different regulatory regimes — IPRs, competition policy (antitrust regulation), environmental protection and core labor standards — with respect to their suitability for inclusion in trade agreements. There are several criteria that economists consider in determining how well regulatory issues may work within the WTO framework. They include how trade-related the areas are, the importance of international externalities that trade rules might overcome, coordination failures of countries to enforce collective interests through stronger standards, and the ability of dispute settlement to deal with them effectively.

Thus, the article offers a qualitative ranking of the four regulatory areas in each of these areas. For example, as noted in Chapter 2, IPRs are strongly trade-related and investment-related, suggesting there is scope for reducing trade distortions with the TRIPS Agreement. Competition policy and environmental regulations also affect trade, though arguably somewhat less in empirical terms, while core labor standards generally have little impact on measured trade flows. In contrast, both IPRs and environmental rules embody significant international externalities that could be resolved, in principle, through trade negotiations. Dispute resolution is complex in all of these areas. The chapter offers a thorough discussion of such issues and compares an overall ranking, in both weighted and unweighted terms. The essential conclusion is that a significant case exists for incorporating IPRs and antitrust rules into the WTO but the case is weak for the other areas. However, this finding is qualitative and cannot be definitively supported by theory and data. It is offered to readers as a means of provoking thought and debate.

Chapter 16 — "Economic Perspectives on a Multilateral Agreement on Open Access to Basic Science and Technology," with John H. Barton, in Simon J. Evenett and Bernard M. Hoekman, editors, Economic Development and Multilateral Trade Cooperation, 2006

The essay in Chapter 16 focuses squarely on the issue of sustaining and improving access to scientific knowledge on a global basis. The authors set out a general proposal for a multilateral agreement on open access to basic science and technology. The chapter discusses the economic foundations for such an accord, called an Agreement for Access to Basic Science and Technology (ABST), and analyzes the principal issues that would emerge in establishing it. The agreement could be structured around open access for inputs (coordination and movement of research projects and scientific personnel), open access to outputs (basic research results) or both, and be founded on basic WTO principles. The agreement could also include provisions for preferential treatment for developing countries. Its central purposes would be to ensure widespread access to essential scientific results and to enhance the transfer of basic technological information to the developing world.

The authors propose such an agreement for three essential reasons. First, there is an increasingly widespread concern that government restrictions on access to data and research results could harm the pace of global scientific advances and the diffusion of knowledge, particularly to the detriment of transition and developing countries. There has been a sharp policy shift toward making knowledge a private commodity, despite its inherent character as a public good, raising fundamental questions for science, education and the diffusion of information.

Second, issues of technology transfer are at the center of the contentious debate about how the TRIPS Agreement might affect developing countries. There is a common perception that not much increase in actual technology transfer has happened since TRIPS was established. At the same time, the exclusive rights offered by TRIPS bear the potential for limiting access of developing countries to even publicly generated basic research that might otherwise enable greater competition and local innovation. Finally, the economics of knowledge creation and the non-rival nature of information implies that global investments in basic science and technology are underfunded in comparison with the global optimum. Knowledge is a prime example of a global public good that can be more effectively provided by cooperative multilateral actions.

Chapter 17 — "Developing and Distributing Essential Medicines to Poor Countries: The DEFEND Proposal," with Mattias Ganslandt and Eina V. Wong, The World Economy, 2001

One critical task facing the global economy is to develop mechanisms that encourage research aimed at finding treatments for diseases, such as HIV, malaria, tuberculosis, and leishmaniasis, which are common in impoverished nations. Equally important is achieving widespread international distribution of these essential medicines at sufficiently low costs to make them effective and affordable. In this chapter, the authors discuss the economics of these questions thoroughly. They offer a review of data on disease burdens and patient costs in poor countries, and describe potential access problems that could arise as patents in new medicines are adopted and strengthened on a global scale. The latter policy could improve innovation incentives but also bears the risk of significantly reducing competition in drugs from generic producers in India, China and elsewhere. Such an outcome could threaten the affordability of essential medicines in the developing world.

As the analysis points out, demand for new drugs to treat such "neglected diseases" is unlikely to be sufficient in private markets, even if protected by stronger patent rights, to stimulate enough R&D and innovation to meet the problem. It follows that public support, perhaps from rich-country governments and charitable foundations, is needed to incentivize drug development. Moreover, subsidies would also be needed to help pay for distribution systems and clinical facilities, plus the costs of preventing the program drugs from being re-exported from poor countries to richer markets. The authors calculate that the total cost for an international strategy would be in the range of \$8.2 billion to \$12.1 billion annually. This would correspond to less than 0.4 percent of total GDP in the OECD countries in 2001. To put this in further perspective, if this amount were fully paid by the United States, the European Union and Japan, it would come to only \$13.50 per person per year in 2001 prices.

PART I

INTELLECTUAL PROPERTY RIGHTS

Chapter 2

How Trade-Related Are Intellectual Property Rights?*

Keith E. Maskus and Mohan Penubarti†

1. Introduction

In the latter half of the 1980s international disputes over the protection of intellectual property rights (IPRs) became common. The USA enacted a special provision in the Omnibus Trade Act of 1988 that elevated the problem of weak foreign protection for US intellectual property to priority status for bilateral negotiations and potential trade retaliation. The European Community has struggled with the particulars of its attempts to harmonize IPRs as part of the establishment of the unified market in 1993. Numerous industrializing and developing economies have engineered substantial strengthening in their IPRs regimes since 1986, both because of foreign pressure to do so and because of the apparent belief that such a policy change would prove inviting to foreign firms interested in investing in those countries. IPRs were a major issue in the negotiations covering NAFTA and the Uruguay Round.

The link between trade and IPRs is important. In principle, patent protection directly affects growth through inducements to

*Reprinted from *Journal of International Economics*, Vol. 39, No. 3/4, pp. 227–248.

†This research was funded partly by NSF grant no. SBR-9410625. We are grateful to James Alm, Elias Dinopoulos, K.C. Fung, David Gould, James Harrigan, James Markusen and Thomas Prusa for comments. We also thank participants at the NBER Universities Conference on Trade Institutions and Rules, December 1993 and seminar participants at Purdue University. Finally, two anonymous referees were most helpful in guiding the revisions. Any remaining errors are our responsibility.

innovation. Returns to innovation could be influenced by variations in international patent laws, with a primary channel being decisions by firms to trade in different markets. Thus, patent regimes could be an additional factor in the relationship between trade and growth (Segerstrom *et al.*, 1990; Grossman and Helpman, 1991; Helpman, 1993; Taylor, 1994). Despite the practical and conceptual importance of this subject, empirical evidence on trade and international patent protection is absent. In the case of the GATT negotiations, this is particularly odd since the issue was always framed rather narrowly, at least in principle. Specifically, the GATT could only be invoked as an appropriate forum for the negotiation of disciplines on international policies to the extent that different levels of IPRs tend to distort trade flows, or, in other words, that IPRs are "trade-related".

The aim of this paper is to provide the first systematic evidence on whether differential patent laws influence international trade. The analysis relies on an empirical version of a static, general-equilibrium trade model (specifically, the Helpman–Krugman model) and uses trade data for a single year. Thus, we address solely the positive question of whether the distribution of bilateral trade across nations depends on the importing country's patent regime. In this regard, the approach is a single component of the larger normative questions concerning IPRs, trade, and growth. We cannot, for example, compute the impacts of patent laws on profits and associated innovative activities and growth. However, results from this initial specification should help provide an understanding of the sensitivity of trade decisions to patent regimes. If there are strong links at this level, it is sensible to pursue additional empirical specifications in the broader sense.

We begin in the next section by describing channels through which variations in patent laws could influence trading decisions. The essential point of that review is the unsurprising prediction that trade volumes could rise or fall with the imposition of stronger patent regimes.

Given this ambiguity, we go on in the third section to specify an econometric model of bilateral trade flows to capture this behavior. The model is based on a reduced-form version of the Helpman–Krugman gross imports equations, with the reduced-form being augmented by variables capturing preferences, measures of trade distortions, and a corrected measure of patent protection in importing countries. Data are discussed in the fourth section, while in the fifth section estimates are presented that suggest clearly that different patent laws can affect trade. Interestingly, within the group of large developing economies, wherein most of the trade disputes over IPRs have come, stronger patent laws seem to attract larger-than-expected flows of imports from the OECD economies. This is also true, though to a lesser extent in terms of statistical significance, in the small developing economies in our data set. Thus, it appears that patents are indeed "trade-related". Concluding remarks are provided in the final section.

2. Trade and Differences in Patent Protection

Patent laws vary markedly across countries in terms of length and breadth of coverage, products that are excluded from patentability, and provisions for issuing compulsory licenses (World Intellectual Property Organization, 1988). Moreover, enforcement of existing laws differs widely.[1] Thus, a firm making export decisions across different markets faces an array of effective levels of patent protection. If the firm takes this array into account, trade will be "distorted" in the sense that policy parameters will influence its international distribution. Note that this is an unusual definition of distortion in that there is no clear standard for defining global optimality against which to assess the trade impacts.

Although there are few theoretical treatments of the impacts of variations in patent laws on trade flows, a number of papers can be cited that, taken together, make the unsurprising point that the predicted effect is ambiguous. That is, there is no clear presumption that, in comparing two otherwise identical nations, the country with the stronger patent law will attract more imports of patentable commodities.

[1] Indeed, enforcement may be so stringent that it acts as a barrier to trade, which has been an issue in the Uruguay Round negotiations.

Consider a price-discriminating firm deciding on the distribution of its exports of a patentable commodity to various nations. The most straightforward reason for the ambiguity is that there is a tradeoff between the enhanced market power for the firm created by stronger patents and the larger effective market size generated by reduced abilities of local firms to imitate the product. The *market-power effect* would reduce the elasticity of demand facing the foreign firm and would ordinarily induce the firm to export less of its patentable product (or products made by a patentable process) to the market with the stronger IPRs. However, the *market-expansion effect* would increase the demand curve facing the firm and attract larger sales. In larger markets we might also imagine a *cost-reduction effect* that would raise exports if a stronger patent law reduces the need of the foreign firm to undertake private expenditures to deter local imitation (Taylor, 1993).[2] This ambiguity exists in all markets. It seems probable, however, that the market-expansion effect would tend to be more dominant in larger countries with highly competitive local imitative firms while the market-power effect would tend to be stronger in smaller countries with limited capacity for imitation.

This fundamental indeterminacy in the effects of patents on trade flows exists in a wide range of imperfectly competitive, static market structures, including a Stackelberg model (Maskus and Penubarti, 1994) and Cournot duopoly models (Taylor, 1993; Maskus and Eby-Konan, 1994) in identical products. Further, if we interpret patents as product-specific tariffs in a monopolistic-competition model there are few unambiguous predictions about their effects on trade (Flam and Helpman, 1987; Brown, 1991).

These conclusions are reinforced by other factors. First, the trade reactions of foreign firms to changes in patent laws cannot be considered in isolation from the structure of import protection. It is possible, for example, that the profit-maximizing reaction of an exporter could be to either increase or decrease the volume of trade with a

[2]Private expenditures by foreign firms on local enforcement of their IPRs are common in the major industrializing economies.

country that strengthens its patent law, depending on the height of the tariff it faces (Maskus and Penubarti, 1994). Similarly, if a binding quota exists on the product the change in the patent law cannot attract more trade. Further, there are potentially complicated interactions between patents and trade barriers affecting trade in intermediate inputs.

Second, decisions by firms owning a new product or process to export to a particular market are codetermined with decisions to service markets through licensing or foreign direct investment (FDI) (Horstmann and Markusen, 1987). Third, there is an important dynamic element to trade if firms find that exporting to economies with strong patent laws raises their global profits and induces additional R&D efforts (Deardorff, 1992; Helpman, 1993).

3. An Empirical Model of Bilateral Trade with Patents and Trade Distortions

This review indicates that no clear prediction may be made in theoretical models about the volume of trade that may emerge in a world of varying patent regimes. Much depends on local market parameters, including demand, the efficiency of local imitative production, and the structure of trade barriers. Also critical are the reactions of foreign firms, which are necessarily imperfectly competitive if they receive patent protection or have some other form of cost or information advantage. Thus, a clear picture can emerge only from empirical work. In this study we opt for an econometric approach to this question based on actual trade flows.

In principle, a fully specified approach would incorporate, for each commodity, estimated parameters for each importing country and would account for market structure issues, such as markups by firms over marginal export costs. Such a specification would place significant and difficult demands on the researcher to measure these characteristics and to account for their endogeneity. In fact, it is not generally possible from available data to identify such market parameters as local imitation costs (including costs from avoiding detection), demand elasticities, and marginal export costs. Thus, a direct approach is infeasible.

We adopt an alternative approach based on an augmented version of the Helpman–Krugman (1985) bilateral gross imports equations. The Helpman–Krugman (HK) model posits monopolistic competition among symmetric firms, which is a sharply defined market structure and cannot capture the full complexity of the possibilities mentioned in the last section.[3] We choose to use the HK trade model on utilitarian grounds along the lines discussed by Harrigan (1993). It is important to estimate trade equations that emanate from a general-equilibrium theory of bilateral trade. Failure to embed the relevant patent and tariff effects in such a model makes interpretation of the resulting coefficients problematic. The HK model predicts what the volume of trade would be in the absence of trade barriers and variations in patent laws. Accordingly, correlating the differences between observed trade flows and those indicated by the model with other influences included bears great promise for providing valid inferences on the trade effects of patents. More ad hoc approaches that fail to control for general-equilibrium trade determinants risk generating spurious or incorrect results.[4]

Moreover, as recently discussed by Harrigan (1992), the HK model may be used as a basis for a flexible examination of bilateral trade flows, where the estimated equations may be interpreted as the second step of a two-stage procedure. The specific prediction of

---

[3]The HK model would correspond most closely to a situation in which each of many symmetric exporting firms from different countries has identical patent protection of, say, different processes for producing similar products and each faces the possibility of local imitation of those products.

[4]One acceptable alternative to the HK model would be gravity models where consumers have a taste for variety and there exist country-specific varieties for technological reasons, as discussed in Bergstrand (1989). Indeed, some prior estimation with that model generated results that are consistent with those in this study (Maskus and Eby-Konan, 1994). Ferrantino (1993) used a gravity approach to suggest that arm's-length trade is unaffected by variations in IPRs, though intra-firm trade is expanded as a response to weaker IPRs. However, he provided no sectoral breakdown in his regressions, used only US trade data, and measured the strength of IPRs by national memberships in international intellectual-property conventions, a notoriously inaccurate measure of the effective strength of local laws.

the model is that under free trade, bilateral imports are proportional to the exporting country's industry output, where the factor of proportionality is the importing country's share in global expenditure. These variables are jointly endogenous and a regression of imports on exporter outputs is difficult to interpret. This forces us into a two-stage approach where the task is estimation of a set of reduced-form import equations. Under additional assumptions, exporting-country outputs are linear functions of exporting-country endowments. Thus, we would expect there to be a strong correlation between exporter output patterns and factor endowments (Harrigan, 1993, p. 94). Accordingly, exporter factor supplies, which are taken here to be exogenous, provide a valid set of instruments for exporter output levels in explaining bilateral imports.

The HK model makes the strong assumption that tastes are identical and homothetic, which implies that each country imports a share of every product that equals its share in world income. This notion is clearly false and calls for allowing some variations in import demand (that is, unequal expenditure shares by sector) across countries. We handle this issue in three ways. First, we estimate intercept terms for each product across nations, allowing for "home market preference" (Harrigan, 1993) and allow additional intercept terms for country-group dummies measuring whether developing-country importers have small or large markets. Second, we incorporate importer per-capita GNP into the imports equations. Per-capita GNP is taken to be exogenous to sectoral imports. Third, we include interaction terms between our measure of patent strength, discussed below, and dummy variables for the size of developing-country importers.

It remains to incorporate trade barriers and measures of the strength of patent laws in the importing nations. Two measures of trade restrictions are included: tariff revenues as a percentage of imports and black-market exchange-rate premia. These measures are not sector-specific, so we consider them to be exogenous and place them directly into the second-stage imports equations. As discussed in the next section, the index of patent strength we use is subject to measurement error and potential endogeneity. Thus, we develop also a first-stage IV equation to explain the cross-country structure of patent laws. The corrected patent indexes are also interrelated with country-group dummies.

Collecting these ideas, the second-stage econometric model is specified as follows:

$$\log(m_{ijk}) = \alpha_{iS}DS + \alpha_{iL}DL + \beta_i \log(Q^*_{ik}) + \gamma_i \log(Y_j) + \tau_i \log(1+T_j) + \theta_i \log\left(1+\frac{B_j}{100}\right) + \pi_i P^*_j + \pi_{iS}P^*_j DS + \pi_{iL}P^*_j DL + \varepsilon_{ijk} \qquad (1)$$

where m_{ijk} is bilateral imports of good i by country j from exporter k, divided by aggregate expenditure in the importer. DS and DL are dummy variables indicating whether an importing developing country has a small or large market. Q^*_{ik} is sectoral exporter output predicted from the first-stage endowments equations and Y_j is importer per-capita income. The variables T_j and B_j are trade-resistance measures for the importing country: tariff revenues as a proportion of total imports and the black-market premia, respectively. Finally, P^*_j is a measure of patent strength in the importer, taken from the first-stage patents equations. In these second-stage equations, the asterisks denote predicted variables from the first-stage estimation.

4. Data Sources and Construction of Variables

The basic unit of analysis is the industry. For example, Eqs. (1) are estimated for each of 28 manufacturing sectors across a set of countries using 1984 data. Data requirements include sectoral bilateral imports, sectoral outputs in exporting countries, factor endowments in exporting countries, and national expenditure, per-capita incomes, measures of trade distortions, and patent strength in importing countries. Consider each of these issues in turn.

The data on trade and production come from the OECD's Compatible Trade and Production (COMTAP) data base.[5] These data are organized according to the 28 3-digit classifications of the International Standard Industrial Classification (ISIC). The database

[5] We are grateful to James Harrigan for providing these data.

includes sectoral outputs by OECD member plus bilateral trade of the OECD countries among themselves and with most countries in the rest of the world. The advantage of these data is that trade and output figures are assembled on a consistent basis, which is difficult to do from other published sources (Maskus, 1991). The disadvantage is that output data are available only for the OECD countries in this source, so that exporters in our model are limited to the 22 OECD member countries (see Appendix for a list of countries and industries). This limitation is not particularly damaging for the model, however, because it is the OECD countries that may be expected to provide the overwhelming share of exports in goods subject to patent protection. Output and trade data are for 1984.

For purposes of computing the regression of sectoral outputs on endowments, we wished to have a larger sample than the OECD nations alone. Thus, data on outputs and factor endowments in 1984 were assembled for the OECD countries and for 25 additional developing countries.[6] Five national factor endowments were included in the output equations, which were estimated separately for each ISIC sector. The first endowment is a measure of the real net national capital stock in millions of US dollars. This is the accumulated, depreciated, and deflated series (15 years, 13.33% depreciation rate) of gross fixed capital formation in each country. Investment deflators were taken from Summers and Heston (1988) and capital formation and exchange rates were from the *International Financial Statistics* of the International Monetary Fund (IMF). The second and third endowments are the skilled and unskilled labor forces in thousands, as reported in the *Yearbook of Labour Statistics* of the International Labour Organization. Skilled labor is defined as occupational categories 0/1 and 2 and low-skilled labor as all other categories. The fourth endowment is the sum of two types of land areas: *LAND1*, the area of arable land and land under permanent crops or permanent pasture and *LAND2*, the area of forests and woodland, both measured in thousands of hectares. These data were taken from the *Production Yearbook* of the Food and Agricultural Organization. The final "factor" is the cif-fob factor reported for each country by

[6]See Maskus (1991) for details on the construction of the endowment variables.

the IMF. It is a measure of the transport costs involved in aggregate commerce with each country. As such it might be better considered a trade distortion, but this treatment is consistent with the view that transport costs represent a "natural" endowment rather than a policy-driven distortion. Sectoral outputs for the non-OECD countries were taken from *Industrial Statistics Yearbook, Volume II* of the United Nations. The results of the first-stage regressions of outputs on endowments by sector are available on request.

Aggregate expenditure in each nation is measured as GNP less the aggregate current account imbalance. These figures, along with per-capita GNP in importing countries in 1984 were taken from The World Bank, *World Tables.* As noted earlier, we argue that the effects of patent laws on trade flows could vary, depending on whether the importing nation is a high-income developed economy, a large developing economy, or a small developing economy.[7] To make this distinction operational, we split our country sample correspondingly into three groups: H, L, and S. High-income countries include the major OECD exporters plus Switzerland. Large developing countries include Greece, Portugal, Turkey, and Yugoslavia among the OECD nations plus 18 others. This group is defined as developing countries with a GDP in excess of $18 billion, which includes both larger poorer economies, such as India, and smaller upper-middle-income economies, such as Singapore. The final group is small developing countries, such as Ecuador, which have both low per-capita incomes and small aggregate GDP figures. Data on GNP per capita ("*INCOME*") and on the classification ("Type") are presented in the Appendix. For estimation purposes, we develop dummy variables for the L and S groups, using them to compute separate group intercepts and patent-interaction terms in each sectoral equation.

It is difficult to measure the extent and effects of trade barriers in specific sectors across countries. Tariff data exist for tariff-line items, but it is difficult to aggregate these into meaningful measures on an ISIC basis and to do so for each trading partner. Moreover, tariffs often represent inconsequential barriers relative to various forms of nontariff barriers (NTBs), especially for developing countries. There

[7]See Maskus and Penubarti (1994) for theoretical details.

is no published categorization of sectoral NTBs in developing countries and, even if there were, assessing their true restrictiveness is an elusive task (Leamer, 1992). We therefore resort to using two national measures of trade restrictions. The first, which was taken from the *Government Finance Statistics Yearbook* of the International Monetary Fund, is simply tariff revenue as a percentage of dutiable imports, a variable often termed the average tariff rate. Problems with this variable as a measure of actual trade restrictiveness are well known, though it seems to accord well with casual beliefs about trade protection rankings across nations. The second is the black-market exchange-rate premium developed by Levine and Renelt (1992).[8] It is a measure of the aggregate "openness" to trade of each importing country, in the sense that a higher premium indicates a more heavily distorted, or closed, economy.

Turning finally to patents, there may be sectoral differences in effective duration or breadth of protection. It is again infeasible to develop a detailed sectoral index by country. However, in principle patent laws are applied uniformly to all sectors, allowing for no sectoral differences within a country. Thus, we consider a measure of the relative strength of patent laws across countries. Such an index was developed by Rapp and Rozek (1990). Based on surveys of business and government officials and an examination of patent laws themselves, the index measures conformity of each nation's patent regulations and enforcement to the minimum standards put forward by the US Chamber of Commerce Intellectual Property Task Force (1987). This index, which is listed as "*PATENT*" in the Appendix, ranks the level of patent protection on a scale from zero to five, where zero reflects a country with no patent law at all and five indicates a country with laws that are fully conforming to the minimum standards. The ranking is valid for 1984 because all relevant legislative matters in each of our sample nations were in place by that time.

To illustrate this index, consider the score of one for Argentina. As of 1984, Argentina had a patent law but did not provide pharmaceutical patents. If a patent were not worked in Argentina for a period of two years from registration, compulsory licenses could be issued on

[8]We are indebted to David Gould for providing these data.

request of local competing firms. The fine for patent infringement was fixed in nominal terms in 1864, meaning that there was no effective deterrent to imitation. Nor were there effective judicial procedures for enjoining against infringement. In contrast, Singapore provided fully enforceable patents, with the exception of issuing compulsory licenses in certain types of inventions and retaining the right to transfer control of pharmaceutical patents for public policy purposes. This system earned Singapore a score of four.

Two econometric problems emerge with regard to this index. First, since the index is subjective, significant potential for measurement error exists. For example, a number of poor nations, such as Ghana and Nigeria, have strong laws on paper because they were British colonies and modeled their regimes on the United Kingdom Patents Act. However, enforcement difficulties considerably reduce the effective strength of patents there. Second, a potential endogeneity problem exists in that levels of economic development and trade flows may influence legislation and enforcement.

For these reasons we adopt an instrumental variables approach to correcting the raw patent index. The instruments include prior indicators of the level of economic development (GDP per capita, primary exports as a share of total exports, infant mortality rate, and secondary enrollment ratios; all for 1965), dummy variables for former British and French colonies, and alternative measures of intellectual property protection (dummy variables for membership in Paris and Berne Conventions and for the existence of legislative provision for pharmaceutical and chemical product patents as of 1984). These variables plausibly are highly correlated with patent strength yet uncorrelated with the reduced-form imports regression error terms. The development indicators were taken from World Bank, *World Development Report, 1988* and the other variables from Siebeck (1990). Detailed results of this IV regression, which registered an adjusted R^2 of 0.58, are available on request. We present the corrected patent measure as "*PATENT**" in the Appendix.

We provide a brief overview of the data. Table 1 lists summary statistics and simple correlations for the national explanatory variables and aggregate merchandise imports in 1984. Per-capita GNP is strongly and positively correlated with imports and the corrected

Table 1. Summary statistics and correlations for importer-specific variables ($n = 77$).

A. Summary statistics

	Mean	St. Dev.	Minimum	Maximum
INCOME	3474	4135	116	15916
IMPORTS	21345	47509	157	338190
*PATENTS**	3.091	1.120	0.902	5.329
*PAT*DS*	1.153	1.376	0.000	3.975
*PAT*DL*	0.833	1.378	0.000	4.880
TARIFF	9.82	8.41	0.000	39.71
BMP($n = 71$)	46	112	0	732

B. Correlations

	INCOME	*IMPORTS*	*PATENT**	*PAT*DS*	*PAT*DL*	*TARIFF*	*BMP*
INCOME	1.000	0.651	0.712	−0.553	−0.128	−0.593	−0.284
IMPORTS		1.000	0.528	−0.359	−0.095	−0.348	−0.177
*PATENT**			1.000	−0.241	−0.005	−0.427	−0.188
*PAT*DS*				1.000	−0.514	0.441	0.350
*PAT*DL*					1.000	−0.055	−0.135
TARIFF						1.000	0.164
BMP($n = 71$)							1.000

index of patent strength. It is negatively correlated with the average tariff rate and the black-market premium. It is also negatively correlated with the interaction terms between patents and the dummy variables for small and large developing countries ("*PAT*DS*" and "*PAT*DL*"), though not very strongly in the latter case. Aggregate imports and patents are negatively correlated with *TARIFF* and *BMP*.

For further perspective, we plot the simple regression line between the corrected patent index and the log of *INCOME* in Figure 1, with t-statistics in parentheses. There is clearly a positive relationship between these variables, though it appears not to be a linear one. Countries with high or low income levels seem to have predicted patent indexes above the regression line, while those in the middle tend to have indexes below the line. From this evidence it appears that effective patent laws differ across nations for reasons other than simply variations in economic development.

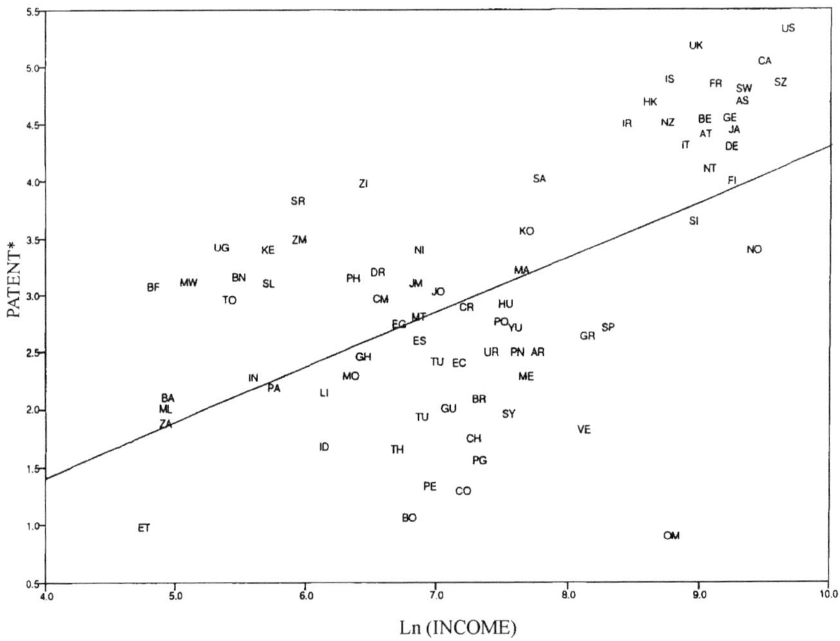

PATENT* = -0.51 + 0.49Ln(INCOME) N=77
(5.73) (6.58) R-SQ = 0.37

Figure 1. Regression line between patents and per-capita GNP.

5. Estimation Results

Results from the reduced-form imports equations are presented in Table 2.[9] Despite the fact that imports are scaled by expenditure, heteroskedasticity could still be a problem in these equations due to differences in country size. Accordingly, the significance levels of the estimates in Table 2 are based on standard errors computed from the White heteroskedastic-consistent matrix. We report first the regression for all 28 sectors pooled, allowing for sectoral intercept dummies.[10] Significance levels are indicated by the superscripts "*" (99% level) and "**" (95% level).

As required by the HK model, the coefficient on exporter output is near unity. Both the tariff rate and the black-market exchange premium register significantly negative impacts on bilateral imports. Per-capita GNP has an insignificant effect on scaled imports, consistent with the notion of homothetic preferences. However, there are market-specific demand effects as suggested by the significant coefficients on the small-country and large-country dummies. The coefficient on the corrected patent index is positive and highly significant, indicating that across all sectors and countries a strengthening in effective patent strength does raise bilateral imports on average. This effect is strengthened considerably in both the small and large developing economies, as evidenced by the significantly positive coefficients on the patent interaction terms. The superscript "+" designates that coefficient sums (on *PATENT* plus the relevant interaction term) are significantly positive. In the pooled sample, then, there is substantial evidence that, other things constant, stronger patent laws attract greater bilateral trade across all nations with this effect being particularly pronounced in the developing countries. In brief, intellectual property rights are indeed "trade-related".

It is clear that such effects could vary across sectors. The remainder of Table 2 provides regression estimates for each ISIC 3-digit industry. We organize the sectors into three groups. The first is a set of nine industries (ignore pharmaceuticals for now) that one

[9]Details of the first-stage regressions are available upon request

[10]Coefficients of the 27 industry dummies were highly significant. We do not report them here.

Table 2. Pooled and sectoral reduced-form imports equations.

Sector	*CONS*	*OUTPUT**	*INC*	*TAR*	*BMP*	*DS*	*DL*	*PATENT**	*PAT*DS*	*PAT*DL*	*N*	$\bar{R}^2$
Pooled	−19.33*	1.15*	−0.03	−1.34*	−0.86*	−0.56*	−2.28*	0.18*	0.14*+	0.50*+	39153	0.47
A PRIORI MOST PATENT-SENSITIVE SECTORS												
Pet. & coal products	−18.39*	0.98*	0.06	−1.25	−0.71*	−2.46*	−3.14*	−0.36	0.79*+	0.77*+	1196	0.26
Food products	−15.81*	0.94*	−0.29*	−0.48	−0.74*	−1.40	−3.71*	0.13	0.03	0.74*+	1514	0.20
Professional goods	−18.76*	1.14*	0.13	0.57	−0.74*	−1.18	−2.06*	−0.06	0.26+	0.57*+	1466	0.58
Metal products	−19.97*	1.23*	−0.03	−0.88	−0.67*	−0.64	−2.26*	0.12	0.23	0.58*+	1508	0.49
Electrical machinery	−18.66*	1.22*	−0.06	0.78	−0.60*	−1.58**	−2.10*	0.00	0.29+	0.53*+	1498	0.53
Plastic products	−20.57*	1.20*	−0.06	−2.55*	−1.00*	−0.10	−2.32*	0.27	0.05+	0.49*+	1443	0.49
Other chemical prod.	−20.18*	1.25*	−0.02	−1.43	−0.66*	0.51	−1.18	0.11	−0.06	0.38**+	1494	0.45
Pharmaceuticals	−13.38*	0.74*	−0.14	−1.74	−0.57*	−0.48	−1.69**	−0.10	0.35**+	0.47*+	653	0.33
Machinery, nec	−19.23*	1.09*	0.14	1.92**	−0.46*	−0.62	−1.23	0.09	0.18+	0.35**+	1515	0.55
Industrial chemicals	−19.67*	1.26*	0.08	−0.42	0.86*	−1.26	−0.88	−0.07	0.39**	0.23	1490	0.45
Pooled	−18.09	1.16*	−0.01	−0.39	−0.71*	−0.95*	−2.08*	0.03	0.23*	0.51*	13124	0.52
A PRIORI LEAST PATENT-SENSITIVE SECTORS												
Leather products	−18.20*	1.03*	−0.07	−6.47*	−1.26*	−1.33	−3.97*	0.19	0.28+	0.95*+	1286	0.36
Wearing apparel	−20.97*	1.10*	0.09	−3.46*	0.91*	0.50	−4.35*	0.40**	−0.02+	0.73*+	1426	0.43
Footwear	−18.50*	0.91*	−0.06	−4.63*	−0.95*	−0.89	−3.58*	0.21	0.34+	0.64*+	1212	0.25
Rubber products	−20.37*	1.39*	−0.11	−1.94**	−0.67*	−0.27	−2.26*	0.22	0.10+	0.39**+	1425	0.49

(*Continued*)

Table 2. (*Continued*)

Sector	*CONS*	*OUTPUT**	*INC*	*TAR*	*BMP*	*DS*	*DL*	*PATENT**	*PAT*DS*	*PAT*DL*	*N*	$\bar{R}^2$
Printing & publishing	-22.13^{*}	1.17^{*}	-0.06	-1.65	-0.48^{*}	0.48	-1.55^{**}	0.37^{**}	-0.10^{+}	0.31^{+}	1441	0.43
Transport equip.	-21.64^{*}	1.37^{*}	-0.01	-0.28	-0.58^{*}	0.01	-1.28	0.19	0.03	0.21^{+}	1472	0.53
Nonferrous metals	-19.66^{*}	0.94^{*}	-0.09	-1.55	-1.07^{*}	0.96	0.01	0.56^{*}	-0.44	0.02^{+}	1397	0.30
Beverages	-18.99	0.87^{*}	-0.31^{*}	-6.89^{*}	-1.02^{*}	1.79	-0.26	0.79^{*}	-0.26^{+}	-0.09^{+}	1377	0.23
Iron & steel	-21.26^{*}	1.18^{*}	-0.05	1.37	-1.01^{*}	0.85	0.27	0.42^{**}	-0.29	-0.12^{+}	1430	0.39
Pooled	-19.53^{*}	1.15^{*}	-0.17^{**}	-2.72^{*}	-0.86^{*}	0.14	-1.87^{*}	0.38^{*}	-0.05^{+}	0.33^{*+}	12466	0.43
OTHER SECTORS												
Tobacco products	-16.09^{*}	1.07^{*}	-0.30^{**}	-4.22^{*}	-1.52^{*}	-3.22^{*}	-6.48^{*}	-0.67^{*}	1.14^{*+}	1.55^{*+}	984	0.27
Textiles	-17.69^{*}	1.10^{*}	-0.06	0.50	-1.15^{*}	-2.06^{*}	-4.37^{*}	0.11	0.28^{+}	0.91^{*+}	1487	0.41
Wood products	-18.73^{*}	0.82^{*}	0.11	-0.78	-0.86^{*}	-1.39	4.02^{*}	0.01	0.14	0.85^{*+}	1387	0.24
Furniture	-21.51^{*}	1.19^{*}	-0.09	-2.26^{**}	-0.99^{*}	-0.40	-2.96^{*}	0.39^{*}	0.15^{+}	0.68^{*+}	1369	0.43
Other manuf.	-22.39^{*}	1.26^{*}	0.23^{*}	-2.17^{*}	-1.02^{*}	0.00	-2.50^{*}	0.28	0.03^{+}	0.63^{*+}	1464	0.51
Paper & products	-19.66^{*}	1.17^{*}	-0.06	0.33	-0.83^{*}	-0.65	-1.64	0.14	0.24^{+}	0.54^{**+}	1419	0.30
Glass & products	-20.32^{*}	1.06^{*}	0.02	-0.61	-1.10^{*}	0.37	-1.94^{**}	0.33	0.02^{+}	0.48^{*+}	1412	0.35
Nonmetal products	-22.51^{*}	1.29^{*}	-0.01	0.94	-0.66^{*}	-0.52	-1.66^{**}	0.24	0.18^{+}	0.47^{**+}	1407	0.36
Pottery & china	-19.54^{*}	1.11^{*}	-0.03	-2.41^{*}	-0.91^{*}	-0.58	-2.05^{*}	0.21	0.30^{+}	0.45^{*+}	1353	0.39
Petroleum refining	-21.63^{*}	1.14^{*}	-0.19	-0.98	-0.81^{*}	-0.07	-1.39	0.36	0.01^{+}	0.09^{+}	1281	0.24
Pooled	-19.10^{*}	1.13^{*}	-0.02	-1.00^{*}	-0.98^{*}	-0.86^{*}	-2.88^{*}	0.14^{*}	0.23^{*+}	0.66^{*+}	13563	0.40

Notes: The pooled regression allows separate intercepts for each sector but constrains the reported coefficients to be equal across sectors. *Denotes coefficient that is significantly different from zero at the 99% confidence level. **Denotes coefficient that is significantly different from zero at the 95% confidence level. The significance levels are computed from heteroskedastic-consistent standard errors. +Denotes a coefficient sum that is significantly positive at the 95% confidence level.

might expect to have the greatest economic sensitivity to variations in national patent laws. This grouping includes sectors whose R&D programs in the USA are evidently most dependent on the promise of attaining US patents (Mansfield, 1986; Levin *et al.*, 1987) and those US sectors that report the most significant damages of all kinds from limited foreign intellectual property protection in a government survey (US International Trade Commission, 1988). The second is a set of nine industries one might expect to have the least sensitivity, based on the same subjective criteria. The third contains all other manufacturing sectors.

In considering these groups, two substantive qualifications should be kept in mind. First, at this level of aggregation patents could affect decisions in all industries. For example, we would not necessarily expect a zero influence of patents on trade, even in sectors such as nonferrous metals and wearing apparel, because they contain categories of commodities that are characterized by substantial technological changes and process improvements. A more sensible hypothesis is that trade flows of the most patent-sensitive group are more positively affected than are those of the least patent-sensitive group on average. However, the second qualification could offset this hypothesis. Specifically, as our earlier discussion noted, the reaction of firms to variations in patent laws is theoretically ambiguous. It is possible that in patent-sensitive sectors the market-expansion effect could be offset by the market-power effect, resulting in an insignificant patent coefficient. Further, firms could react to weak patent laws more by engaging in FDI, in order to retain control over proprietary information, than by reducing or shifting exports.

Within each group we arrange sectors according to the magnitude of the coefficient on the patent interaction with the large-country dummy, because the policies of the large developing countries have been the source of greatest policy concern. In all sectoral regressions the coefficient on exporter output is near unity, as expected. The coefficient on *INCOME* is zero except for food products, tobacco products, and beverages, where it is significantly negative (suggesting income-inelastic demands) and other manufactures, where it is significantly positive. The effect of average tariff rates is predominantly negative on bilateral imports, except in machinery. This category consists mainly of industrial machinery and this finding likely reflects the fact that high-tariff developing economies have disproportionately high imports of such goods for growth and import-substitution reasons. On the other hand, the coefficients on our variable capturing the closed nature of national markets, the black-market exchange premium, are uniformly and significantly negative. On this evidence, trade distortions are important determinants of bilateral trade flows, a finding consistent with that of Harrigan (1993).

Consider the impacts of $PATENT^*$, or variations in strength of national patent laws. The coefficients reflect the average impact, which could be positive or negative in theory, of patent laws on bilateral imports across all countries in each sector. In the first group of industries all coefficients are essentially zero, suggesting an insignificant impact overall. In the second group, however, all coefficients are positive, five of them significantly so. Thus, it appears that, across all countries, bilateral trade flows are more influenced by patent laws in sectors with presumably less sensitivity (at least in terms of their R&D efforts) to patent availability. This finding supports the view that trade reductions through the exercise of enhanced market power are more important in patent-sensitive sectors than in patent-insensitive sectors.[11] It might also be that FDI is more significant in the more patent-sensitive sectors as we have explained. For example, Bayard and Elliott (1994) cited figures that in 1985 US pharmaceutical firms had investments totaling $700 million in Brazil, a country frequently noted for its lax patent protection, accounting for $600 million in sales there. In contrast, US exports of drugs to Brazil by these firms amounted to only $50 million.[12] In the group of other sectors (those with presumably intermediate patent sensitivity) the patent coefficients were largely positive though insignificant, with one

[11]It is conceivable that our patent index serves as a proxy for the strength of copyright and trademark laws. In the low-patent-sensitivity group there are some sectors, such as wearing apparel, beverages, and printing and publishing, that would be more sensitive to these IPRs.

[12]Additional suggestive evidence supports this view. In chemicals, machinery, and electrical machinery, domestic sales of US foreign affiliates in 1989, scaled by host-country GDP, were more than twice those for the average of all sectors, while the number of foreign affiliates per US parent was highest in chemicals (and even larger in the subcategory drugs; see US Department of Commerce, 1992).

positively significant case (furniture) and one negatively significant case (tobacco products).

Turning to the interaction terms, their coefficients measure the additional effect of patent protection in small and large developing countries relative to the average impact. In this regard, the greatest sensitivity to patents comes in the most patent-sensitive sectors. In the small countries, the coefficients on PAT^*DS are significantly positive in the cases of petroleum and coal products and industrial chemicals. None of these coefficients is significant in the least patent-sensitive group and one case, tobacco products, is significantly positive in the intermediate group. In the case of the large developing importers, the effect of stronger patents is strongly positive in eight of nine cases in the most patent-sensitive group (excluding pharmaceuticals), in four of nine cases in the least patent-sensitive group, and in nine of ten cases in the intermediate group. Moreover, the coefficients for the large nations are nearly always markedly higher than the coefficients for the small nations. These conclusions are strengthened when we consider the coefficient sums (not shown) on $PATENT^*$ and PAT^*DS, and $PATENT^*$ and PAT^*DL. In only six of 28 sectoral cases is this sum not significantly positive in the small developing economies and in only one case is it insignificant in the large developing economies. In the pooled regressions within each group, these coefficient sums are also significantly positive. Interestingly, the coefficient sums are smallest in the a priori most patent-sensitive group, again suggesting that market-power effects most restrain the trade impacts in this group.[13]

We noted earlier the possibility that these results at the aggregate industry level may mask variations in subcategories. For example, it may be surprising that the "other chemical products" industry has an insignificant coefficient overall and in small countries because it contains pharmaceuticals, the prototypical patent-sensitive industry. To examine this issue, we concorded bilateral trade data in pharmaceuticals with production data for the 4-digit drugs sector.[14] The

[13] Standard F-tests reject the hypothesis of joint equality of these patent and interaction-term coefficients in each group.

[14] Trade data are from the United Nations *Commodity Trade Statistics* volumes. The production data for the 4-digit ISIC sector are from the COMTAP data.

associated regression, listed in the first part of Table 2, indicates that patent strength indeed is a significantly positive determinant of trade flows into both small and large developing countries.

Thus, the evidence supports the hypothesis that stronger patent laws induce greater bilateral trade flows, an effect that is pervasive across manufacturing sectors. This result is robust in several dimensions. First, regressions in a gravity-equation framework provided similar results.[15] Second, all our equations perform sensibly in that the output coefficient was near unity, as required by the underlying model. Third, we have accounted for measurement error in the patent index by instrumenting it. However, the results were little changed when we used the uncorrected patent index. Further, using alternative measures of trade restrictions and openness does not alter the patent results appreciably (Penubarti, 1994).

6. Concluding Remarks

We reviewed models suggesting that the impact of patent laws on trade is ambiguous. Exporting firms with a choice of export markets may base their decisions in part on the strength of local patent laws. However, such firms face a fundamental tradeoff between the market power generated by local patents and the ability to expand sales in the event that local patents raise residual market demand and lower marginal exporting costs. Issues of FDI also make the theoretical impact ambiguous. Thus, the question is ultimately empirical.

The empirical results strongly suggest that exporting firms discriminate in their sales decisions across export markets, taking account of local patent laws. Across the range of developing countries, bilateral imports are significantly larger than levels predicted by the HK model for countries with stronger patent laws. The impact is stronger in the larger countries. It is interesting that this group includes most of the countries with which disputes over intellectual property have been most prominent, such as Mexico, Korea, Brazil, and Argentina. That variations in international patent laws have

[15] For example, a regression for all manufacturers (ISIC 3) resulted in a *PATENT* coefficient sum for the large developing economies of 0.37, which was significantly positive.

detectable influences on international trade indicates that IPRs are indeed "trade-related".

We emphasize that this interpretation is specific to the econometric model at hand. The model used here is explicitly static. More work of a dynamic nature could help place these results into better perspective. For example, though trade flows seem to be affected by variations in patent protection, we cannot conclude that this finding means that stronger and more harmonized global levels of patents would generate more innovation without considering also other determinants of profits and international technology diffusion (Segerstrom *et al.*, 1990).

References

Bayard, T. O. and Elliott, K. A., 1994. *Reciprocity and Retaliation in US Trade Policy* (Institute for International Economics, Washington, DC).

Bergstrand, J., 1989. The generalized gravity equation, monopolistic competition, and the factor-proportions theory in international trade. *Review of Economics and Statistics* 71, 143–153.

Brown, D. K., 1991. Tariffs and capacity utilization by monopolistically competitive firms. *Journal of International Economics* 30, 371–381.

Deardorff, A.V., 1992. Welfare effects of global patent protection. *Economica* 59, 33–51.

Ferrantino, M. J., 1993. The effect of intellectual property rights on international trade and investment. *Weltwirtschaftliches Archiv* 129, 300–331.

Flam, H. and Helpman, E., 1987. Industrial policy under monopolistic competition. *Journal of International Economics* 22, 79–102.

Grossman, G. M. and Helpman, E., 1991. *Innovation and Growth in the Global Economy* (MIT Press, Cambridge, MA).

Harrigan, J., 1992. Openness to trade in manufactures in the OECD, manuscript, University of Pittsburgh.

Harrigan, J., 1993. OECD imports and trade barriers in 1983. *Journal of International Economics* 35, 91–111.

Helpman, E., 1993. Innovation, imitation, and intellectual property rights. *Econometrica* 61, 1247–1280.

Helpman, E. and Krugman, P. R., 1985. *Market Structure and Foreign Trade* (MIT Press, Cambridge, MA).

Horstmann, I. and Markusen, J. R., 1987. Licensing versus direct foreign investment: A model of internalization by the multinational enterprise. *Canadian Journal of Economics* 20, 464–481.

Leamer, E. E., 1992. The structure and effects of tariff and nontariff barriers in 1983, in: R. Jones and A. O. Krueger, eds., *The Political Economy of International Trade* (Blackwell, London), 224–260.

Levin, R. C., Klevorick, A. K., Nelson, R. R., and Winter, S. G., 1987. Appropriating the returns from industrial R&D. *Brookings Papers on Economic Activity* 3, 783–820.

Levine, R. and Renelt, D., 1992. A sensitivity analysis of cross-country growth regressions. *American Economic Review* 82, 942–963.

Mansfield, E., 1986. Patents and innovation: An empirical study. *Management Science* 32, 173–181.

Maskus, K. E., 1991. Comparing international trade data and product and national characteristics data for the analysis of trade models, in: P. Hooper and J. D. Richardson, eds., *International Economic Transactions: Issues in Measurement and Empirical Research* (University of Chicago Press, Chicago, IL), 17–56.

Maskus, K. E. and Eby-Konan, D., 1994. Trade-related intellectual property rights: Issues and exploratory results, in: A. V. Deardorff and R. M. Stern, eds., *Analytical and Negotiating Issues in the Global Trading System* (University of Michigan Press, Ann Arbor, MI), 401–446.

Maskus, K. E. and Penubarti, M., 1994. Patents and international trade, manuscript, University of Colorado, Boulder, CO.

Penubarti, M., 1994. Market openness and US–Japan trade conflict, Ph.D. dissertation, University of Colorado, Boulder, CO.

Rapp, R. and Rozek, R., 1990. Benefits and costs of intellectual property protection in developing countries. Working paper no. 3, National Economic Research Associates, Inc., Washington, DC.

Segerstrom, P. S., Anant, T. C. A., and Dinopoulos, E., 1990. A Schumpeterian model of the product life cycle. *American Economic Review* 80, 1077–1091.

Siebeck, W. E. (ed.), 1990. Strengthening protection of intellectual property in developing countries: A survey of the literature (World Bank, Washington, DC).

Summers, R. and Heston, A., 1988. A new set of international comparisons of real product and price levels: Estimates for 130 countries, 1950–1985 *Review of Income and Wealth* 34, 1–25.

Taylor, M. S., 1993. TRIPs, trade, and technology transfer. *Canadian Journal of Economics* 26, 625–638.

Taylor, M. S., 1994. TRIPs, trade, and growth *International Economic Review* 35, 361–382.

United States Chamber of Commerce, 1987. *Guidelines for Standards for the Protection and Enforcement of Intellectual Property Rights* (Washington, DC).

United States Department of Commerce, 1992. *US Direct Investment Abroad: 1989 Benchmark Survey, Final Results* (Washington, DC).

United States International Trade Commission, 1988. *Foreign Protection of Intellectual Property Rights and the Effect on US Industry and Trade* (Washington, DC).

World Intellectual Property Organization, 1988. *Background Reading Material on Intellectual Property* (WIPO, Geneva).

Appendix: Countries and Basic Data

Case	Country	Symbol	Type	*PATENT*	*PATENT**	*INCOME*	*TARIFF*	*BMP*
Exporters and Importers								
1.	Australia[a]	AS	H	4	4.641	11165	7.90	0
2.	Austria[a]	AT	H	4	4.372	8393	1.58	0
3.	Belgium[a]	BE	H	5	4.537	8393	0.003	0
4.	Canada[a]	CA	H	4	5.036	13240	3.78	0
5.	Denmark	DE	H	5	4.375	10226	0.10	0
6.	Finland[a]	FI	H	4	4.332	10240	1.41	0
7.	France[a]	FR	H	5	4.844	9035	0.07	0
8.	Germany[a]	GE	H	5	4.549	10165	0.02	0
9.	Ireland[a]	IR	H	4	4.495	4592	5.66	0
10.	Italy[a]	IT	H	5	4.309	7216	0.05	0
11.	Japan[a]	JA	H	4	4.444	10471	2.17	0
12.	Netherlands[a]	NT	H	5	4.345	8640	0	0
13.	New Zealand[a]	NZ	H	4	4.505	6296	5.84	0
14.	Norway[a]	NO	H	4	3.392	12342	7.31	0
15.	Spain[a]	SP	H	4	2.711	4002	5.78	0
16.	Sweden[a]	SW	H	5	4.805	11229	0.76	0
17.	UK[a]	UK	H	5	5.180	7713	0.02	0
18.	US[a]	US	H	5	5.329	15916	3.39	0
19.	Greece[a]	GR	L	4	2.639	3413	1.14	7
20.	Portugal[a]	PO	L	3	2.766	1786	2.53	10
21.	Turkey[a]	TU	L	1	1.927	981	6.06	13
Exporter only								
22.	Yugoslavia[a]	YU	L	2	2.684	1964	10.56	NA
Importers only								
23.	Switzerland[a]	SZ	H	5	4.852	14935	4.87	0
24.	Argentina[a]	AR	L	1	2.506	2355	10.69	80
25.	Brazil[a]	BR	L	1	2.086	1500	5.33	29
26.	Colombia[a]	CO	L	2	1.520	1336	9.90	6
27.	Egypt[a]	EG	L	2	2.751	820	26.99	79
28.	Hong Kong[a]	HK	L	3	4.685	5498	8.55	0
29.	India[a]	IN	L	1	2.287	269	39.71	19
30.	Indonesia[a]	ID	L	0	1.681	464	2.83	3
31.	Israel[a]	IS	L	5	4.880	6359	4.75	16
32.	Korea[a]	KO	L	3	3.565	2146	7.79	9
33.	Malaysia[a]	MA	L	3	3.222	2073	7.75	2
34.	Mexico[a]	ME	L	2	2.289	2159	6.47	15
35.	Nigeria[a]	NI	L	4	3.276	954	14.27	78
36.	Pakistan[a]	PA	L	3	2.185	316	25.55	22
37.	Philippines[a]	PH	L	4	3.149	580	13.48	11

(Continued)

(*Continued*)

Case	Country	Symbol	Type	*PATENT*	*PATENT**	*INCOME*	*TARIFF*	*BMP*
38.	Singapore[a]	SI	L	4	3.654	7566	0.85	1
39.	South Africa[a]	SA	L	5	4.027	2357	2.66	20
40.	Thailand	TH	L	1	1.653	802	12.34	3
41.	Venezuela[a]	VE	L	2	1.828	3345	8.22	44
42.	Bangladesh	BA	S	2	2.059	140	15.01	109
43.	Benin	BN	S	2	3.157	240	27.85	1
44.	Bolivia[a]	BO	S	1	1.060	890	6.49	178
45.	Burkina Faso	BF	S	2	3.069	124	11.74	1
46.	Cameroon	CM	S	2	2.969	709	23.63	1
47.	Chile	CH	S	2	1.757	1455	13.21	19
48.	Costa Rica	CR	S	3	2.890	1378	8.77	19
49.	Dom. Rep.	DR	S	2	3.196	700	9.92	43
50.	Ecuador[a]	EC	S	1	2.372	1298	17.66	31
51.	El Salvador	ES	S	3	2.463	956	5.12	54
52.	Ethiopia	ET	S	0	0.981	116	15.77	111
53.	Ghana	GH	S	4	2.458	621	13.97	732
54.	Guatemala	GU	S	3	2.354	1197	10.43	28
55.	Jamaica	JM	S	3	3.216	931	3.79	42
56.	Jordan	JO	S	4	3.031	1102	12.91	3
57.	Kenya	KE	S	4	3.395	301	16.65	19
58.	Liberia	LI	S	4	2.141	463	17.77	19
59.	Malawi	MW	S	4	3.114	163	17.46	70
60.	Morocco[a]	MO	S	4	2.291	565	12.84	10
61.	Panama	PN	S	2	2.651	2010	8.33	0
62.	Paraguay	PG	S	1	1.563	1511	6.14	54
63.	Peru[a]	PE	S	1	1.338	1041	23.18	30
64.	Sri Lanka[a]	SR	S	4	3.834	376	14.19	132
65.	Syria	SY	S	2	1.962	1891	10.55	99
66.	Togo	TO	S	2	2.951	224	12.70	1
67.	Tunisia[a]	TN	S	3	2.415	1098	26.18	24
68.	Uganda	UG	S	4	3.418	210	7.53	553
69.	Uruguay[a]	UR	S	3	2.501	1667	12.12	9
70.	Zaire	ZA	S	4	1.883	136	26.74	134
71.	Zambia	ZM	S	3	3.490	383	6.34	110
72.	Zimbabwe	ZI	S	4	3.975	618	25.34	156
Other countries								
73.	Hungary[a]	HU	L	3	2.918	1850	7.06	NA
74.	Mali	ML	S	2	2.011	137	4.85	NA
75.	Mauritius	MT	S	4	2.802	959	17.88	NA
76.	Oman	OM	S	0	0.902	6550	3.47	NA
77.	Sierra Leone	SL	S	4	3.104	300	17.98	NA

Note: [a]Denotes country for which data on factor endowments were assembled.

Chapter 3

Intellectual Property Rights, Licensing and Innovation in an Endogenous Product-Cycle Model*

Guifang Yang and Keith E. Maskus†

1. Introduction

Controversy persists over the role of intellectual property rights (IPRs) in encouraging firms in developed countries to innovate and in helping developing countries gain access to knowledge on the global frontier. An important means of technology acquisition is licensing, which has not been studied formally in this context. In this paper we develop a dynamic general-equilibrium product-cycle model to analyze the effects of Southern IPRs on incentives of Northern firms to innovate and to license state-of-the-art technologies to the South.

Helpman (1993) first modeled the dynamic effects of IPRs in a general-equilibrium framework with an innovative North and an imitative South. He finds that stronger IPRs would diminish both the Northern rate of innovation and Southern welfare when imitation is the only channel of technology transfer. Thus, in the absence of licensing, strengthened IPRs would raise imitation costs, restrict technology diffusion, and reduce long-run incentives to innovate. This pessimistic conclusion was reinforced by Glass and Saggi (1995), who allow both imitation and foreign direct investment (FDI) as channels

*Reprinted from *Journal of International Economics*, Vol. 53, No. 1, pp. 169–187. Copyright 2001, with permission from Elsevier.

†We are grateful to Yongmin Chen, participants at the Midwest International Economic Society Meetings in Evanston, Illinois, and two anonymous referees for their comments. This research was supported in part by NSF Grant 9410625.

of technology transfer. They show that a strengthening of Southern IPRs would reduce rates of innovation and imitation and also would decrease the flow of technology transfer. In a more positive vein, Lai (1998) finds that the effects depend crucially on the channel of technology transfer. Stronger IPRs in the South would expand technology transfer and innovation under FDI but would have opposite effects if production were transferred through imitation.[1]

In this literature imitation and FDI are the primary sources through which the South may gain access to the North's advanced technology. Licensing has been largely ignored. In this paper we focus on licensing as the means by which the South gets advanced technology. Licensing is conceptually and practically important for a number of reasons. First, in some situations, imitation may be sufficiently difficult that it is not profitable (Bagchi *et al.*, 1984). Second, recent empirical evidence suggests that the volume of arm's-length technology transfers is rising (Mansfield, 1995). Furthermore, government policies in some technology-importing countries prefer licensing to equity investment as the mode of technology transfer (Contractor and Sagafi-Nejad, 1981).

In our model the Northern innovative firm first chooses the intensity of effort it devotes to innovation. Once the innovation is successful, it chooses whether to license. The advantage of licensing technology to the South is higher instantaneous profits due to lower wages there. However, licensing incurs costs, including those involved in contract negotiation and in performing various activities to ensure the transfer of the necessary technology know-how. Furthermore, the licensor has to give up some rents to the licensee in order to deter imitation. Therefore, the Northern firm needs to strike a balance between saving labor costs, sacrificing rents, and incurring licensing-transfer costs.

We argue that stronger intellectual property rights in the South both would reduce the imitation risk faced by innovative Northern

[1]In a framework of strategic interaction between firms, Taylor (1994) shows that failure by a country to provide patent protection for foreign inventions forces innovative firms to employ "less than best-practice technologies" and reduces worldwide research and development (R&D) and growth.

firms and would create an improved legal framework for the enforcement of licensing contracts. The former process would reduce the imitation risk from the licensee, allowing the licensor to deter imitation by giving up a smaller share of the licensing rents to the licensee. The latter process would lower the legal costs associated with establishing and policing arm's-length licensing relationships. Put simply, the degree of IPRs protection determines not only the total economic returns from licensing (the size effect) but also the distribution of the returns between two parties (the distribution effect).

Through providing a higher rent share and reducing licensing costs, more rigorous IPRs would alter Northern firms' incentives to license and, in turn, to innovate. Results from the model show that stronger IPRs in the South would generate a higher rate of innovation in the North and a higher rate of licensing to the South. Licensing would be encouraged in order to take advantage of lower labor costs. The intuition for these results is as follows. First, when economic returns to licensing increase due to decreased licensing costs, there would be a direct rise in the economic returns to innovation as well, providing firms more incentive to license and to innovate. Second, stronger IPRs protection in the South would enhance the licensor's share of licensing rents, thereby encouraging licensing and innovation for the same level of licensing costs. Third, since licensing would require fewer resources, more labor would be freed up for innovation in the North. Furthermore, when more production is transferred to the South, additional resources would be freed in the North for innovation.

However, stronger IPRs in the South would have ambiguous effects on the relative wage between regions. If stronger IPRs led to a large magnitude of technology transfer, the Southern relative wage would rise. If a small amount of production moved, the Southern relative wage would fall.

Our model is similar in spirit to product-cycle models in which innovation and technology transfer are the key processes for understanding the spread of production and technological advancement. Its inspiration is Vernon (1966), in which these mechanisms are described but not formally modeled. It is also related to Krugman (1979) and Dollar (1986) in which technical change and technology transfer are exogenous processes. However, we present a quality-ladders model with endogenous innovation and licensing, thereby integrating licensing into the theory of endogenous product cycles (Grossman and Helpman, 1991).

The remainder of this paper is organized as follows. In Section 2 we describe the model and derive the steady-state equilibrium. In Section 3 we study the steady-state comparative statics of IPRs. In Section 4 we decompose the results into size and distribution effects and study the impact on the relative wage. We conclude in Section 5.

2. The Model

2.1. *Consumption*

The consumption side of the model is essentially identical to Grossman and Helpman (1991), so we only highlight its features. Consider an economy with a continuum of goods indexed by $\omega \epsilon [0, 1]$. Each good potentially may be improved a countably infinite number of times, indexed by qualities $j = 0, 1, 2, \ldots$. The increments to quality are common to all products and exogenously given by a parameter $\lambda > 1$. Each good may be supplied in all discovered quality levels.

Consumers live forever and share identical preferences. The intertemporal utility function for the representative consumer is given by

$$U = \int_0^\infty e^{-pt} u(t) dt \qquad (1)$$

where ρ is the subjective discount rate, and $u(t)$ represents instantaneous utility at time t. We specify

$$u(t) = \int_0^1 \ln \left[\sum_{j=0}^{\infty} \lambda^j d_{jt}(\omega) \right] \mathrm{d}\omega \qquad (2)$$

where $djt(\omega)$ denotes consumption of quality j of good ω at time t. Every consumer maximizes discounted utility subject to an intertemporal budget constraint

$$\int_0^\infty e^{-R(t)} E(t) \mathrm{d}t = A(0) \qquad (3)$$

where $R(t) = \int_0^t r(s)ds$ is the cumulative interest factor up to time t and $A(0)$ is the value of initial asset holdings plus the present value of factor income. The term $E(t)$ is the consumer's expenditure flow at time t, given by

$$E(t) = \int_0^1 \sum_{j=0}^{\infty} p_{jt}(\omega)d_{jt}(\omega)d\omega \tag{4}$$

where $p_{jt}(\omega)$ is the price of a product ω of quality j at time t.

The consumer's utility-maximization problem may be broken into two stages. In the first stage the consumer evenly spreads lifetime spending across time. In the second stage the consumer allocates an equal expenditure share to every product ω and chooses the single variety that offers the lowest quality-adjusted price. At any point of time, only the highest quality level available is sold as it offers the lowest quality-adjusted price.

2.2. *Market Structure*

Consider a dynamic world economy of two regions: North and South. North and South are different in their abilities to conduct "state of the art" research and development. The North is more productive in conducting innovation so all innovation takes place in the North in a steady-state equilibrium. In the absence of licensing, the South would not have the technology to produce the top quality level by itself, but it may license technology from the North.

To focus on licensing we make the following assumptions. First, inward FDI is banned or infeasible. Indeed, in some developing countries government policies traditionally have preferred licensing to equity investment as the mode of technology trade. Second, imitation from direct examination of imported goods is costly and is not economically feasible. Imitation by "inventing around" the patent is not economically advantageous, given only the information revealed in patent applications.[2] In short, Southern firms must buy the technology first and then engage in imitation. Therefore, licensing is

[2]Levin *et al.* (1987) find that for most industries patent disclosure does not significantly limit the effectiveness of patents in protecting technology.

the only means by which the South can acquire Northern top-level technologies.

There are many firms in the North. We define "leaders" as those firms that have innovated the current highest quality level of any good and other firms as "followers". Following Grossman and Helpman (1991), we assume that a Northern producer with an existing technology lead would not conduct research to improve the quality of its own product. Therefore, R&D for product improvement would be conducted by followers.

After successful innovation, the innovative firm chooses the location of production. However, in deciding whether to license technology to the South, the Northern licensor has to find a balance between lower labor costs in the South, additional costs incurred in transferring technology, and rents given up to the licensee to deter imitation.

The literature on technological change recognizes that it takes substantial resources to establish and police an arm's-length relationship in order to make transferring a new process or product feasible (e.g., Teece, 1976; Rugman, 1986). These observations suggest that it is costly for a licensor both to design and enforce an acceptable contract. An important problem is that a licensor faces imitation risk when it licenses technology to the South. Either the licensee could imitate directly and defect from its contract or other local firms could imitate the initial licensee's technology by hiring away its skilled workers.[3]

Regarding the first problem, we assume that commitment-license contracts are feasible with close monitoring and that the licensor uses the rent share to deter imitation by the licensee. This means that the licensee commits to payments for all future time even if it imitates the licensor's product and the licensor commits to refund part of the payments should it choose to switch licensees. Therefore, it is in the licensor's interest to commit to an exclusive licensing arrangement in order to avoid incurring transfer costs to another licensee and to

[3]Imitation could happen under FDI as well. Markusen (2001) develops a strategic model in which local managers learn a multinational firm's technology and may choose to organize rival production.

maintain plant-level scale economies. However, in order to induce the licensee to commit not to imitate, the licensor must share some rents. Under commitment contracts, licensees have no incentive to imitate the licensor's product because the expected gain from imitation is the same as that under the contract.

Rent sharing is commonly observed in licensing contracts. Caves *et al.* (1983) indicate that, on average, licensors earn only 40% of the rents from innovation. The division of rents between the licensor and licensee depends on many factors, such as each party's bargaining power, the degree of market competition, and government intervention (Contractor and Sagafi-Nejad, 1981). One important factor is that license rents are used to deter imitation. Gallini and Wright (1990) show that when imitation is possible and there is asymmetric information, the licensor would sacrifice rents though its share would rise with imitation costs. Thus, stronger IPRs should be associated with higher licensor rent shares.

For our purposes we consider solely the effect of IPRs on the rent split. Stronger Southern protection would render it harder for the licensee to imitate the licensor's product. Thus, the licensee would commit not to imitate at a lower rent share. It is reasonable to assume that the licensor's rent share is a positive function of the effectiveness of Southern IPRs.[4] Letting δ be the licensor's share, we stipulate that $0 < \delta < 1$, $\delta = \delta(\kappa)$, and $\partial\delta/\partial\kappa > 0$, where κ is the degree of strength in the Southern IPRs regime.

Furthermore, stronger IPRs would reduce the monitoring costs imposed on licensors, raising the return to transferring technology. Put differently, tightened property rights would reduce the imitation risk faced by the Northern firm by improving the legal framework

[4]The model would be more satisfactory if the rent share were endogenized. However, endogenization of the rent share would require designing strategic licensing contacts between the licensor and the licensee in which both of them face uncertainty about the stream of licensing rents. In Gallini and Wright (1990), for example, imitation risk and asymmetric information about quality levels of licensed technologies force licensors to share rents with licensees as a signal of true quality. In their approach, the licensor's rent share is an increasing function of imitation costs. In our case, because such costs would rise with stronger Southern intellectual property rights, the licensor's rent share would increase with stronger IPRs.

for the enforcement of licensing and diminishing the legal costs associated with establishing and policing an arm's-length relationship. Thus, we model the cost of licensing as a negative function of the degree of Southern protection.

Regarding imitation by Southern followers, we follow Helpman (1993) and Segerstrom (1991) in positing Bertrand competition between licensees and potential imitators. Both firm types operate in the South and have the same marginal costs. If a follower firm were to imitate the licensee's product successfully, the licensee would choose to price at the imitator's marginal cost and neither firm would earn positive profits. This means that after the licensee buys the technology other firms would have no incentive to imitate because of positive imitation costs. Therefore, there is no imitation from Southern followers.

In short, neither contract defection nor imitation occurs in equilibrium in our model but the need to deter such activities affects both costs and rent shares. A tightening of Southern IPRs would raise payoffs to licensing for both reasons.

Returning to the model, there are two possible market types for each good. In the first type of market (S) the highest-quality goods are produced in the South through licensing. We call this the licensed South technology market. In the second type of market (N) the highest-technology goods are produced in the North. We label this case the original North technology market. Innovation targets both markets. When innovation targeting the S market is successful, the market would become an N and the newly innovated highest quality would be produced in the North. When innovation targeting the N market is successful, it would remain an N. Production in the North would continue until a successful licensing contract were concluded and output shifted to the South.

In addition to licensing costs there may be technical reasons why production of highest-quality goods remains in the North for a period of time before transfer to the South. For example, Vernon (1966) suggests that production is likely to remain in innovative countries during the early standardization period of a product life cycle. Moreover, some time period may be needed for adapting the technology to the Southern economic environment. However, if all licensing decisions were due to these technical factors, the decision to license would be exogenous. In our model the licensing decision is endogenous and a random process, as discussed in the next sub-section. There is a positive probability that the new technology would be licensed to the South in the next instant. Successful licensing of a high-quality product would shift its production to the South and the N market would become an S.

2.3. *Innovation and Licensing*

Firms spend resources to innovate "state-of-the-art" products. Following Grossman and Helpman (1991), we assume that individual research success is a continuous Poisson process. The probability of success during any time interval does not depend upon the resources that have been spent in previous unsuccessful periods. Thus, this probability is proportional to the intensity of effort during the current interval. A firm that engages in innovation at intensity I for an interval of time length dt succeeds with probability $I dt$. This effort requires $a_I \cdot I$ units of labor per unit of time. The variable I is the Poisson arrival rate at which "state-of-the-art" technology will be innovated in the next instant and is endogenous here.

After a firm succeeds in innovating a technology for producing a higher quality product, it optimally chooses how much effort to devote to licensing its production technique to the South. Licensing incurs costs and entails uncertainty for the individual enterprises. A successful technology transfer means that the transferee plant achieves the proper design and performance specifications. Before that point, considerable resource costs are typically incurred. It is reasonable to model technology transfer in a fashion similar to that in modeling innovation because the costs and processes involved are not much different. In his case study of 26 investment projects involving international technology transfer, Teece (1976) finds that the total transfer costs ranged from two percent to 59 percent of the licensee's total project costs. Further, in the case of licensing, protracted bargaining periods could pertain.

Because of these costs, we model successful licensing as a random process. Assume the duration τ between the time of innovation and the time of licensing has an exponential distribution with cumulative density $Pr(\tau \leq t) = 1 - e^{-\iota t}$. Thus, $\iota \Delta t$ gives the probability that licensing takes place in the time interval $(t, t + \Delta t)$, given that the product is produced in the North after innovation up to time t. The variable ι is the Poisson arrival rate at which the high-quality technology will be licensed to the South in the next instant. This licensing rate is endogenously determined in steady-state equilibrium. A licensee that engages in licensing at intensity ι for an interval of time length dt succeeds in adopting and absorbing the technology with probability ιdt.

To incorporate IPRs, we take licensing to be costly and assume that it requires $(a_L \iota / \kappa)$ units of labor per unit of time. Recall that the variable κ represents the degree of IPRs protection. A higher κ means that firms must spend fewer resources for a given licensing intensity to enforce the licensing contract. Furthermore, the licensor's rent share is a positive function of κ.

With this background, first consider the Northern firm's effort to innovate. Successful innovators attain a market value of V_I (I stands for "innovating"). Each firm may achieve an expected gain of $V_I I$dt at cost $wa_I I$dt, by undertaking R&D at intensity I for an interval dt. The Northern wage rate is w. By the zero-profit condition due to free entry and exit in innovation, we have

$$V_I = wa_I, \quad I > 0. \tag{5}$$

Next, consider the decision of Northern leading firms to license. Successful licensors among those who undertake such efforts attain a stock market value of $\delta(\kappa)V_L$ (L stands for "licensing"). Each such firm may achieve an expected gain of $\delta(\kappa)V_L \iota$dt, at cost $V_I \iota$d$t + (wa_L/\kappa)\iota$dt by undertaking licensing-related R&D at intensity ι for an interval dt. The term $V_I \iota$dt reflects the opportunity cost of licensing by giving up production in the North. Here, we normalize the wage rate in the South to be 1. By similar logic, the costs of adaptation must equal expected rewards in order for the licensing intensity to be finite and positive. Therefore we have

$$\delta(\kappa)V_L = V_I + \frac{wa_L}{\kappa}. \tag{6}$$

We have defined V_I and V_L as firm values, which in steady-state must be constant. The value of a firm must equal its present value of lifetime profits. Firms in the North technology market face the risk of innovation and the risk of the highest-quality technology being licensed to the South. When either of these factors happen, firm values would change. For the leading firm in the N market, V_I is given by

$$V_I = \frac{\pi_N + \iota\delta(\kappa)V_L}{\rho + I + \iota}. \tag{7}$$

Here π_N is the instantaneous profit that the leading firm gets when the follower's innovation fails and its own licensing does not occur. If rival innovation were to happen the leading firm in the N market would be driven out of the market. If licensing happened the firm would become the leading firm in the S market and would earn $\delta(\kappa)V_L$.

Next, consider the incentive of a firm to license a technology. A licensee that succeeds in adopting and absorbing a state-of-the-art technology from the North earns the following reward in steady state:

$$V_L = \frac{\pi_L}{\rho + I}. \tag{8}$$

Here π_L is the instantaneous profit that the licensee achieves when the Northern follower's innovation does not occur. If innovation happened, the leading firm in the S market would be driven out of the market. The licensor would share the rents with the licensee and would earn $\delta(\kappa)V_L$.

Now turn to the instantaneous profits earned by each kind of firm. For the leading firm in the N market, its closest competitor is the Southern firm that can produce the second-level quality product. Thus, the firm sets a quality-adjusted price equal to the marginal cost of production of the competitor. We assume that production exhibits constant returns to scale, which means that one unit of output requires one unit of labor. Thus, $P_N/\lambda = 1$ and $P_N = \lambda$. The firm captures the entire market and makes sales of E/λ where E

represents aggregate spending. The firm earns instantaneous profits:

$$\pi_N = E\left(1 - \frac{w}{\lambda}\right). \tag{9}$$

For firms with licenses in the S market, the quality of their products is still at a one-level lead. Licensees price against those Southern firms that are able to produce the second-level products and therefore set $P_L = \lambda$ as well. Their sales will be E/λ and each licensee earns instantaneous profits:

$$\pi_L = E\left(1 - \frac{1}{\lambda}\right). \tag{10}$$

2.4. *Resource-Market Clearance*

In equilibrium, all resources are fully used for production, innovation, and licensing in both the North and South. We only have one input, labor. Let the labor supply in the North be L_N and that in the South be L_S, where both are exogenously given. We denote the measure of the North technology market as n_N and the measure of the South licensing market as n_S. The labor-market clearance condition in the North is as follows:

$$\frac{E}{\lambda}n_N + Ia_I n_N + Ia_I n_S + \iota\left(\frac{a_L}{\kappa}\right) n_N = L_N. \tag{11}$$

The first term represents labor resources used in production, the second term and the third term represent labor used in innovation, and the fourth term represents labor used in licensing. The labor market clearance condition for the South is given by:

$$\frac{E}{\lambda}n_S = L_S. \tag{12}$$

The left-hand side is labor used in Southern production.

2.5. *Constant Steady-State Market Shares*

In steady state, measures of products produced in the N and S markets are constant. Recall that N denotes the proportion of industries with high-quality products produced in the North and S denotes the proportion of industries with high-quality products licensed to the

South. Thus, the flow of production out of the N market must be the same as that into the N market, and the flow of production out of the S market must be the same as that into the S market, as indicated in Eq. (13):

$$In_S = \iota n_N. \tag{13}$$

The flow out of the N market is $In_N\mathrm{d}t + \iota n_N \mathrm{d}t$ for an interval $\mathrm{d}t$ and the flow into the N market is $In_N\mathrm{d}t + In_S\mathrm{d}t$. Further,

$$n_N + n_S = 1. \tag{14}$$

This completes the description of the steady-state equilibrium.

3. Steady-State Equilibrium and the Effects of IPRs

3.1. *Steady-State Equilibrium*

From Eqs. (13) and (14), we have

$$n_N = 1 - n_S \quad \text{and} \quad \iota = \frac{In_S}{1 - n_S}. \tag{15}$$

Recall that n_S is the market measure of the proportion of industries which license their technology to the South and therefore we can interpret it as the extent of licensing. The remaining task is to solve for four endogenous variables (I, n_S, E, and w) in terms of the exogenous variables.

Combining Eqs. (11), (12), and (15) and eliminating E, we derive an equation for the joint resource constraint in the North and South.

$$\frac{(1 - n_S)}{n_S} L_S + \frac{a_L}{\kappa} In_S + Ia_I = L_N. \tag{16}$$

Equation (16) gives the relationship between I and n_S when resource constraints in both the North and the South are satisfied. We wish to graph this joint resource constraint in terms of innovation I versus the extent of high-quality licensing n_S. Taking total derivatives of Eq. (16) shows that

$$\frac{\mathrm{d}n_S}{\mathrm{d}I} > 0 \quad \text{and} \quad \frac{\mathrm{d}^2 n_S}{\mathrm{d}I^2} > 0 \quad \text{if} \ \frac{L_S}{n_S} > \frac{a_L I n_S}{\kappa}.$$

Since $n_S < 1$, we may conclude that $L_S/n_S > L_S$. The term $a_L I n_S/\kappa$ is the labor used in the North for transferring technology. It is reasonable to assume that the labor force in the South is larger than the labor required in licensing technology from the North to South. Therefore, I and n_S are positively related, meaning that along this constraint as licensing intensity increases so does innovation effort. The intuition behind this positive relationship is straightforward. If n_S were to increase more production would be transferred to the South, making more resources in the North available for innovation.

From Eqs. (5) through (10) we derive the following expressions:

$$E\left(1-\frac{w}{\lambda}\right) = w\left[a_I(\rho+I) - \frac{a_L}{\kappa}\left(\frac{I n_S}{1-n_S}\right)\right] \tag{17}$$

and

$$\delta(\kappa)E\left(1-\frac{1}{\lambda}\right) = w\left(a_1 + \frac{a_L}{\kappa}\right)(\rho+I). \tag{18}$$

Equation (17) gives the zero-profit condition in innovation and Eq. (18) gives the zero-profit condition in licensing. Combining Eqs. (17) and (18) and using Eqs. (12) and (15) to eliminate E and w, we next derive a joint valuation equation:

$$\left(a_I + \frac{a_L}{\kappa}\right)(\rho+I) = \delta(\kappa)\left(1-\frac{1}{\lambda}\right)\left[a_I(\rho+I) - \frac{a_L I}{\kappa}\frac{n_S}{1-n_S} + \frac{L_S}{n_S}\right]. \tag{19}$$

Equation (19) gives the relationship between I and n_S when the zero-profit conditions in both licensing and innovation are satisfied. Taking total derivatives of this equation shows that $\mathrm{d}n_S/\mathrm{d}I < 0$. Therefore the rate of innovation and the extent of licensing are negatively related. The intuition is as follows. Suppose that innovation effort by Northern follower firms rises. This would cause the current leading firm to face a higher risk of losing its market no matter whether its production takes place in the North or in the South through licensing. Therefore the leading firm would earn its instantaneous profits for a shorter period. Licensing incurs costs, entails uncertainty, and requires rent sharing. Thus, it would become less

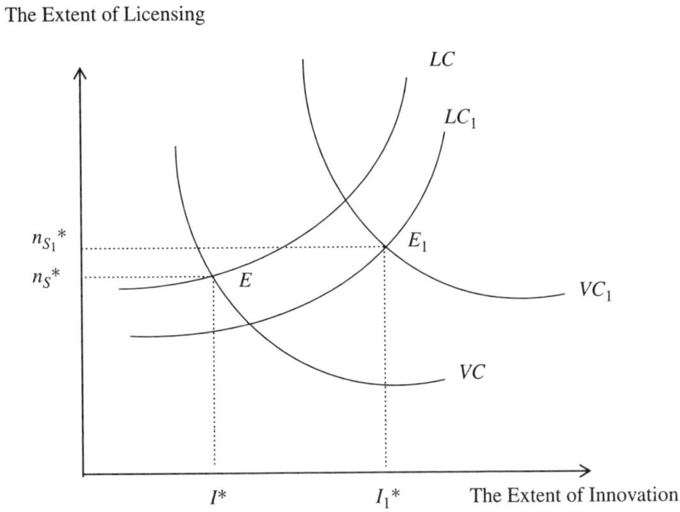

Figure 1. Licensing and innovation in general equilibrium.

attractive for the leading firm to transfer technology to the South if it knew the life cycle of its product were shortened, providing it with less time and smaller returns to cover its transfer costs.

We depict the combinations of I and n_S that satisfy the joint resource constraint (Eq. (16)) by the curve labeled LC in Fig. 1. It is positively sloped and convex. The curve labeled VC in Fig. 1 shows the combinations of I and n_S that satisfy the joint valuation Eq. (19). It is negatively sloped. The intersection of LC and VC, at point E, gives the steady-state equilibrium rates of innovation and licensing.

3.2. *Steady-State Comparative Statics*

Consider first the effects of tightened IPRs on the innovation effort and the extent of licensing. Stronger IPRs in the South increase κ, causing the LC curve to shift. Holding n_S constant in Eq. (16) and taking the derivative of I with respect to κ shows that $\mathrm{d}I/\mathrm{d}\kappa > 0$. Therefore, when stronger IPRs protection is adopted in the South, the LC curve would shift to the right. When κ increases, licensing costs would decrease and labor resources used in transferring technology would be reduced in the North, making more labor available for innovation.

Stronger IPRs protection in the South also would shift up the *VC* curve. Holding n_S constant in Eq. (19) and taking the derivative of I with respect to κ shows that $\mathrm{d}I/\mathrm{d}\kappa > 0$. When the cost of transferring technology decreases, satisfaction of the zero-profit condition in licensing would require firms to earn the instantaneous rents for shorter periods. In turn, this would require more innovation from the followers in general equilibrium. Furthermore, tighter IPRs in the South would afford a higher rent share to the licensor, which would expand the incentives both to license and to innovate. As a result, the *VC* curve would shift to the right.

Accordingly, in Fig. 1 when stronger IPRs are implemented in the South both the *LC* curve and the *VC* curve would shift to the right. The new equilibrium is at point E_1. Thus, stronger IPRs would lead unambiguously to a higher rate of innovation. However, the effect on the extent of licensing is unclear and would depend on which curve shifts further. In the Appendix we show that the *VC* curve would shift further in response to a higher κ. As a result, the extent of licensing would also increase. We therefore can state:

Proposition 1. *If IPRs are strengthened in the South, both the rate of innovation in the North and the extent of licensing to the South would increase.*

The intuition behind Proposition 1 is as follows. Stronger IPRs not only would create a better legal framework for the enforcement of licensing contracts, reducing the costs associated with establishing and policing an arm's-length relationship, but also would raise the rent share the licensor receives. Therefore, the degree of IPRs protection not only would determine the size of the pie (the size effect) but also the distribution of the pie between the two licensing parties (the distribution effect). Firms would be encouraged to license technology to the South in order to take advantage of lower labor costs there and earn higher instantaneous economic returns.

Thus, several impacts are relevant here. First, when economic returns to licensing increase due to lower licensing costs, there would be a direct increase in the economic returns to innovation as well, providing firms more incentives both to license and to innovate. Second, a strengthening of IPRs protection in the South would increase the licensor's rent share, thereby encouraging more licensing and innovation for the same level of licensing costs. Third, since licensing would require fewer resources with stronger intellectual property protection, more labor would be made available for innovation in the North. Furthermore, when additional production is transferred to the South, more resources would be freed in the North for innovation.

4. Special Cases and the Relative Wage Effect

4.1. *Isolation of the Size Effect and the Distribution Effect*

In this section, we isolate the size effect and the distribution effect in order to gain a better understanding of the basic forces in the model. First, consider the case with the size effect only. That is, the rent share does not depend on κ and we assume $\delta = 1$ without loss of generality. It is straightforward to show that both the LC and VC curves would shift to the right, but that the VC curve would shift farther, in response to tighter IPRs protection in the South. Therefore, with stronger protection both licensing and innovation would expand through higher returns to licensing and innovation and greater availability of effective resources.

Second, consider the case with the distribution effect only ($\kappa = 1$). In this case, the joint resource constraint does not depend on κ and the LC curve would not shift in response to tighter IPRs. However, the VC curve would shift to the right, which would induce both licensing and innovation to increase in the new equilibrium. Although there would be no direct effect on resources available for other activities, stronger IPRs would give the licensor a larger rent share and encourage more licensing and more innovation. Thus, through freeing resources formerly used in Northern production, the policy change would indirectly increase the resources available for innovation.

Therefore, in this model the size effect and the distribution effect would enhance each other in encouraging more licensing and more innovation in response to stronger Southern protection of intellectual property.

4.2. *The Relative Wage Effect*

We next study impacts on the relative wage between North and South. For simplicity, we use the case with the size effect only ($\delta = 1$) to illustrate the results without loss of generality.[5] From Eqs. (5), (6), (8), and (10), we have:

$$w = \frac{E\left(1 - \frac{1}{\lambda}\right)}{(a_I + \frac{a_L}{\kappa})(\rho + I)}. \tag{20}$$

From Eqs. (16) and (12), we can derive I as a function of n_S:

$$I = \frac{L_N + L_S - \frac{L_S}{n_S}}{\left(a_I + \frac{a_L n_S}{\kappa}\right)}. \tag{21}$$

Substituting E from Eq. (12) and I from Eq. (21) into Eq. (20), we can derive an equation for w as a function of n_S:

$$w = \frac{\frac{\lambda L_S}{n_S}\left(1 - \frac{1}{\lambda}\right)}{\left(a_I + \frac{a_L}{\kappa}\right)\left(\rho + \frac{L_N + \left(L_S - \frac{L_S}{n_S}\right)}{\left(a_I + \frac{a_L n_S}{\kappa}\right)}\right)}. \tag{22}$$

Taking the total derivative of Eq. (22), we can write

$$\frac{\mathrm{d}w}{\mathrm{d}\kappa} = C - D\frac{\mathrm{d}n_S}{\mathrm{d}\kappa} \tag{23}$$

where $C > 0$, $D > 0$, and $\mathrm{d}n_S/\mathrm{d}\kappa > 0$ (see the Appendix). Therefore, the sign of $\mathrm{d}w/\mathrm{d}\kappa$ could be positive or negative. If stronger IPRs caused a large increase in the extent of licensing ($\mathrm{d}n_S/\mathrm{d}\kappa$ is large), then the sign of $\mathrm{d}w/\mathrm{d}\kappa$ could be negative. However, if stronger IPRs induced a small increase in the extent of licensing ($\mathrm{d}n_S/\mathrm{d}\kappa$ is small), then the sign of $\mathrm{d}w/\mathrm{d}\kappa$ could be positive. We therefore have:

Proposition 2. *The effect of stronger IPRs on the Southern relative wage* ($1/w$) *is ambiguous. If stronger IPRs were to induce a large*

[5] The proof for the general case will be provided upon request.

increase in the extent of licensing, the Southern relative wage could increase. However, if stronger IPRs were to lead to a small increase in the extent of licensing, the Southern relative wage could decrease.

The intuition underlying Proposition 2 is straightforward. When Southern IPRs are strengthened both the rate of innovation and the extent of licensing would rise. These factors would engender opposing effects on the relative Southern wage. In the South, when the extent of licensing increases the demand for Southern labor would rise and the Southern wage would go up unambiguously. However, the Northern wage could increase or decrease. On the one hand, stronger IPRs would reduce labor demand in both licensing and production. On the other hand, both higher returns and a larger rent share would encourage innovation, which would increase the demand for labor.

Thus, if the Southern wage increases and the Northern wage decreases, the Southern relative wage would rise. However, if both the Southern and Northern wages rise, the Southern relative wage could increase or decrease depending on the increased extent of licensing to the South due to the policy change. If stronger IPRs caused a sufficiently large amount of licensing to the South, the Southern relative wage would go up.

5. Conclusion

In this paper we develop a simple dynamic general-equilibrium version of the product-cycle model to study the effects of stronger IPRs in the South on the incentives of firms to innovate and license new technologies. The cost of technology transfer depends on the degree of IPRs protection in the South. By reducing licensing costs, stronger IPRs would lead to higher economic returns to the licensor and would liberate resources for innovation, thereby leading to both higher innovation and licensing. Furthermore, stronger IPRs protection in the South would raise the rent share of the licensor and therefore would encourage additional licensing and innovation. The net effect of stronger IPRs on the relative wage would be ambiguous because innovation and licensing have potentially opposing impacts on demand for labor in North and South.

Our results are more optimistic about the possible impacts of stronger intellectual property rights in developing countries than are those in Helpman (1993) and Glass and Saggi (1995). Helpman shows that in the long run the South could suffer from stronger IPRs because of higher imitation costs and the consequent reduction in technology availability. Glass and Saggi similarly show that more protective IPRs could decrease the flow of technology transferred to the South. Although our model is not directly comparable with theirs, we show that tighter IPRs could actually increase the rate of innovation and licensing. This result stems from the plausible notion that stronger property rights would reduce the costs of making and enforcing licensing contracts and would increase the rent share of the licensor. In turn, the returns to both licensing and innovation would rise, while additional labor would be made available to effect these processes. The Southern relative wage could rise or fall.

We believe that these outcomes are more consistent with the prevailing notion among policy makers that inward technology transfer would rise in nations that adopt stronger IPRs. They also help explain survey findings demonstrating that licensing responds positively to patent rights in developing countries (Mansfield, 1994). Our results bolster Lai's (1998) finding that tighter IPRs would increase innovation and technology transfer. However, his paper employs a product-variety model with FDI as the channel through which transfers are made. Our endogenous product cycle model introduces licensing as the transfer mechanism.

References

Bagchi, A. K., Banerjee, P., and Bhattacharya, U. K., 1984. Indian patents act and its relation to technological development in India: a preliminary investigation. *Economic and Political Weekly*, 287–304.

Caves, R. E., Crookell, H., and Killing, J. P., 1983. The imperfect market for technology licenses. *Oxford Bulletin of Economics and Statistics* 45, 249–267.

Contractor, F. J. and Sagafi-Nejad, T., 1981. International technology transfer: major issues and policy responses. *Journal of International Business Studies* 12, 113–135.

Dollar, D., 1986. Technological innovation, capital mobility, and the product cycle in North-South trade. *American Economic Review* 76, 177–190.

Gallini, N. T. and Wright, B. D., 1990. Technology transfer under asymmetric information. *RAND Journal of Economics* 21, 147–160.

Glass, A. J. and Saggi, K., 1995. Intellectual property rights, foreign direct investment, and innovation. Working Paper 95-06, Ohio State University, Columbus, Ohio.

Grossman, G. M. and Helpman, E., 1991. Quality ladders and product cycles. *Quarterly Journal of Economics* 106, 557–586.

Helpman, E., 1993. Innovation, imitation, and intellectual property rights. *Econometrica* 61, 1247–1280.

Krugman, P., 1979. A model of innovation, technology transfer, and the world distribution of income. *Journal of Political Economy* 87, 253–266.

Lai, E. L. C., 1998. International intellectual property rights protection and the rate of product innovation. *Journal of Development Economics* 55, 115–130.

Levin, R., Klevorick, A., Nelson, R., and Winter, S., 1987. Appropriating the returns from industrial research and development. *Brookings Papers on Economic Activity* 20, 783–820.

Mansfield, E., 1994. Intellectual property protection, foreign direct investment, and technology transfer. Discussion Paper 19, International Finance Corporation, Washington DC.

Mansfield, E., 1995. *Innovation, Technology and the Economy: The Selected Essays of Edwin Mansfield*, volume II (London: Edward Elgar).

Markusen, J. R., 2001. Contracts, intellectual property rights, and multinational investment in developing countries. *Journal of International Economics* 53, 189–204.

Rugman, A. M., 1986. New theories of the multinational enterprise: an assessment of internalization theory. *Bulletin of Economic Research* 38, 101–118.

Segerstrom, P. S., 1991. Innovation, imitation, and economic growth. *Journal of Political Economy* 99, 807–827.

Taylor, M. S., 1994. Trips, trade, and technology transfer. *International Economic Review* 35, 361–381.

Teece, D. J., 1976. *The Multinational Corporation and the Resource Cost of International Technology Transfer* (Ballinger Publishing, Cambridge, MA).

Vernon, R., 1966. International investment and international trade in the product cycle. *Quarterly Journal of Economics* 80, 190–207.

Appendix

Result 1: The VC curve would shift further than the LC curve in response to a higher κ.

Holding n_S constant and taking the derivative of I with respect to κ in Eq. (16), we have

$$\frac{\mathrm{d}I}{\mathrm{d}\kappa} = \frac{\frac{a_L}{\kappa^2} I n_S}{\left(a_I + \frac{a_L n_S}{\kappa}\right)} > 0. \tag{A.1}$$

Holding n_S constant and taking the derivative of I with respect to κ in Eq. (19), we have

$$\frac{\mathrm{d}I}{\mathrm{d}\kappa} = \frac{(\rho+I)\left[\frac{a_L}{\kappa^2} + \frac{\delta'(\kappa)}{\delta(\kappa)}\left(a_I + \frac{a_L}{\kappa}\right)\right] + \delta(\kappa)\left(1-\frac{1}{\lambda}\right)\frac{n_S}{1-n_S}\frac{a_L I}{\kappa^2}}{a_I\left[1-\delta(\kappa)\left(1-\frac{1}{\lambda}\right)\right] + \frac{a_L}{\kappa}\left(1+\delta(\kappa)\left(1-\frac{1}{\lambda}\right)\frac{n_S}{(1-n_S)}\right)} > 0. \tag{A.2}$$

It is straightforward to show that the right-hand side of Eq. (A.2) is greater than the right hand side of Eq. (A.1). Therefore, in Fig. 1, the *VC* curve shifts further to the right than the *LC* curve.

Result 2: The sign of $\mathrm{d}w/\mathrm{d}\kappa$ *is ambiguous.*

Taking the derivative of w with respect to κ in Eq. (22), we get

$$\frac{\mathrm{d}w}{\mathrm{d}\kappa} = C - D\frac{\mathrm{d}n_S}{\mathrm{d}\kappa} \tag{A.3}$$

where

$$C = \frac{a_L}{\kappa^2}L_S(\lambda-1)\left[\frac{(\rho+I)}{n_S} - \left(a_I + \frac{a_L}{\kappa}\right)\frac{\left(L_N + L_S - \frac{L_S}{n_S}\right)}{\left(a_I + \frac{a_L n_S}{\kappa}\right)^2}\right]$$
$$\times\left[\frac{1}{\left(a_I + \frac{a_L}{\kappa}\right)^2(\rho+1)^2}\right]$$

and

$$D = \frac{L_S}{n_S}(\lambda-1)\left(a_I + \frac{a_L}{\kappa}\right)$$
$$\times\left[\frac{(\rho+I)}{n_S} - \frac{\left(L_N + L_S - \frac{L_S}{n_S}\right)\frac{a_L}{\kappa} - \frac{L_S}{n_S^2}\left(a_I + \frac{a_L n_S}{\kappa}\right)}{\left(a_I + \frac{a_L n_S}{\kappa}\right)^2}\right]$$
$$\times\left[\frac{1}{\left(a_I + \frac{a_L}{\kappa}\right)^2(\rho+I)^2}\right].$$

Substituting I in Eq. (21) into terms C and D, it is readily shown that $D > 0$ and $C > 0$. Therefore the sign of $\mathrm{d}w/\mathrm{d}\kappa$ is indeterminate and depends on the magnitude of $\mathrm{d}n_S/\mathrm{d}\kappa$.

Chapter 4

Vertical Distribution, Parallel Trade and Price Divergence in Integrated Markets*

Mattias Ganslandt and Keith E. Maskus†

1. Introduction

The European Union has adopted a rigorous regime of regional exhaustion, which states that parallel trade — i.e. cross-border arbitrage — is impermissible from outside the community but the first sale within its territory exhausts the rights of the original manufacturing firm to restrict further trade of its product in the common market.[1] Both the European Commission and the European Court of Justice have consistently defended the presumed importance of parallel trade in support of the single market (Ganslandt and Maskus, 2003). The essential justification for this policy is a belief that parallel imports generate competition at the retail level, inducing a tendency

*Reprinted from *European Economic Review*, Vol. 51, No. 4, pp. 943–970. Copyright 2007, with permission from Elsevier.
†We thank the Swedish Competition Authority and the Jan Wallander and Tom Hedelius Foundation for financial support. We are also grateful for comments from Sofronis Clerides, Wilfred Ethier, Jim Markusen, Lars Persson and two referees. We also thank seminar participants at Lund University, University of Kiel, IUI, NOITS 2004, the CEPR International Research and Policy Symposium, and the 2005 spring Midwest International Economics meeting.
[1]The policy is essentially equivalent to the U.S. "first-sale doctrine" under which initial sale of a good inside the territorial United States exhausts further rights to control distribution. Thus, our analysis of PI and exclusive territories applies also to "gray-market" unauthorized trade within the United States.

toward retail price convergence and pro-competitive gains from price integration.[2]

Despite these long-standing efforts to integrate markets, there remains considerable retail price divergence within the European Union, which continues to puzzle observers. For example, the European Central Bank (2002, p. 39) recently stated that, "The data available suggest that price level dispersion for many tradeable goods and services remains higher between euro area countries than within individual countries, implying that further improvements in the functioning of the Internal Market are possible."

One explanation for this failure of prices to move together could be that imperfectly competitive firms may be capable of sustaining divergent retail prices in reaction to reductions in trade barriers. In this paper we set out such a model and explore fully the price impacts of PI in a setting where an original manufacturer — i.e. the holder of the intellectual property rights — sells its product through independent distributors in two national markets.[3] Our point of departure is the observation that the bulk of PI actually takes place at the wholesale level, rather than the retail level (Maskus and Chen, 2002; National Economic Research Associates [NERA], 1999).[4] Because

---

[2]There exists evidence that this is indeed relevant in some markets. For instance, Ganslandt and Maskus (2004) present empirical evidence that competition from European PI firms in the pharmaceutical sector has a significantly negative effect on original manufacturers' prices in Sweden.

[3]Friberg and Martensen (2001) offer a model with similar conclusions but in a very different analytical context.

[4]The study by National Economic Research Associates (1999) reported survey evidence of significant flows of parallel trade within the European Union in the early 1990s. While the report tended to focus on retail price differences, it pointed out that the bulk of parallel trade happens at the wholesale or distributor level. Maskus and Chen (2002) performed econometric analysis with detailed U.S. export-price data and discovered that the international distribution of wholesale export prices did follow the predicted U-shaped relationship in U.S. tariff rates and that the level of such prices across destination regions depended on the legal treatment of parallel trade in those regions. Thus, empirical evidence points to the vertical-control problem as being central to PI. This line of inquiry was pursued extensively within the EU context by Ganslandt and Maskus (2003). Their empirical evidence supported the view that there are multiple causes for PI. Econometric analysis of European prices suggested that both horizontal arbitrage and vertical-control problems are important practical explanations for such trade. Pricing behavior by exporters from high-price markets — such as Denmark and the United Kingdom — indicated that such firms increase export prices in countries that are in close proximity (and therefore have low trade costs) in an attempt to deter PI.

manufacturing firms place goods on the market initially through vertical contracts with local distributors, it is important to study the implications of such contracts for PI volumes and the consequent scope for retail price integration. In this context, EU competition policy also rigorously supports parallel trade at the distributor level as a means of reducing market power conveyed by exclusive territories.

In fact, we find that the claim that parallel imports unambiguously bring down retail prices in expensive locations is misleading. We show that in some market circumstances, for an important and empirically relevant range of trade costs, the existence of parallel imports can cause retail prices both to diverge between markets and to increase in high-cost locations, precisely opposite to the conventional intuition.[5] The essential reason is that original manufacturing firms that own intellectual property rights undertake vertical pricing decisions to manage the threat and costs of PI, including consolidation of the number of distributors as trade costs fall. Thus, a reorientation of competition policy within the EU toward understanding vertical restraints seems in order.

In addition, we provide a welfare analysis of permitting parallel trade and compare it with market segmentation based on, for instance, exclusive territories or legal restraints.[6] The analysis advanced here builds on the insights in a paper by Chen and Maskus (2005) in several important ways. They noted three essential efficiency tradeoffs in a model where a licensed distributor in one market can trade goods back to the home market of the original manufacturer. The manufacturer has to balance the losses from a pro-competitive price effect of PI into its own market, on the one hand,

[5] Arbitrage against differences in retail prices was the focus of early literature (Tarr, 1985; Hilke, 1988), which found that rapid and large dollar appreciation generated rising volumes of parallel imports into the United States.

[6] The welfare effects of price-integrating horizontal arbitrage have been studied in several papers. In an important theoretical paper, Malueg and Schwartz (1994) argued that a regime of uniform retail pricing would be globally inferior to one in which firms could price-discriminate on the basis of countries grouped by demand elasticity. Also arguing from a standard pricing model, Richardson (2002) argued that neither a global policy of uniform pricing (international exhaustion) nor of full segmentation (national exhaustion) could be supported as a Nash equilibrium, suggesting that negotiations to achieve a consistent global exhaustion regime at the World Trading Organization would be frustrated.

and the resource costs wasted in the activity of parallel trade and the double-markup problem in inducing a profit-maximizing retail price in the foreign market, on the other. The authors discovered a U-shaped welfare curve in the cost of trading PI goods. If trade costs are low, the optimal regulatory policy is to permit free parallel trade, for the pro-competitive gains dominate the other efficiency losses. However, at intermediate-to-high trade costs it is efficient to ban PI to avoid inefficient vertical pricing.

Our model is different and more general than that in Maskus and Chen (2002) and Chen and Maskus (2005) as we permit vertical distributors in two markets, an approach that generates a richer menu of potential price effects even within the framework of Cournot competition. With two independent distributors, permission of parallel trade may harm consumers due to the possibility that such trade could result in retail price divergence as trade costs fall. Consequently, consumers may gain if policy permitted exclusive territories. Maskus and Chen (2002) found a similar result but only for high PI costs.[7]

We believe that there is substantial scope for understanding better the implications of parallel trade by focusing on models of vertical pricing behavior in markets integrated at the wholesale level. It is conceivable that wholesale prices may be set in a way that offsets or even counteracts the anticipated impacts of an open PI regime. This is the point of departure for the analysis that follows.

The paper is organized as follows. In the next section we set out a basic model of parallel imports in two vertically controlled markets that are linked by the possibility of PI. The following section provides an analysis of three different cases that emerge as equilibria for different ranges of trade costs. In the fourth section we take a closer look at the equilibrium retail prices, profits and consumer welfare. In the fifth section we separately analyze the implications of introducing two important modifications of the model: retail arbitrage and product differentiation. We offer concluding remarks in the final section.

[7]In two related papers, Raff and Schmidt (2005a, 2005b) analyzed the implications of exclusive territories used by firms in international trade. They concluded that trade liberalization may lead manufacturers to offset the loss of tariff barriers with contracts imposing exclusive territories, which may decrease competition and welfare.

2. A Theoretical Model of Parallel Imports

We develop a model of vertical pricing with parallel imports. The model demonstrates the three deleterious impacts of PI on the manufacturing firm's profits. First, competition between distributors in the PI-recipient market reduces profits through a "pro-competitive effect" even as it raises consumer welfare there. Second, the act of engaging in PI wastes resources in transport, which reduces profits through a "trade-cost effect". Finally, the need to manage PI could generate an inability to set the optimal wholesale price of zero (equal to marginal production costs) in the PI-source market. Specifically, this "double-markup effect" reduces profitability the greater the gap between wholesale price and marginal cost. The original manufacturer must strike a balance among these impacts in choosing its prices.

2.1. *Model Assumptions*

Here we set out the basic model assumptions and offer a justification for them. Consider the situation in which an original manufacturer, M, sells its product in two countries, A and B. Firm M sells its product through an independent exclusive distributor in country A and another independent exclusive distributor in country B. The demand in A is $Q_A = 1 - p$, and that in B is $Q_B = S(1 - bp)$, where S is the (relative) population of market B, letting the population in A be normalized to 1. Parameter $b > 1$ governs (relative) demand elasticity. Manufacturer M has a constant marginal cost of production normalized to zero, and the marginal cost of retailing in both countries is normalized to zero as well.

In our primary model, goods sold in both markets are homogeneous, which is a reasonable assumption in the PI context because the ultimate origin of all products is the single manufacturing firm. However, the NERA (1999) survey suggests that PI goods may be considered slightly inferior to goods sourced domestically and we extend the analysis to a situation of quality differentiation in the final analytical section.

Initially, we shall assume that consumer markets are segmented. Neither consumers nor firms can buy the product at the retail level in one country and ship it to the other in quantities sufficient to affect prices. Segmentation of consumer markets at the retail level can be motivated on several grounds. In some industries arbitrage is illegal, while parallel trade at the wholesale level is permitted. This is the case with prescription drugs in the European Union and alcoholic beverages in the United States. In other industries, physical products and non-tradeable services are tied, causing effective market segmentation at the retail level. Local warranties bundled with capital goods and local calling plans bundled with cell phones are examples in this category. Finally, retail arbitrage can be prohibitively costly, while available margins could support parallel trade at the wholesale level. Near the end of the paper we shall relax this assumption and illustrate the effects of arbitrage based on retail price differentials.

Suppose that the manufacturing firm M can offer the distributor in market i ($i = A, B$) any contract in the form (w_i, T_i), where w_i is the wholesale price at which the distributor purchases from M, and T_i is a transfer payment (franchise fee) from the distributor to M. Thus, we assume that the manufacturing firm can only control supply with regular two-part tariffs of the kind envisioned here. There is complete information, permitting the manufacturer to extract all profits from distributors with these contracts.[8] We rule out contracts that incorporate an agreement to limit parallel trade directly or indirectly. The manufacturing firm can neither control quantities with supply curbs nor target parallel trade with a "dual-pricing" scheme (i.e., charging higher prices for products that are exported than for those sold domestically). We consequently focus on the effects of parallel imports for distributor-level competition when the manufacturing firm uses a standard strategy to solve the vertical control problem (differentiating wholesale prices while avoiding double marginalization) rather than trying to target parallel trade as such. More sophisticated strategies may also be

[8]Under incomplete information there would be a principal–agent problem, with two agents competing. This problem would be more difficult and would offer a greater incentive for vertical integration. However, our framework permits a tractable investigation of the tradeoffs arising in the common situations where vertical integration is incomplete.

illegal under competition law in jurisdictions such as the European Union.

Regarding competition, we assume that M cannot prevent the distributor in B from selling the product in market A, either directly or through such intermediaries as firms specializing in parallel trade. That is, either the manufacturer cannot legally limit the distributor's territory of sales or it is too costly to enforce any such constraint. If the distributor in B sells in market A it incurs an additional marginal cost of $t \geq 0$ and competes with the distributor in A in a Cournot fashion. We also assume that only the licensed distributor in market B can sell in that market. Thus, the situation we have in mind describes one market protected from PI competition but a vertical relationship that is vulnerable to wholesale-level competition in the other market.

There are several practical justifications for this assumption of one-way trade. First, if there are fixed costs (not explicitly modeled in this paper) and market B is small while market A is large, an asymmetry is possible, with products flowing from the small to the large country but not the other way around. Portugal and the United Kingdom within the European Union would exemplify this case. Second, there may be asymmetric product standards between the two countries. Third, the countries may vary in their legal treatment of PI, with international exhaustion the rule in A and national exhaustion in B. Examples of the former are Australia and Hong Kong, which are open to PI in copyright goods, and examples of the latter are Japan and the United States, which are not (Maskus, 2000).

Another important assumption is that the manufacturing firm sells the products to its distributors prior to the opening of the retail markets. The wholesale price can only be charged based on the quantity sold by the manufacturing firm to the distributor, but not on the quantities sold by the distributors to consumers. This is consistent with the manufacturing firm's inability to directly control or monitor quantities sold by the distributors. Consequently, the distributor contracts cannot be based on subsequent retail supplies and the manufacturing firm must therefore offer a contract with a non-negative wholesale price, i.e. $w_i \geq 0$. If the wholesale price could be negative, it would pay the distributor to order an infinite quantity from the manufacturing firm and sell only a limited quantity to consumers, a case we rule out on the grounds of realism.[9]

The situation we describe rules out other strategies or market circumstances that might mitigate the PI problem. For example, the manufacturer might acquire the B distributor and, through this vertical integration, prevent products from escaping to the A market. For another, if the manufacturer used competitive wholesale dealers in market B, it could preclude PI and avoid double marginalization by setting the monopoly price there, knowing that the dealers could not charge a markup themselves.[10] Our justification for the specific case we consider is based on two general observations. First, in many products PI exist at the wholesale level, which would not be possible under vertical integration or competitive distribution, suggesting that some market power arises at that level. Second, it is common for original manufacturers to establish single, but independent, distributors in each national market. For example, the Coca-Cola company has separate general distributors for several countries in the EU. More specifically, Coca-Cola Enterprise SA is the sole licensed bottler in France, though it has five bottling plants and two distribution warehouses.[11] Automobiles also are commonly marketed through single wholesale dealers in the EU (NERA, 1999). To illustrate, Ferrari Great Britain Ltd. is the exclusive dealer for the company in the United Kingdom.[12] The recorded music industry may best exemplify our assumptions. The Sony-BMG company sustains single national

[9]It is in this context that our assumption of zero marginal production costs is significant. As we discuss later, under some circumstances involving positive PI the manufacturing firm would like to subsidize sales in market A, charging a wholesale price below the marginal cost of production. However, the only way to avoid the distributor placing excessive order amounts is to have a strictly positive wholesale price. It is consequently impossible to subsidize the distributor unless the distributor can be forced to sell everything it has ordered from the manufacturing firm. This inefficiency in vertical pricing generates the outcome in our model that our intermediate pro-competitive range (where the non-negative wholesale price in A is a binding condition) would gradually diminish for higher marginal costs and at some point would vanish. Ganslandt and Maskus (2006) consider the case of positive marginal production costs.

[10]We are grateful to a referee for making this point.

[11]See http://www2.coca-cola.com/ourcompany/cfs/cfs_france.htm, last visited 15 January 2006.

[12]See http://www.ferrari.co.uk/home.php, last visited 15 January 2006.

distributors in most European countries, including such relatively low-income locations as Hungary, the Czech Republic, and Spain.[13] At the same time, the volume of PI in recorded music within the EU overall was between five and ten percent of the aggregate market in 1998, with some releases experiencing up to 25 percent of sales arriving through parallel trade, in part because of low transport costs (NERA, 1999). The NERA report (p. 81) further noted about this industry that "vertical restraints ... are not so tight that they would prevent parallel trade."

The strategy of having exclusive national dealers arises largely because of the need to support market-specific services, advertising and guarantees, which would be difficult to sustain under the free-riding that would exist with competitive distribution. Further, it limits reductions in quality or dilution of trademark value that could occur under open competition. Finally, there may be scale economies in distribution. Indeed, these are major pro-competitive justifications for permitting exclusive territories.

2.2. *Basic Analysis*

Under the Cournot assumption, let the quantities sold in A by the two distributors be q_A^* and q_B^*, respectively, and the quantity sold in B by the sole distributor be q_B. A subgame-perfect Nash equilibrium is a pair (q_A^*, q_B^*) that constitute a Nash equilibrium for any (w_i, T_i) for $i = A, B$, together with an optimal choice of q_B by the distributor in market B for any (w_B, T_B) and an optimal choice of (w_i, T_i) for $i = A, B$ by the manufacturing firm M. Let w denote the vector (w_A, w_B) and T denote the vector (T_A, T_B).

Our main objective is to analyze how the manufacturing firm sets the wholesale prices and the transfer payments to maximize its profit. The manufacturing firm's profit is equal to the total revenues in equilibrium minus real costs incurred. More precisely, the objective of the manufacturing firm is to maximize:

$$\Pi(w) = p_A(q_A^* + q_B^*) - tq_B^* + p_B q_B \qquad (1)$$

[13]See http://www.sonybmg.com, last visited 29 January 2006.

where the first term on the right hand side is the total revenue in market A, the second term is the real cost of parallel trade between the markets and the third term is the total revenue in market B. Note that with two-part tariffs and complete information the manufacturer is able to extract all economic profits from both distributors.

We assume that at least one distributor can sell in market A, for any w and T accepted by the distributors. The equilibrium quantity for distributor A is[14]

$$q_A^*(w) = \begin{cases} \dfrac{1-2w_A^*+w_B^*+t}{3} & \text{if } 2\left(w_B^*+t\right)-1 < w_A^* < \dfrac{1+w_B^*+t}{2} \\ \dfrac{1-w_A^*}{2} & \text{if } w_A^* \leq 2(w_B^*+t)-1 \\ 0 & \text{if } \dfrac{1+w_B^*+t}{2} \leq w_A^* \end{cases} \tag{2}$$

and the corresponding quantity for distributor B in market A, i.e. the volume of PI, is

$$q_B^*(w) = \begin{cases} \dfrac{1+w_A^*-2w_B^*-2t}{3} & \text{if } 2w_A^*-1 < w_B^*+t < \dfrac{1+w_A^*}{2} \\ \dfrac{1-w_B^*-t}{2} & \text{if } w_B^*+t \leq 2w_A^*-1 \\ 0 & \text{if } \dfrac{1+w_A^*}{2} \leq w_B^*+t \end{cases} \tag{3}$$

and in market B the unique subgame equilibrium is the monopoly quantity

$$q_B(w) = S\left(\frac{1-bw_B}{2}\right) \quad \text{if } w_B < \frac{1}{b}, \tag{4}$$

where the condition for the wholesale price ensures that the distributor sells a positive quantity in that market. The quantities given in equations (2), (3) and (4) characterize all subgame equilibria and

[14]Details are provided in the Appendix.

we can proceed to analyze the decision of the manufacturing firm. More precisely, the firm chooses optimal wholesale prices for the distributors.

The optimal choices depend on trade costs. Four possible outcomes emerge as equilibria for different ranges. First, in the range of low trade costs the manufacturing firm is able to set optimal wholesale prices in both markets and PI exist. Second, for intermediate trade costs the firm is constrained to set a non-negative price in A but can choose an optimal price in B and PI occur in equilibrium. Third, for high trade costs the manufacturer sets wholesale prices to deter PI. Finally, sufficiently high trade costs are prohibitive and markets will be segmented.

For both low and intermediate trade costs the manufacturing firm sets wholesale prices to accommodate the effects of arbitrage. Accordingly, parallel imports occur in equilibrium. The quantities chosen by distributors are all positive. The equilibrium profit is denoted $\Pi^{PI}(w)$, where superscript PI refers to "parallel imports".

For high trade costs, on the other hand, parallel trade is deterred and no arbitrage occurs. The margin between the retail price in market A and the wholesale price in market B must be smaller than the unit trade cost, i.e.,

$$p_A(w) - w_B \leq t. \tag{5}$$

In this case, the quantity sold in market A by distributor B is zero and we refer to this outcome as the arbitrage-free equilibrium. The profit is denoted $\Pi^{NA}(w)$, where NA refers to "no arbitrage".

Finally, for prohibitive trade costs each distributor would only sell in its designated market.[15] In the segmented equilibrium the manufacturing firm can maximize its profit without a constraining arbitrage condition. It follows immediately that in the segmented equilibrium the manufacturing firm set wholesale prices equal to zero (marginal production costs) in both markets (i.e. $w_A = 0$ and $w_B = 0$) to eliminate the double mark-up problem and the

[15] More generally, segmentation may also be caused by contractual restrictions and legal barriers.

corresponding segmented retail prices are

$$p_A = \frac{1}{2} \quad \text{and} \quad p_B = \frac{1}{2b}. \tag{6}$$

We now proceed to analyze the wholesale prices set by the manufacturing firm in the PI equilibrium and the arbitrage-free equilibrium and make a comparison with the segmented equilibrium. We then proceed to analyze consumer welfare.

3. Optimal Pricing and Parallel Trade

3.1. *Retail Price Divergence*

It is natural to start with the analysis in the range of low trade costs. In this situation the distributor in market B finds it profitable to sell a positive quantity in market A, implying that parallel imports occur in equilibrium. The analysis in this section shows that parallel trade does not unambiguously result in retail price convergence.

To find the optimal wholesale prices we differentiate the accommodation profit $\Pi^{PI}(w)$ with respect to wholesale prices in the two markets. The first-order condition with respect to the wholesale price in market A, w_A, is

$$\frac{d\Pi^{PI}(w)}{dw_A} = \left[\frac{1}{9} - \frac{2}{9}w_A - \frac{2}{9}w_B - \frac{2}{9}t\right] - \left[\frac{1}{3}t\right] = 0 \tag{7}$$

and the first-order condition with respect to the wholesale price in market B, w_B, is

$$\frac{d\Pi^{PI}(w)}{dw_B} = \left[\frac{1}{9} - \frac{2}{9}w_A - \frac{2}{9}w_B - \frac{2}{9}t\right] + \left[\frac{2}{3}t\right] - \left[\frac{Sbw_B}{2}\right] = 0. \tag{8}$$

The first-order condition in market A illustrates two of the three problems facing the manufacturer. The first term in square brackets is the pro-competitive effect of parallel imports in that market and it characterizes the manufacturer's incentive to control the total supply there. Competition between distributors lowers the manufacturer's profits. Note that the pro-competitive effect is positive when w_B and the unit trade cost are low. In this situation the firm would moderate the pro-competitive effect of arbitrage by raising its wholesale price in market A. In contrast, the effect is negative when w_B and the trade cost are high and the firm would choose to reduce w_A. The incentive to reduce the volume of arbitrage and charge a low wholesale price in market A is therefore stronger when the trade cost is high. The second term is the trade-cost effect, which captures the incentive to save resources wasted in parallel trade. Clearly, the higher is t, the greater the incentive to limit PI by reducing the wholesale price in A.

In the first-order condition for market B, the first term in square brackets is the pro-competitive effect of parallel imports in market A. High trade costs permit the firm to reduce w_B and still limit this competition. Note that the pro competitive effect is identical in both (7) and (8). In other words, the manufacturing firm has two instruments with which to control the total supply in market A: the wholesale price in A and the wholesale price in B. The second term in square brackets also reflects the problem with resources used in trade costs. Here, the trade-cost effect is positive and, to moderate the volume of arbitrage, the manufacturing firm would charge a high wholesale price in market B, while simultaneously charging a low wholesale price in market A.

Note that a third term appears in the latter first-order condition, reflecting the double-markup effect. This is the profit-reducing impact of double marginalization in market B, capturing the inability of the manufacturer to set the efficient wholesale price of zero while simultaneously limiting the volume of parallel trade. The higher is w_B, other things equal, the greater the incentive to reduce that price toward vertical pricing efficiency. If the manufacturing firm is forced by PI to charge a positive wholesale price in B, its revenues there would be reduced because the local monopoly distributor there would set a retail price above the optimal (revenue-maximizing) level for the IPR holder.[16] The manufacturing firm thus has an incentive to keep the wholesale price in B low in order to minimize the double-markup problem.

[16]This insight explains why there is no double-markup problem in A, where the distributor must compete with imports and is not a monopolist. Increases in w_A accordingly reflect conditions of competition and generate profit-maximizing retail prices there.

Having analyzed the incentives of the manufacturing firm we can solve the system of equations consisting of the two first-order conditions (7) and (8) to find the optimal wholesale prices. The system has a unique solution:

$$w_A = \frac{1}{2} - \frac{5}{2}t - \frac{2}{Sb}t, \tag{9}$$

$$w_B = \frac{2}{Sb}t, \tag{10}$$

where both prices are continuous and linear functions in the unit trade cost t. The wholesale price in market $A(B)$ is a decreasing (increasing) function of the trade cost. Note that the wholesale price in market A is non-negative for $t \in [0, \underline{t}]$, which we term the range of low trade costs. The upper bound for this interval is determined by $w_A(t) = 0$ and the relevant threshold level is

$$\underline{t} = \frac{Sb}{4 + 5Sb}. \tag{11}$$

Our results support two observations. First, the lower are trade costs, the stronger is the incentive to reduce the pro-competitive effect by raising the wholesale price in A (note that w_A rises as t falls). Moreover, few resources are wasted by PI in this range of costs. Second, declining trade costs permit the firm to decrease the wholesale price in market B and reduce the double-markup distortion. Thus, parallel trade has an effect on vertical price control that is opposite to the typical effect of arbitrage in price-discrimination models. The wholesale price in the importing country increases as the unit trade cost falls, despite an increasing volume of parallel imports. Correspondingly, the wholesale price in the export market decreases when the unit trade cost falls.

Readers may prefer to think in terms of rising trade costs. Thus, put another way, as trade charges increase in this low range the manufacturer would face diminished PI and lower competitive pressure, permitting it profitably to decrease the wholesale price in A. In B the distributor price would rise, which would exacerbate the double-markup problem. Rising trade costs would also increase resources wasted in PI activities.

We insert the optimal wholesale prices into (2) and (3) to obtain the subgame-perfect equilibrium quantities. We then insert the equilibrium outputs into the inverse demand functions to obtain the retail prices in markets A and B:

$$p_A = \frac{1}{2} - \frac{1}{2}t, \tag{12}$$

$$p_B = \frac{1}{2b} + \frac{t}{Sb}, \tag{13}$$

where both functions are continuous and linear in trade costs. Thus, in the range of low trade costs a decline (increase) in t generates a divergence (convergence) in retail prices.[17] We can summarize our result for the range of low trade costs as follows:

Proposition 1. *Assume that retail markets are segmented and the unit trade cost of parallel trade is $t \in [0, \underline{t})$. If the trade cost increases in this range, the retail price in market A decreases, the retail price in market B increases and the volume of PI decreases.*

Proof. The proof of the first part of the proposition follows immediately from the optimal retail prices in (12) and (13). The volume of arbitrage is equivalent to the quantity chosen by distributor B in market A. We insert the equilibrium wholesale prices to obtain the equilibrium volume of parallel imports:

$$q_B^* = \frac{1}{2} - \frac{2}{Sb}t - \frac{3}{2}t \quad \text{if } t \in [0, \underline{t}] \tag{14}$$

which is a continuous and declining function in the trade cost. □

It is also worth noting that the volume of arbitrage is an increasing function in the size parameter S as well as the price-elasticity (captured by parameter b) in market B. The manufacturing firm has an incentive to keep the wholesale price in that market low when it is large and when the consumers are price-sensitive. In this case the

---

[17]Strictly speaking, because these prices depend on the values of S and b, it is possible to have $p_b > p_a$ at $\underline{t}$, in which case as t falls these prices would come together before diverging as in the text. This situation is consistent with PI flowing from the high-retail price nation to the low-retail price nation, a result found in Chen and Maskus (2005).

manufacturing firm is willing to accept a larger volume of arbitrage to avoid a serious double-markup problem in B.

One interesting aspect of the parallel trade equilibrium is that at a unit trade cost of zero the retail prices are identical to those that would be set by a vertically integrated monopolist. In other words, the manufacturing firm can solve the vertical control problem perfectly if no real resources are wasted in arbitrage and consumer markets are segmented.[18] The wholesale price in market B would be set to zero and the wholesale price in market A would be set at a prohibitive level for the distributor there, pushing it out of business. Distributor B would sell the revenue-maximizing quantity in both markets and A would be supplied from B but no real resources would be used in transportation. In addition, there would be no double-markup problem in market B. Consequently, the retail prices in an integrated equilibrium would be identical to the prices set in a completely segmented equilibrium. This result stems from the fact that as trade costs disappear it is profit-maximizing to shift the source of supply from the distributor in A to the distributor in B. Once distributor B becomes the sole supplier at zero trade cost, the manufacturer would choose to authorize exports from B to A, implying that PI would be replaced by standard trade. In effect, the result would be a market integrated at the wholesale level but segmented at the retail level by a single distributor with two territories.

3.2. *A Pro-Competitive Effect of Parallel Imports*

For sufficiently high trade costs, i.e. a unit cost above the critical threshold $\underline{t}$, the optimal wholesale price in market A would be negative, meaning that the manufacturing firm would like to sponsor the supply of distributor A in that market through a variable subsidy and recoup this subsidy by charging a higher fixed fee. As discussed earlier we have ruled out this possibility on grounds that effective

[18] In principle, the manufacturer could solve the problem even with positive trade costs by vertically integrating with the distributor in B and not shipping any goods to A. As noted earlier, we rule this case out in order to focus on the situation of independent distributors and the vertical control problem. Vertical integration is not feasible in a variety of circumstances and independent wholesalers are common.

monitoring may be impossible. Therefore, the firm must determine the profit-maximizing wholesale price in market B subject to the condition that the wholesale price in market A is zero.

This constraint adds an additional distortion and further complicates the vertical control problem by limiting the manufacturer to a single price instrument. The maximization problem is now reduced to finding the solution to the first-order condition for the manufacturing firm with respect to the wholesale price in market B:

$$\frac{d\Pi^{PI}(w)}{dw_B} = \left[\frac{1}{9} - \frac{2}{9}w_B - \frac{2}{9}t\right] + \left[\frac{2}{3}t\right] - \left[\frac{1}{2}Sbw_B\right] = 0. \tag{15}$$

As before, the first term reflects the pro-competitive effect in A, the second term is the trade-cost effect and the third term reflects the double-markup problem in B.

The important difference compared to the first-order condition when the manufacturing firm could charge an optimal wholesale price in market A resides in the first term. The manufacturing firm has an incentive to charge a lower wholesale price in market B to avoid generating a double-markup problem in market A.[19] Hence, the wholesale price in B is lower than it would be if w_A could be negative. We solve the first order condition (15) to obtain the equilibrium wholesale price

$$w_B = 2\left(\frac{1+4t}{4+9Sb}\right). \tag{16}$$

In this case, as trade cost goes up the manufacturer would reduce the volume of PI by raising the wholesale price in market B. However, it is not profitable for the manufacturing firm to block PI completely by this strategy because it would create an excessively large double-markup effect in B. The higher the trade cost, the stronger is the incentive to reduce the volume of PI. This incentive is a combination of two effects: a higher trade cost wastes more resources in PI and also protects market A from competition. These effects reduce the net cost of limiting PI with an inefficient wholesale price in market B.

[19]In contrast to the unconstrained case, the possibility of a double-markup problem in A arises because of the minimum pricing constraint, which might prevent establishment of the optimal retail price.

Accordingly, the wholesale price in B increases with t while the wholesale price in A would be kept at the minimum level of zero. As previously we insert the subgame-perfect equilibrium quantities from (2) and (3) in the inverse demand functions to obtain the retail prices in markets A and B. The equilibrium prices are given by

$$p_A = \frac{1}{3} + \frac{2}{3}\left(\frac{1+4t}{4+9Sb}\right) + \frac{t}{3}, \tag{17}$$

$$p_B = \frac{1}{2b} + \left(\frac{1+4t}{4+9Sb}\right), \tag{18}$$

where both functions are continuous and linear in t. It is interesting to note that both retail prices would increase with higher trade costs.

For sufficiently high trade costs, parallel trade is unprofitable and no arbitrage occurs, whereas for low and intermediate trade costs, i.e. $t \in [0, \bar{t})$, the volume of parallel trade is positive. The critical threshold for eliminating arbitrage is given by

$$p_A - w_B = t, \tag{19}$$

where p_A is given by (17) and w_B is given by (16). Accordingly, the upper threshold at which PI vanish is

$$\bar{t} = \frac{3}{2}\left(\frac{Sb}{4+3Sb}\right) \tag{20}$$

which takes values on the open interval $(0, 1/2)$. The threshold level is consequently strictly positive and less than the prohibitive unit trade cost ($t = 1/2$) for relevant values of S and b. We have the following result:

Proposition 2. *Assume that retail markets are segmented and the unit trade cost of parallel trade is* $t \in (\underline{t}, \bar{t})$. *If the trade cost increases in this range, the retail prices in markets A and B both increase and the volume of PI decreases.*

Proof. For the purpose of proving the proposition, note that the optimal retail prices (17) and (18) are linear and increasing functions in t. Second, insert the equilibrium wholesale prices to obtain the equilibrium volumes of parallel imports:

$$q_B^* = \frac{1}{3} - \frac{4}{3}\left(\frac{1+4t}{4+9Sb}\right) - \frac{2t}{3} \quad \text{if } t \in (\underline{t}, \overline{t}) \tag{21}$$

which is a continuous and declining function of trade costs. □

The evolution of the retail price differential between markets A and B is also interesting in the intermediate range of trade costs. Both retail prices rise with costs in this interval, but they actually diverge. Using the equilibrium prices (17) and (18) we compute the difference between retail prices, i.e. $p_A - p_B$. The derivative of the difference with respect to the unit trade cost is

$$\frac{d(p_A - p_B)}{dt} = \frac{3Sb}{4+9Sb}, \tag{22}$$

which is strictly positive. In other words, retail prices differ more for high trade costs than for low trade costs in this range. This result is the outcome of the manufacturing firm's incentive to set a wholesale price in market B to reduce the volume of PI while keeping the wholesale price in market A at zero. Put differently, suppose unit trade cost declines in this range. The manufacturer would accommodate a larger volume of PI while also choosing a lower wholesale price in market B. As a result, the pro-competitive effect of PI in market A gets stronger and the double-markup effect in market B gets weaker. The retail prices in both markets consequently decrease and tend to converge as trade costs fall.

The essence of the analysis in this subsection is that, because the manufacturer no longer has the ability to set the wholesale price in A optimally, it is limited to using only the wholesale price in B to manage these various tradeoffs. Accordingly, the firm must permit a higher PI volume, magnifying the pro-competitive impact in market A.

3.3. *A Double-Markup Problem with Arbitrage-Free Prices*

For high trade costs, i.e. $t \in [\overline{t}, 1/2]$, the manufacturing firm sets wholesale prices to block PI and the equilibrium is, consequently, arbitrage-free. In this equilibrium the distributor in B finds it profitable to sell a positive quantity in her own market but does not ship goods to market A.

Noting that the wholesale price in A must be non-negative, i.e. $w_A \geq 0$, we differentiate the arbitrage-free profit function $\Pi^{NA}(w)$ with respect to wholesale prices in the two markets, subject to the condition in inequality (5) to find the optimal wholesale prices. These prices in the arbitrage-free equilibrium are

$$w_A = 0, \tag{23}$$

$$w_B = \frac{1}{2} - t. \tag{24}$$

This result has a straightforward interpretation. The higher the trade cost, the more protected is market A by the natural barrier and, further, the wholesale price in market B can be lowered to reduce the double-markup effect in B. This price would be set both to deter PI and limit the price distortion in B. When t reaches its maximum level (i.e. $t = 1/2$), the optimal wholesale price in B becomes zero and the price distortion disappears. The corresponding equilibrium retail prices in A and B are given by

$$p_A = \frac{1}{2}, \tag{25}$$

$$p_B = \frac{1}{2b} + \frac{1}{4} - \frac{t}{2}. \tag{26}$$

We have the following result:

Proposition 3. *Assume that retail markets are segmented and the unit trade cost of parallel trade is $t \in (\bar{t}, 1/2)$. In this range of trade costs, the retail price in market A is equivalent to its value in the segmented equilibrium. Moreover, as trade cost increases, the retail price in market B declines.*

Proof. The proposition follows from (25) and (26). □

The retail price in B declines with trade costs because the manufacturing firm sets a wholesale price in this market to deter PI. This wholesale price reflects a double-markup distortion that diminishes with higher t. Thus, for higher trade costs the manufacturing firm would eliminate PI with a lower wholesale price and experience a correspondingly weaker double-marginalization problem. In this case equilibrium retail prices diverge as trade cost increases, which is more in line with the usual intuition. The potential competition from arbitrage is strong when the trade cost is low and weak when the trade cost is high.

Note that $p_B = \frac{1}{2b}$ at $t = 1/2$. For prohibitive trade costs, i.e. $t > 1/2$, markets are fully segmented.

4. Prices, Profits and Consumer Welfare

The analysis of prices in the previous section characterizes the equilibrium for all trade costs and other parameter values. In this section we take a closer look at retail prices, profits and consumer welfare. We are particularly interested in comparing the equilibrium in a market subject to actual or potential PI competition with the segmented equilibrium.

4.1. *Retail Prices*

We first consider retail prices for different trade costs. Equilibrium retail prices in the two markets are illustrated in Figure 1 for parameter values $S = 1$, $b = 2$. The upper line corresponds to p_A and the lower line corresponds to p_B. The horizontal lines indicate fully segmented price levels. The first vertical line demarcates the lower

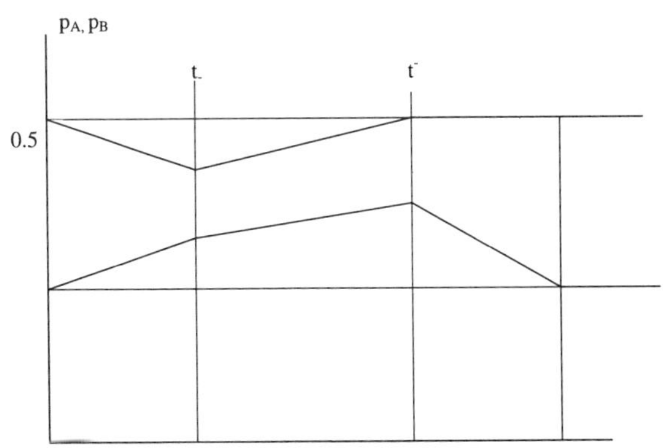

Figure 1. Equilibrium retail prices in $A, B(S = 1, b = 2)$.

threshold, below which the non-negativity constraint does not bind and retail prices diverge with a decreasing trade cost. This range corresponds to accommodation equilibrium conditions and PI are positive. Proceeding to intermediate trade costs, prices increase but grow apart as t increases in the second range, which corresponds also to positive volumes of PI. The second vertical line demarcates the upper threshold of t for which PI exist. In the third range, corresponding to the arbitrage-free case, retail price is at its monopoly level in market A but falls in market B. Above that range of trade cost the markets become fully segmented.

It is straightforward to see that the arbitrage-free equilibrium is Pareto-dominated by a completely segmented equilibrium. This follows directly from the fact that the retail price in market B is higher than the segmented retail price (because of the double-markup problem), the retail price in market A equals the segmented retail price and the profit of the manufacturing firm is lower than it would be under full segmentation.

4.2. *Profits*

Next, we compute the manufacturing firm's profit in equilibrium (see the Appendix for details). It is illustrated as a function of the variable trade cost in Figure 2 for the same parameter values. For low trade costs ($t < \underline{t}$), the profit decreases as trade cost rises, both because of more resources lost in parallel trade activities and an

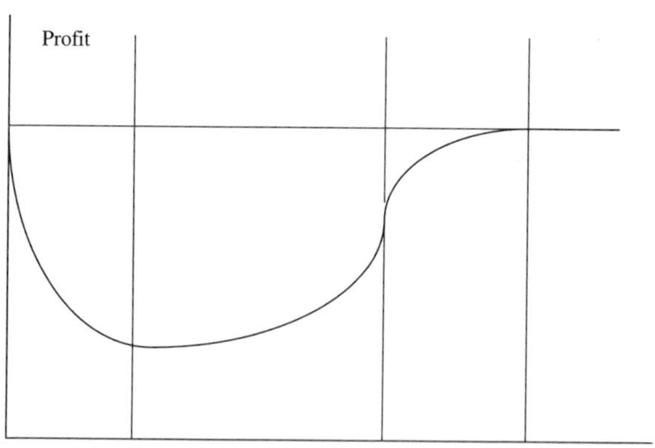

Figure 2. Manufacturing firm's profit ($S = 1, b = 2$).

increasing inability to induce optimal retail prices in both markets. For intermediate trade costs ($\underline{t} \leq t < \bar{t}$), the profit goes up as trade costs increase. The importance of the trade cost as a natural barrier increases; parallel imports vanish, fewer resources are wasted and the double-markup problem in market B is moderated. For high trade costs ($t \geq \bar{t}$), the profit also increases with t. The equilibrium is arbitrage-free and no resources are wasted in trade. The double-marginalization problem in the export market is gradually reduced as the markets move towards segmentation.

4.3. *Consumer Welfare*

Finally, we analyze consumer welfare with PI and exclusive territories. Both competition policy and market integration policy typically have the objective of enhancing consumer welfare and maximizing consumer surplus. It is therefore of interest to evaluate the effects of legalizing PI on consumer welfare when the manufacturing firm employs two-part tariffs for vertical control purposes. We compare these effects to the welfare outcome when the manufacturing firm is permitted to segment the wholesale market with exclusive territories and discriminate with different prices in the two countries.

Aggregate (two-country) consumer surplus with an integrated wholesale market is

$$CS = \left[\left(Q_A - \frac{Q_A^2}{2}\right) - p_A Q_A\right] + \left[\frac{1}{b}\left(Q_B - \frac{Q_B^2}{2S}\right) - p_B Q_B\right] \quad (27)$$

where the first term is consumer surplus in A and the second term is consumer surplus in B. We insert the demand functions in markets A and B to get consumer surplus as a function of retail prices:

$$CS(p) = \frac{1}{2}(1 - p_A)^2 + \frac{S}{2b}(1 - bp_B)^2 \quad (28)$$

which is continuous, concave and decreasing in both p_A and p_B.

Differentiating consumer surplus with respect to t in the accommodation range of low trade costs ($t < \underline{t}$) gives the following result

$$\frac{dCS}{dt} = -\frac{1}{2b} + \frac{1}{4} + \frac{t}{4Sb} + \frac{t}{4}, \quad (29)$$

and in the accommodation range of intermediate trade costs ($\underline{t} \leq t < \bar{t}$),

$$\frac{dCS}{dt} = -\frac{2S(b+1)+2}{4+9Sb} + \frac{(4+Sb)t}{4+9Sb} \tag{30}$$

and, finally, in the arbitrage-free range of high trade costs ($t \geq \bar{t}$):

$$\frac{dCS}{dt} = \frac{2S - Sb + 2Sbt}{8}. \tag{31}$$

It is straightforward to show that combined consumer surplus in markets A and B has its unique global minimum at $t = \bar{t}$. This has an intuitive explanation. In any arbitrage-free equilibrium, the retail price in market A is at its maximum and identical to the segmented price. At the same time, the retail price in B has its maximum at $\bar{t}$ and is strictly higher than the segmented price. Here, the manufacturing firm charges the highest wholesale price in that market to ensure that prices are arbitrage-free, since the trade cost is only a partial barrier for parallel imports. In other words, for high trade costs both combined consumer surplus and producer profits are lower when parallel trade is permitted than when the manufacturing firm can segment the wholesale market with exclusive territories.

In addition, for two sufficiently similar markets, i.e. $b < \underline{b}(S)$, the policy of permitting parallel trade has a negative effect on aggregated consumer surplus at low and intermediate trade costs. The unique potential local maximum for consumer welfare at interior trade costs, $t \in (0, 1/2)$, is at the lower critical threshold $\underline{t}$. Accordingly, market segmentation dominates PI as long as consumer welfare at this trade cost is lower than the consumer welfare in a segmented equilibrium. We insert equilibrium prices from the PI equilibrium at $\underline{t}$ and the segmented equilibrium into the consumer surplus function. We then set the two levels equal and solve for b. The critical level is

$$\underline{b}(S) = \frac{10}{11} - \frac{6}{11S} + \frac{2}{11S}\sqrt{9 + 14S + 25S^2} \tag{32}$$

and we have the following result:

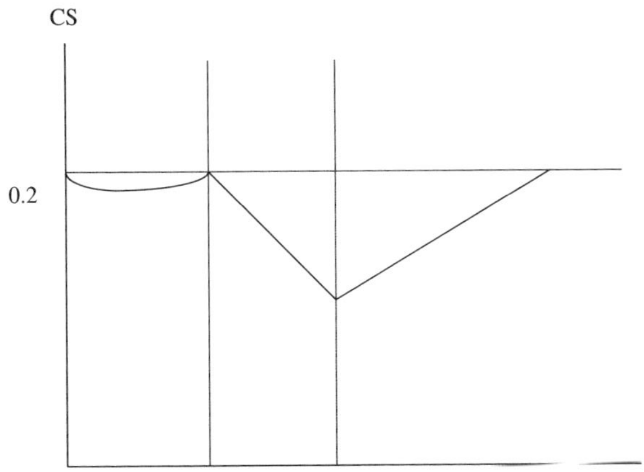

Figure 3. **Joint CS ($S = 1, b = b^*$).**

Proposition 4. *For sufficiently similar markets, i.e.* $b < \underline{b}$, *joint consumer surplus has its global maxima at* $t = 0$ *and* $t = \frac{1}{2}$ *(see Figure 3).*

Proof. First, note that retail prices are equal to the segmented prices at $t = 0$ and $t = \frac{1}{2}$. Consumer surplus is consequently the same at these two points. Second, consumer surplus in the arbitrage-free range of high trade costs ($t \geq \bar{t}$) is strictly less than consumer surplus in the segmented equilibrium as p_A is identical to the segmented price and p_B is strictly higher. Third, in the accommodation range of intermediate trade costs ($\underline{t} \leq t < \bar{t}$), the derivative of the consumer surplus function is strictly negative and consumer surplus reaches its maximum at $\underline{t}$ in this interval (both retail prices increase in t in this range). Finally, consumer surplus in a segmented equilibrium is constant and consumer surplus in the range of low trade costs is quadratic and strictly convex in t (the derivative is linear and increasing in t). The two consumer surplus levels, consequently, can only be equal at two trade cost levels and consumer surplus must be lower in any interior PI equilibrium than in the segmented equilibrium. The first trade cost is $t = 0$. The second is

$$t = \frac{S(4 - 2b)}{Sb + 4} \tag{33}$$

and this threshold is larger than $\underline{t}$ if

$$b < \frac{10}{11} - \frac{6}{11S} + \frac{2}{11S}\sqrt{9 + 14S + 25S^2} \tag{34}$$

which is sufficient for the consumer surplus to be lower in the PI equilibrium than in the segmented equilibrium. Consumer surplus is consequently lower in an equilibrium open to PI than in a segmented equilibrium for any $t \in (0, 1/2)$, if $b < \underline{b}$. □

While open parallel trade is beneficial for consumers in the import market for low and intermediate trade costs, it is not in the interest of the consumers of both countries taken as a group and it also damages producers. On the contrary, if the manufacturing firm uses two-part tariffs to solve vertical control problems, exclusive distribution territories can benefit consumers.

On the other hand, for sufficiently different countries parallel trade may increase consumer welfare (see Figure 4). In the range of low trade costs consumer surplus is an increasing function in t for $b \geq 2$. In this case, combined consumer surplus has its maximum at $\underline{t}$. Consumers can jointly benefit when the manufacturing firm sets prices to minimize the volume of parallel trade. The difference in retail prices reaches its minimum at $\underline{t}$, where the aggregated consumer surplus is highest. This is akin to the classical welfare effect of uniform pricing: price convergence increases aggregate consumer

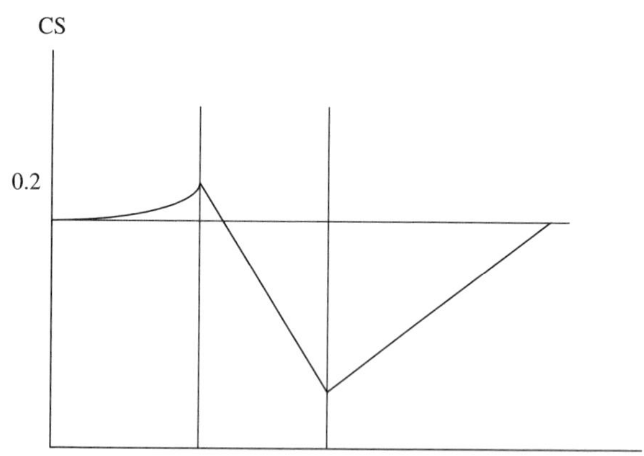

Figure 4. Joint CS ($S = 1, b = 2$).

surplus as long as total output is unchanged or higher compared to a situation with price discrimination.

This result shows that consumers in both markets may jointly benefit from parallel imports for low and intermediate trade costs as long as the difference in the price elasticity of demand is sufficiently large between the two consumer groups. The intuition for this result is that the pro-competitive effect of PI in A would dominate the double-markup effect in B since the optimal price in the latter market would be sufficiently low to induce a significant price reduction in the former market. This outcome would pertain to situations where consumers in the PI-source country were substantially more price-sensitive than in the PI-recipient country. Indeed, licensee prices of original manufacturer's goods generally seem to follow this pattern across countries, at least to the degree that per-capita income differences are negatively correlated with demand elasticities.[20] In that regard, where trade costs are low permitting PI among countries at significantly different income levels could raise overall consumer welfare, even as it would harm consumers in the lower-income countries.

As a final observation in this section we note that the two most important parameters for reaching welfare conclusions are b and t, which are the price sensitivity in market B and the trade cost, respectively. Parameter S, the size of market B, is of minor importance. Our main conclusion, that permitting parallel trade can reduce combined consumer surplus for positive trade costs, is true independent of the size of market B as long as the price-sensitivity in that market is below the limit of the critical threshold as S goes to zero, which requires only that $(b < 4/3)$.[21]

5. Model Extensions

In this section we consider two important relaxations of our model assumptions: the possibility of arbitrage at the retail level and the existence of quality-differentiated products.

[20] Ganslandt and Maskus (2004) found this for pharmaceuticals. See also NERA (1999).
[21] We demonstrate this assertion in a separate appendix available on request.

5.1. *Retail Arbitrage*

We initially assumed that retail markets are segmented, justifying the realism of this assumption on the grounds of varying policies, product standards, and differential market sizes. While we think these factors are relevant in a broad class of goods, readers may wonder about the implications of retail arbitrage in the model. Thus, here we relax the assumption of retail segmentation.

First, consider the possibility that consumers could engage in perfectly elastic retail arbitrage. Consumer arbitrage is not profitable if the retail price differential is less than the unit trade cost t. Formally, the no-arbitrage condition is

$$p_A - p_B \leq t \tag{35}$$

and we insert the retail prices and solve for the critical trade cost. A sufficient condition for retail markets to be segmented in this case is $t \geq \widetilde{t}$, where

$$\widetilde{t} = \frac{S(b-1)}{2+3Sb}. \tag{36}$$

Note that this threshold is lower than $\underline{t}$, meaning that this form of arbitrage limits the range of price divergence, for given market parameters, more than does the first form. Nevertheless, a range of price divergence continues as trade costs fall. Observe that the threshold is close to zero for b close to 1. Markets remain segmented at the retail level by the natural barrier t for similar markets or high trade costs. For dissimilar markets and low trade costs, arbitrage at the retail level could limit the scope for price differentiation and prices would converge. Figure 5 illustrates the consumer-level retail arbitrage conditions. The solid lines depict retail prices in markets A and B. The dotted vertical line corresponds to the threshold level $\widetilde{t}$ at which consumer arbitrage would kick in. For trade costs below that level, prices would tend to converge along the dashed lines making up the retail price cone (p'_A and p'_B), as shown.

Next, consider the possibility that the retail services provided in market A are necessary complements for consumption there. In other words, the distributor in market A (or an agent) can import the

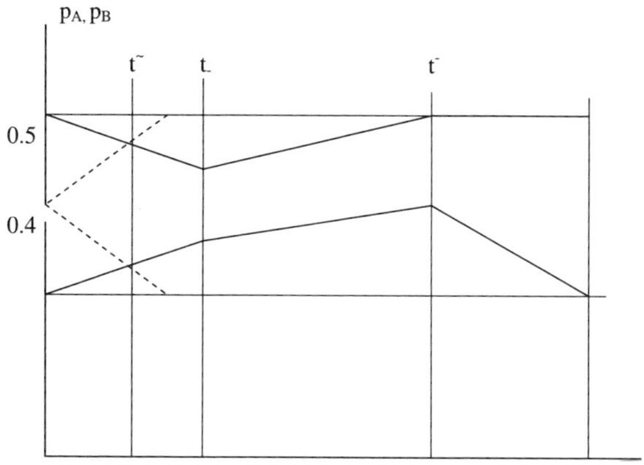

Figure 5. Retail prices with retail arbitrage ($S = 1, b = 1.5$).

product from B but consumers in A cannot buy the product at the retail level in B. Arbitrage by distributor A (in which he buys at the retail level abroad in order to find a cheaper source of supply than the available wholesale price) is not profitable if the margin between the wholesale price in market A and the retail price in market B is lower than the unit trade cost. Formally, the no-arbitrage condition in this case is

$$w_A - p_B \leq t \tag{37}$$

and in order to determine whether this condition is slack in equilibrium we insert the accommodation wholesale price in market A and the retail price in market B. A sufficient condition for retail markets to be segmented in this case is

$$t > \frac{S(b-1)}{6+7Sb} \tag{38}$$

where the right hand side is close to zero for b close to 1. Markets are consequently segmented at the retail level by the natural barrier t for similar markets or high trade costs. For dissimilar markets and low trade costs, however, arbitrage at the retail level could limit the scope for price differentiation and prices would converge.

We reiterate that perfect retail arbitrage is a strong assumption and inconsistent both with the fact that the bulk of PI occurs at the distributor level and with persistent differences in retail prices within the EU. There are good reasons to expect limited arbitrage of this kind even without restrictions on parallel trade. First, as noted above there are likely to be complementarities in retail services that cannot be provided by arbitrageurs. Second, there may be significant fixed costs in organizing cross-border retail trade. Third, there may be large information costs for consumers in determining product prices and characteristics for purposes of organizing arbitrage. Thus, we think our analysis of distributor-level PI is valid in many realistic circumstances.

5.2. *Product Differentiation*

In the previous sections we modeled parallel imports and domestic products as homogeneous goods. In some situations, however, it is more realistic to assume that the PI product is differentiated from the good sold by the local distributor. More specifically, the PI variety can be of slightly inferior quality, perhaps due to the absence of pre-sale and post-sale services, limited warranties, longer distribution lags or a need for domestic adaptation. In this subsection we explore the effects of quality differentiation within our framework of one-way parallel trade. We restrict our attention to the range of low trade costs, with a strictly positive volume of PI, since differentiation in the absence of PI competition is not interesting. Further, we consider both quantity competition and price competition.

5.2.1. *Quantity Competition*

Assume that demand in market B is given by $Q_b = S(1 - p_b)$ as before, where p_b denotes the price of the product sold there. Next, consider market A. Because the PI good is inferior, it offers utility at a fraction $\lambda < 1$, of the product sold by the local distributor.[22] The demand for the product sold by the local distributor in A and

[22]We treat the parameter λ as fixed and exogenous. For some products and markets, product differentiation may be endogenous. For example, the manufacturing firm could increase product differentiation by local adaptation or limits on the geographical scope of warranties. This is an interesting topic for future research.

the demand for the PI product are

$$q_A = 1 - \frac{p_A - p_{PI}}{1-\lambda} \tag{39}$$

$$q_{PI} = \frac{\lambda p_A - p_{PI}}{(1-\lambda)\lambda} \tag{40}$$

where p_A is the price of the variety sold by the local distributor and p_{PI} is the price of the PI variety. We solve the equilibrium in the usual backward fashion. For a sufficiently low trade cost, the equilibrium quantities in market A are

$$q_A(w) = \frac{2 - \lambda + w_B + t - 2w_A}{4-\lambda} \tag{41}$$

$$q_{PI}(w) = \frac{\lambda + w_A\lambda - 2w_B - 2t}{\lambda(4-\lambda)} \tag{42}$$

while the subgame in market B, for a given w_B, is the same as before. Working backwards, the manufacturing firm chooses wholesale prices to maximize aggregated profit. The optimization problem has a unique solution in which equilibrium wholesale prices are

$$w_A = \frac{S\lambda^3 b - t(4S\lambda b + 4 + S\lambda^2 b)}{2(4(1-\lambda) + Sb\lambda(4-3\lambda))} \tag{43}$$

$$w_B = \frac{2(1-\lambda)\lambda + t\lambda}{4(1-\lambda) + Sb\lambda(4-3\lambda)}. \tag{44}$$

It follows from these expressions that the wholesale price in A decreases and that in B increases in t. This is qualitatively the same result we found with homogeneous products. We insert the equilibrium wholesale prices in the demand functions to obtain the following retail prices

$$p_A = \frac{4(1-\lambda) + Sb\lambda(2-\lambda)^2 - tSb\lambda^2}{2(4(1-\lambda) + Sb\lambda(4-3\lambda))} \tag{45}$$

$$p_{PI} = \frac{4\lambda(1-\lambda) + Sb\lambda^2(2-\lambda) + (4(1-\lambda) + Sb\lambda(4-5\lambda))t}{2(4(1-\lambda) + Sb\lambda(4-3\lambda))} \tag{46}$$

$$p_B = \frac{1}{2b} + \frac{\lambda(1-\lambda) + t\lambda}{4(1-\lambda) + Sb\lambda(4-3\lambda)} \tag{47}$$

where the denominators are strictly positive for $\lambda < 1$.

It follows that the retail price for the locally distributed product in A decreases and the retail price in B increases, if the trade cost increases. This result corresponds qualitatively to our result for homogeneous products.

Interestingly, the result for the PI variety depends on the degree of product differentiation. For sufficiently similar products (high λ), the retail price for the PI product decreases in t. If the PI product is a perfect substitute for the local variety both products must sell for the same price. However, if the PI product is very inferior compared to the local variety, the retail price for the PI product increases in t since the distributor can pass on the higher marginal cost to consumers.

5.2.2. *Price Competition*

Assuming that products are differentiated we can analyze the equilibrium when distributors set prices rather than quantities. Let both distributors choose profit-maximizing prices in market A while distributor B also sets the price in her market. The optimal retail prices as functions of wholesale prices are

$$p_A(w) = \frac{2 - 2\lambda + 2w_A + w_b + t}{4 - \lambda} \tag{48}$$

$$p_{PI}(w) = \frac{\lambda(1 - \lambda) + 2w_B + w_A\lambda + 2t}{4 - \lambda} \tag{49}$$

and the volume of PI is

$$q_{PI}(w) = \frac{\lambda(1 - \lambda) + w_A\lambda - (2 - \lambda)(w_B + t)}{(4 - \lambda)(1 - \lambda)\lambda}. \tag{50}$$

In other words, if the wholesale price in A is sufficiently high and that in B is sufficiently low, parallel trade is strictly positive for low unit trade costs. Working backwards, we insert the retail prices in the demand functions and solve the manufacturing firm's optimization problem to obtain the unique equilibrium wholesale prices

$$w_A = \frac{2(S\lambda b + 1)\lambda - 2S\lambda^3 b}{4 - 3S\lambda^2 b + 4S\lambda b} - \frac{1}{2}t \tag{51}$$

$$w_B = \frac{2\lambda}{4 - 3S\lambda^2 b + 4S\lambda b} \tag{52}$$

and it follows that the wholesale price for distributor A decreases in t, while the wholesale price for distributor B is constant. The result is qualitatively similar, but not identical, to quantity competition. The manufacturing firm has three major problems to solve in both cases. It has to moderate the pro-competitive effect in market A, reduce the waste of resources and avoid a double-markup problem in market B. The difference between quantity and price competition is that the problem of resource waste in PI activities is less severe in the latter case. The trade cost also has a more direct effect on competition in market A. In addition, product differentiation has a positive value in market A, which adds a fourth effect of legal PI.

We can use the equilibrium wholesale prices to obtain the corresponding equilibrium retail prices

$$p_A = \frac{2(1 - S\lambda^2 b + S\lambda b)}{4 - 3S\lambda^2 b + 4S\lambda b} \tag{53}$$

$$p_{PI} = \frac{\lambda(2 - S\lambda^2 b + S\lambda b)}{4 - 3S\lambda^2 b + 4S\lambda b} + \frac{1}{2}t \tag{54}$$

$$p_B = \frac{1}{2b} + \frac{\lambda}{4 - 3S\lambda^2 b + 4S\lambda b} \tag{55}$$

and the equilibrium PI quantity is

$$q_{PI} = \frac{S\lambda b}{4 - 3S\lambda^2 b + 4S\lambda b} - \frac{t}{2(1-\lambda)\lambda}. \tag{56}$$

If distributors compete in prices, retail prices for the locally distributed products remain constant if the unit trade cost is below the critical level establishing a positive volume of PI. Within this range, a higher unit trade cost would generate a lower PI volume and less competition in the recipient market, but retail prices would not diverge as conventional intuition would suggest. Instead, these prices would be unaffected by changes in t. The retail price for the PI variety in market A, however, would be increased by a rise in the trade cost, in line with conventional intuition.

The main difference between quantity and price competition is that retail prices of the locally distributed goods have a tendency to converge with quantity competition if the trade cost increases within the low range, while the price differential is constant in the case of price competition. The main intuition for this difference is that in the former case the manufacturing firm must push retail prices together to avoid wasteful PI activities, while the competitive disadvantage of a higher trade cost would be sufficient to reduce the volume of PI in the latter case.

6. Concluding Remarks

We developed a model in which a manufacturing firm owns an intellectual property right in two markets but its ability to limit parallel imports from one market to the other is exhausted. In this environment, the firm has the ability to set differential wholesale prices to its independent distributors in the two locations. It will use these instruments to maximize profits within the vertical-control framework. There are three essential tradeoffs for the manufacturer. It wishes to restrict the extent of competition from PI in the A market, limit the amount of PI because it wastes real resources in transport costs, and avoid the double-markup problem in market B arising from the inability to set an efficient (zero) wholesale price.

Our analysis turned up some interesting results. Because of the cross-cutting effects of PI on wholesale prices in the two markets, it is possible to observe a divergence in wholesale prices as trade costs are reduced within a low range. As a result, retail prices may diverge as well. Arguably, the EU is in a situation of low and declining internal trade restrictions. In this context, paradoxically, the policy of free parallel trade among member states may be a force for retail price divergence.

For intermediate and high trade costs, the firm might wish to set a negative wholesale price in A but could be constrained to a minimum price of zero. As trade costs increase, the volume of PI declines and the double-markup problem in B diminishes. Retail prices rise in both locations as transport costs increase within the intermediate range, then diverge at high trade costs as these prices achieve levels expected with wholesale-market segmentation.

Perhaps the most interesting implication of our analysis is that there is a substantial difference between integrating retail markets and wholesale markets. In the former case, as trade costs fall due to declining trade barriers and transport costs, straightforward arbitrage would push markets toward price convergence, which is the intuitive result and the outcome anticipated by integration policy. However, in the latter case, where PI are prevalent, declining trade costs could integrate wholesale markets even as they push retail markets toward greater segmentation. Again, this possibility suggests that the favorable view of parallel imports in EU competition policy, stemming from a tradition that does not consider vertical distribution arrangements, may be due for reconsideration.

It could be argued that the double-markup problem in the export market is partly due to seller concentration at the retail level. The manufacturing firm perhaps could introduce several competing retailers to reduce the retail margin and moderate the double-marginalization problem. This benefit of having more than one retail distributor could, however, be offset by other disadvantages such as free-riding and duplicated fixed costs in retailing. Hence, extending the analysis to study the effects of parallel trade with an endogenous retail structure is an interesting subject for future research.

References

Chen, Yongmin and Keith E. Maskus. 2005. Vertical Pricing and Parallel Imports. *Journal of International Trade and Economic Development* 14: 1–18.

European Central Bank. 2002. Price Level Convergence and Competition in the Euro Area. *Monthly Bulletin*, August: 39–50.

Friberg, R. and K. Martensen. 2001. Endogenous Market Segmentation and the Law of One Price. Stockholm School of Economics, SSE/EFI Working Paper No. 471.

Ganslandt, Mattias and Keith E. Maskus. 2003. Vertical Restraints, Distribution, and the Price Impact of Parallel Imports: Implications for the European Union and Sweden. In K. Lundvall, ed., *High Prices in Sweden — A Result of Poor Competition?* (Stockholm: Swedish Competition Authority), 160–223.

Ganslandt, Mattias and Keith E. Maskus. 2004. The Price Impact of Parallel Trade in Pharmaceuticals: Evidence from the European Union. *Journal of Health Economics* 23: 1035–1057.

Hilke, J. C. 1988. Free Trading or Free Riding: an Examination of the Theories and Available Evidence on Gray Market Imports. *World Competition* 32: 75–92.

Malueg, David A. and Marius Schwartz. 1994. Parallel Imports, Demand Dispersion, and International Price Discrimination. *Journal of International Economics* 37: 187–196.

Maskus, Keith E. 2000. Parallel Imports. *The World Economy* 23: 1269–1284.
Maskus, Keith E. and Yongmin Chen. 2002. Parallel Imports in a Model of Vertical Distribution: Theory, Evidence, and Policy. *Pacific Economic Review* 7: 319–334.
National Economic Research Associates. 1999. *The Economic Consequences of the Choice of Regime in the Area of Trademarks* (London: National Economic Research Associates).
Raff, Horst and Nicholas Schmitt. 2005a. Endogenous Vertical Restraints in International Trade. *European Economic Review* 49: 1877–1889.
Raff, Horst and Nicholas Schmitt. 2005b. Exclusive Dealing and Common Agency in International Markets. CES-IFO working paper no. 1168.
Richardson, Martin. 2002. An Elementary Proposition Concerning Parallel Imports. *Journal of International Economics* 56: 233–245.
Tarr, David G. 1985. An Economic Analysis of Gray Market Imports. U.S. Federal Trade Commission, manuscript.

Appendix

The Subgame Equilibrium

It is assumed that at least one distributor finds it profitable to supply market A. Accordingly, we have to consider three different cases: both distributors supply that market, only distributor A supplies it or, finally, only distributor B does so. Consider duopoly first, where the distributors compete in quantities in market A. The profit functions of the two distributors are

$$\pi_A^* = (1 - (q_A^* + q_B^*) - w_A^*)q_A^*, \tag{A.1}$$

$$\pi_B^* = (1 - (q_A^* + q_B^*) - w_B^* - t)q_B^*, \tag{A.2}$$

and the corresponding first-order conditions are

$$1 - 2q_A^* - q_B^* - w_A^* = 0, \tag{A.3}$$

$$1 - q_A^* - 2q_B^* - w_B^* - t = 0, \tag{A.4}$$

and the unique solution to this system of equations is

$$q_A^* = \frac{1 - 2w_A^* + w_B^* + t}{3}, \tag{A.5}$$

$$q_B^* = \frac{1 + w_A^* - 2w_B^* - 2t}{3}. \tag{A.6}$$

Distributor A's quantity is positive if

$$w_A^* < \frac{1 + w_B^* + t}{2} \tag{A.7}$$

and distributor B's quantity is positive if

$$w_B^* < \frac{1 + w_A^* - 2t}{2}. \tag{A.8}$$

These two conditions must simultaneously hold in duopoly. We combine them and rewrite the expression to obtain

$$2w_B^* + 2t - 1 < w_A^* < \frac{1 + w_B^* + t}{2} \tag{A.9}$$

and

$$2w_A^* - t - 1 < w_B^* < \frac{1 + w_A^* - 2t}{2}. \tag{A.10}$$

Next, consider a deterrent wholesale cost for distributor B,

$$w_B^* \geq \frac{1 + w_A^* - 2t}{2} \Longleftrightarrow w_A^* \leq 2(w_B^* + t) - 1,$$

for which $q_B^* = 0$ in equilibrium. Then distributor A is the only supplier in his market. It maximizes

$$\pi_A^* = (1 - q_A^* - w_A^*)q_A^* \tag{A.11}$$

and the first-order condition is

$$1 - 2q_A^* - w_A^* = 0 \tag{A.12}$$

with a unique solution

$$q_A^* = \frac{1 - w_A^*}{2} \tag{A.13}$$

which is positive for $w_A^* < 1$. Correspondingly, if the wholesale cost for distributor A is a deterrent, i.e.,

$$w_A^* \geq \frac{1 + w_B^* + t}{2} \Leftrightarrow w_B^* + t \leq 2w_A^* - 1, \tag{A.14}$$

then $q_A^* = 0$ in equilibrium and distributor B is the only supplier in A. It maximizes

$$\pi_B^* = (1 - q_B^* - w_B^* - t)q_B^* \tag{A.15}$$

and the first-order condition is

$$1 - 2q_B^* - w_B^* - t = 0 \tag{A.16}$$

with a unique solution

$$q_B^* = \frac{1 - w_B^* - t}{2} \tag{A.17}$$

which is positive for $w_B^* < 1 - t$.

The Manufacturing Firm's Profit

Insert segmented prices in the profit function to obtain the profit in the segmented equilibrium:

$$\Pi^S = \frac{S + b}{4b} \tag{A.18}$$

and correspondingly retail prices and the PI quantity for low trade costs to obtain the profit in the PI equilibrium for $t < \underline{t}$

$$\Pi_{t<\underline{t}}^{PI} = \frac{Sb + 5Sbt^2 - 2tSb + 4t^2 + S^2}{4Sb}. \tag{A.19}$$

Next insert retail prices and the PI quantity for intermediate trade costs to obtain the profit in the PI equilibrium for $t \geq \underline{t}$

$$\Pi_{t\geq\underline{t}}^{PI} = \frac{8Sb^2 + 20Sb^2t^2 - 8tSb^2 + 9S^2b + 16bt^2 + 4b + 4S}{4b(4 + 9Sb)} \tag{A.20}$$

and, finally, retail prices for high trade costs to obtain the profit in the no-arbitrage equilibrium

$$\Pi^{NA} = \frac{4b + 4S - Sb^2 + 4tSb^2 - 4Sb^2t^2}{16b}. \tag{A.21}$$

Chapter 5

Parallel Imports and the Pricing of Pharmaceutical Products: Evidence from the European Union*

Mattias Ganslandt and Keith E. Maskus†

1. Introduction

Parallel imports (PIs) are legitimately produced goods imported legally into a country without the authorization of a trademark, copyright, or patent holder. The legal doctrine governing the permissibility of parallel imports is exhaustion, or the point of distribution at which rights to control further distribution are ended. Under national exhaustion the rights holder may prevent such importation. However, under international exhaustion PIs are legal. The essential purpose of such trade is arbitrage between countries with different prices.

For several years, parallel trade of prescription drugs has been an important issue for the pharmaceutical industry and numerous policy institutions in Europe, including the European Commission, the European Court of Justice, and member states of the European Union. Manufacturers prefer restraints on such trade in order to

*Reprinted from *Journal of Health Economics*, Vol. 23, No. 5, pp. 1035–1057. Copyright 2004, with permission from Elsevier.

†The research for this paper was begun while Maskus was a lead economist at the World Bank. We thank seminar participants at the World Bank, ETSG in Glasgow, European Economic Association, IUI, University of Colorado, University of Oregon, New York University, University of Hawaii and Stockholm University for valuable comments. Ganslandt is grateful to STINT and The Jan Wallander and Tom Hedelius Foundation for financial support. Special thanks to LIF and Kajsa Savfstrom for data and Eina Wong for research assistance. We are also grateful to the editor and two referees for comments.

support higher prices in markets with weaker price controls. However, PIs exist and recent industry estimates suggest that lost sales in the EU currently amount to some $3 billion per year.[1]

Parallel trade in prescription medicines has become a prominent issue in other parts of the world as well. For example, American consumers and policy makers have grown increasingly concerned about the relatively high prices of patented and brand-name drugs in the United States. Several measures have been proposed by U.S. policy makers, including direct regulation of drug prices and a policy to admit PIs. In an effort to reduce drug costs for consumers both the House and the Senate approved a measure in July 2000 that would have permitted pharmacists and wholesalers to import cheaper drugs from other countries.[2] The Clinton Administration refused to implement the bill in December 2000, claiming that it could not guarantee the safety of such products. Nevertheless, high prices of patented and brand-name drugs in the United States remain an important policy issue and new legislation was introduced in 2003 to deregulate restrictions on parallel trade from Canada.

At the global level some analysts argue for establishing restraints on parallel trade in order to support low drug prices in poor countries (World Health Organization, 2001). It is argued that if tight segmentation could be maintained between rich markets and poor markets, drug companies might be willing to supply the latter with large volumes at marginal production costs. In August 2003, members of the World Trade Organization signed a waiver to the Agreement on Trade-Related Aspects of Intellectual Property Rights (TRIPS) to permit developing countries to import generic substitutes of patented medicines in order to cope with national emergencies, such as HIV, tuberculosis or malaria.[3] A key provision of this waiver is that beneficiary countries agree to control re-exportation of drugs they import

[1]See "Drug Makers Try to Curtail Cheap Imports," *The Wall Street Journal*, 11 April 2002.
[2]See "Senate Approves Measure on Import of Cheap Drugs," *The Wall Street Journal*, 20 July 2000.
[3]See "Implementation of Paragraph 6 of the Doha Declaration on the TRIPS Agreement and Public Health," 30 August 2003, WT/L/40 at http://www.wto.org/english/tratop_e/trips_e/implem_para6_e.htm.

in this fashion, for even generic drugs could be subject to parallel trade given the anticipated price differentials.

Parallel imports of pharmaceuticals are controversial because their welfare effects are generally ambiguous. First, there is a tension between two major public-policy objectives: innovation and development of new drugs, on the one hand, and short-run cost-containment strategies for the health care system and broad access to existing medicines, on the other.

The research-intensive pharmaceutical industry relies heavily on patents, the value of which depends in part on the scope for price differentiation. This scope depends critically on the existence of barriers to arbitrage. The value of a patent is, therefore, partly determined by the definition of the geographical area within which the product can be freely circulated and re-sold without the original manufacturing firm's consent. In the terminology of intellectual property law, patent distribution rights are "exhausted" over a pre-defined area upon first sale. Once these rights are exhausted the patent holder can no longer restrict the circulation of the product. In other words, parallel trade is permitted over the geographical area where the rights to control distribution have been exhausted but not from regions or countries outside the area. One should, therefore, expect arbitrage to limit the scope for price discrimination within the area of exhaustion. The narrower the area of exhaustion, the greater is the scope for price differentiation. Consequently, countries with national exhaustion policies provide stronger incentives to innovate at the expense of higher consumer costs. In this context, permission of PIs could reduce incentives to innovate, even as consumers in high-price countries would gain due to lower prices for existing drugs.[4]

Second, there is an important conflict between divergent price regulations in different countries, on the one hand, and the consequences of parallel trade, on the other. Where PIs are permitted, arbitrageurs can exploit regulated price differentials, leaving the original manufacturer with fewer alternatives to adjust its behavior. More specifically, patent holders may have an incentive to accommodate

[4]Danzon (1998) argues that a segmented equilibrium with price-discriminating monopolies can be optimal from a welfare perspective.

their pricing strategies to PIs while also taking measures to raise the trade costs of arbitrageurs. This accommodation would result in some price convergence but could expend considerable real resources on trade activities.

The tension between these multiple policy objectives is evident in EU policy.[5] Under current EU law a member state has an exclusive right to define its health care policy, including price regulations and benefits, while the principle of free movement of goods allows individuals or firms within the EU to trade goods across borders without the consent of the producer.

The conflict between national price regulations, on the one hand, and mandates for EU-wide circulation of products, on the other, is recognized by the EU authorities. The Commission notes in its *Communication on the Single Market in Pharmaceuticals* (1998) that, "Unless parallel trade can operate dynamically on prices, it creates inefficiencies because most, but not all, of the financial benefits accrue to the parallel trader rather than the health care system or patient." Even so the Commission concludes that "...parallel trade must equally be seen as an important driving force for market integration and, consequently, for achieving the Single Market."

Thus, while there are numerous issues concerned in the welfare tradeoffs, a central question for policymakers is the potential for PIs to generate significant price competition in higher-priced countries. Despite the importance of this question, there are no direct estimates of the price impacts of parallel trade in drugs. The aim of this paper is, therefore, to provide the first systematic economic investigation of these effects. More specifically, our focus here is on the price-integrating impact of PIs in pharmaceutical products within the European Union.

For this purpose we first develop a simple two-country model of parallel trade in which an original manufacturer maximizes profits in the high-price nation against a demand curve residual to import competition. In order to capture the effects of differing price controls, we assume that PIs come from a country in which price is capped.

[5]See Ganslandt and Maskus (2000) for further details.

The number of parallel-importing firms is endogenous to market conditions and fixed costs under free entry, generating a prediction for price as a function of the number of entrants. In equilibrium the manufacturer sets a price in the high-price market that accommodates PIs, generating partial price convergence with the exporting country. We estimate this model using biweekly product-level data from the Swedish market from the first quarter of 1994 to the third quarter of 1999. We have compiled a unique data set consisting of prices for 50 major pharmaceutical products, the sales of patent holders, the wholesale prices of their drugs, the identity and time of entry of parallel importers, and the sources of PIs from within the EU.

The Swedish market provides a natural test for our theoretical hypotheses. Before 1995 Sweden prohibited PIs of pharmaceutical products. However, Sweden's entry into the European Union on January 1, 1995 required that country to adopt the EU-wide exhaustion principle and permit such trade. Thus, we analyze the impacts of entry after an exogenous shock to the patented market. We find that prices in Sweden of drugs subject to competition from PIs declined relative to other pharmaceutical prices, with the effect concentrated at the end of the period. The econometric analysis shows that competition from the entry of parallel traders had a significant effect on the prices of original manufacturers in Sweden. We estimate the effect of such entry to be a reduction of between 12 and 19 percent of manufacturers' relative prices.

This result may be compared to other findings in the literature regarding competition from generic producers. Caves *et al.* (1991) found that in the United States the original innovator's price declined as the number of generic entrants rose, but the rate of price decline was small. Wiggins and Maness (1994) also reported that brand-name drug prices declined as generic entry expanded. Lu and Comanor (1998) found that the number of branded substitutes had a significantly negative effect on launch prices of new drugs in the United States. However, such findings are not universal. Grabowski and Vernon (1992) observed an increase in brand-name prices in response to entry. Frank and Salkever (1992, 1997) discovered that generic entry reduced the own-price elasticity of branded pharmaceutical products in the United States after patent expiration, permitting those prices to rise while generic producers targeted the higher-elasticity demand segments at considerably lower prices.

This issue remains unsettled and our results provide evidence for the hypothesis that entry through parallel imports exerts a significant downward influence on price by directly competing with the original producer in the same product. It should be noted that PIs are different from generic products, for the former exist while the original product is on patent and the latter must await patent expiration. Further, because PIs are sourced from licensees or distributors of original manufacturers, they are identical to the patented product save for differences in packaging. Thus, they should be highly substitutable with the original goods.

The rest of this paper is organized as follows. We provide a brief overview of EU case law in Section 2. We present the theoretical model of PIs in Section 3. In Section 4 we develop the econometric approach and estimate the price effects of PIs from other countries in the EU to Sweden. We make concluding remarks in Section 5.

2. Current EU Case Law on Parallel Imports of Pharmaceutical Products

The European Court of Justice has held that free circulation of goods takes precedence over protection of intellectual property rights. In *Merck v Stephar* (C 187/80) the European Court of Justice held that a patent holder marketing its product in two different member states cannot prevent arbitrage between the two local markets, despite differences in intellectual property protection in the two countries. Thus, exhaustion applies upon first sale anywhere in the EU. Moreover, varying degrees of price control across countries do not justify prevention of parallel imports from countries with more rigorous regulations to markets with less rigorous regulations, as found in *Merck v Primecrown* (joined cases C-267/95 and C-268/95). Furthermore, parallel importers have limited rights to use original trademarks in marketing their products (*Dior v Evora*, C-337/95, and *BMS and Others v Paranova*, joined cases C-427/93, C-429/93, and C-436/93). Finally, manufacturers cannot partition the single market by introducing a new variety in member states, which could have the effect of replacing market authorization for the prior variety, where its product is subject to competition from parallel imports (*Rhone-Poulenc Rorer*, case C-94/98).

However, exhaustion in the European Union has important limitations. Most importantly, it does not extend to countries outside the common market (*EMI v CBS*, case C-51/75 and *Silhouette v Hartlauer*, case C-355/96). Thus, the ECJ has established a principle of "community exhaustion" but rejected the idea of international exhaustion. However, the principle of community exhaustion does not extend to cases where the goods are sold in a member state under a compulsory license, as established in *Pharmon v Hoechst* (C-19/84). To summarize, the EU system essentially mandates free parallel imports within its territory, despite the existence of national intellectual property regimes and price controls, so long as the manufacturer has placed the good voluntarily on the market.

In an important modification of EC policy, the European Court of First Instance ruled in 2000 that original manufacturers could impose supply quotas for foreign wholesalers so long as those quotas did not constitute contractual agreements prohibiting export of a product (*Bayer AG v Commission of the European Communities*, Case T-41/96, 26 October 2000). A literal interpretation of this ruling would be that it permits restraints on sales from manufacturers to licensed wholesalers but does not impede the ability of parallel importers to acquire drugs and ship them abroad. However, evidence suggests that manufacturers do limit supplies available for licensees within the EU that could escape into other markets in the region (see footnote 1). As a consequence, parallel traders may encounter limits on the quantities available for their activity.

3. A Model of Parallel Imports

We develop a simple model of price arbitrage between two countries. The most important feature of this model is that parallel trading firms jointly choose the maximum amount of drugs they acquire for shipment to higher-priced markets. The notion that parallel traders do not acquire unlimited volumes is consistent with evidence that the share of such trade in total pharmaceutical sales in high-price markets, such as Sweden, Germany and the UK, rarely exceeds ten percent except in a few major products. This situation occurs despite the existence of marked price differentials between these markets and EU member countries where prices are substantially lower (see footnote 1). This may be because parallel importers find it difficult to locate distributors who are willing to supply their needs. Further, as just discussed, original manufacturers tend to limit their sales to foreign wholesalers, or to cap production levels permitted to foreign licensees, in order to restrict the volume of parallel trade.[6] However, rather than relying on such constraints, in the model we derive the profit-maximizing quantity chosen by the PI firms.

The assumption of a quantity limit can be justified analytically on two grounds. First and foremost, our model is consistent with market conditions in Sweden. The model takes into account that PI firms are few and act strategically. The PI firms must order and ship the product from abroad prior to sale in Sweden and the quantity of PI firms is consequently pre-determined at the time when the manufacturing firm sets its price and markets clear.[7] The ordering and shipping of PI products may therefore be considered a commitment mechanism for PI firms and the sequence of decisions in our model is consistent with these real-world conditions. Second, the quantity limit for PI firms is a convenient way to take into account the increasing marginal cost of PIs. As limited volumes of pharmaceutical products are available for parallel trade, PI firms use low-cost suppliers first and more costly distributors only second.[8] In addition, PI firms would take costly

[6]The President of the European Association of Pharmaceutical Full-line Wholesalers (GIRP) recently stressed "....the full-line wholesalers' concern that they will no longer be able to guarantee the reliable and timely distribution of all pharmaceutical products due to the practice of supply restrictions by several manufacturers" (GIRP Press release, 17 December 2003).

[7]Pharmacy managers in Sweden report that there is often a shortage of PI products. Such products are frequently out-of-stock and wholesalers are unable to meet the demand from pharmacies. Quantity limits are consequently of significance for the Swedish market (Holmberg *et al.*, 2003).

[8]Recent evidence from Canada suggests that it is increasingly costly for cross-border retailers to find large volumes available for arbitrage. For instance, the supply control systems used by major pharmaceutical companies have changed the behavior of Canadian online pharmacies selling prescription drugs to US citizens. According to some sources online pharmacies increasingly turn to wholesalers unaffected by the supply control systems. However, unaffected wholesalers charge 30 percent higher prices due to the risk of supply cutoffs. See "Drugmakers Curb Canadian Outlets," *Detroit Free Press*, 6 September 2003.

actions to circumvent supply restrictions imposed by manufacturing firms.[9] Accordingly, we expect an increasing marginal cost of engaging in parallel trade.

Consider a model with two markets (home and foreign) denoted h and f. We assume throughout that the home market is the high-income country and the foreign market is the low-income country. Demand for a specific pharmaceutical product in the home market is

$$D_h(p) = a - bp \tag{1}$$

where b is proportional to the marginal utility of money.[10] We make the simplifying assumption that no substitute therapies exist so that only the own-price appears in this demand function.

The product is patented in both countries and produced by a single manufacturing firm at marginal cost c, which for simplicity we set equal to zero in markets h and f. There is an autonomous government in the foreign country that is able to set a cap on the price charged to retail outlets (hospitals and pharmacies) in its own market. The price cap is denoted ($\bar{p_f}$) and is always (strictly) binding in that market. The price in the home market is unregulated. The manufacturing firm consequently sells the product in the home and foreign market at prices p and $\bar{p_f}$, respectively.

Assume that there is a small number of symmetric parallel-importing firms permitted to engage in commercial arbitrage, while arbitrage by individual patients between the two markets is prohibited. The marginal cost of engaging in parallel trade is t per unit and the fixed cost is T. The PI product is a perfect substitute for the product sold by the manufacturing firm to retail outlets in the h market. It is assumed that the PI price is below the price set by the manufacturing firm, which ensures that the entire PI quantity shipped to market h is sold. Denoting this quantity by X, the

[9]Tactics used by Canadian on-line pharmacies to circumvent supply restrictions include payments to storefront pharmacies in exchange for supplies as well as the establishment of own retail outlets, permitting on-line drugstores to move supplies between channels. See "Pfizer Moves to Stop Drugs from Canada," *The New York Times*, 14 January 2004.

[10]This system of demand functions can be obtained approximately from a linear-quadratic utility function (i.e., $\alpha x - \beta x^2/2 + v(y)$) so long as the expenditure on x is a relatively small share of the consumer's budget and we substitute $a = \alpha/\beta$ and $b = v'(m)/\beta$, assuming that $v'(m) > 0$ and $v''(m) < 0$.

manufacturing firm's residual demand in the home market is

$$D_h^m(p) = D_h(p) - X. \tag{2}$$

The strategic interaction is modeled as a multi-stage game. In the first stage, n symmetric parallel importers enter the market, applying for an approval from the authorities in h to import the product from market f.[11] PI firms will only enter the market if the expected profit is non-negative. In the second stage, each PI firm (indexed by i) simultaneously ships a quantity x_i from the wholesaler in the foreign market to the home market, incurring the variable trade cost t. The shipped quantity x_i is known to all firms. In the third stage, the manufacturing firm sets its price (p) in the home market and markets clear. Throughout the game the manufacturing firm takes the price in the foreign market ($\bar{p_f}$) as given.[12] Furthermore, it is assumed that both markets will be served in equilibrium.

In the theoretical analysis we look for the unique sub-game perfect Nash equilibrium. We solve the game in the usual fashion, starting in the last stage and working backwards to the first stage. This procedure also allows us to discuss different aspects of parallel trade.

In the last stage, when solving for an optimal price for the manufacturing firm subject to a particular quantity of PIs, we identify a relationship between the volume of PIs and the price in the home market. Working backwards to the second stage, we solve the non-cooperative sub-game among the PI firms and find the equilibrium volume of imports. This yields a relationship among the volume of PIs and the number of PI firms, the size of the h market, and the price in the f market. Proceeding to the first stage we solve for the

---

[11] This assumption is motivated by the fact that in the Swedish market a parallel importer must have an approval by the Medical Products Agency before it can commence to import a pharmaceutical product.

[12] Pecorino (2002) presents a model in which a monopolistic pharmaceutical firm is unconstrained in price in its home market but bargains with the price authority in the foreign market. In his setup the permission of parallel imports (or "reimports") into the home market generates a higher price in the foreign market due to a change in Nash bargaining positions, raising overall monopoly profits, while prices in both countries are integrated by trade. The result is driven both by the assumption that the countries share similar incomes and by the existence of a single firm. In our model there are different incomes and free entry of PI firms.

equilibrium number of PI entrants. This shows how the number of PI firms in equilibrium depends on the size of the home market, prices, as well as variable and fixed costs.

3.1. *Unlimited Parallel Imports Lead to Price Convergence*

Before we proceed to our analysis of the multi-stage game, it is worthwhile looking at a benchmark in which arbitrage is perfectly elastic.[13] In this case PI firms import a product to the home market as long as the retail price difference is larger than the trade cost. If parallel imports were actually to happen, the manufacturing firm would achieve no sales in h. Thus, the only way it can sell in the home market is by setting a price in the home market that makes such trade unprofitable. We shall refer to this as an "arbitrage-free" price.

When potential arbitrage is unlimited and perfectly elastic, blocking PIs with an arbitrage-free price in the home market is more profitable than accommodating the inflow. This strategy generates home sales at per-unit revenue equal to the foreign price plus trade cost, while under accommodation the manufacturer only sells in the foreign market with per-unit revenue equal to foreign price.

Compared to the segmented equilibrium, the threat of PIs results in price convergence (up to the level of trade cost) by virtue of a price decline in the home market. Thus, potential arbitrage is enough to induce price convergence but no actual PIs exist in equilibrium. It is worth noting that from a welfare standpoint, potential competition from PIs is more efficient than actual PI competition because in the former case no real resources are devoted to arbitrage (no trade costs are incurred), while they are in the latter.[14]

3.2. *The Effects of a Quantity Limit*

The impacts of PIs are sensitive to the existence of a maximum volume constraint, i.e. the supply elasticity of parallel imports.

[13]Ganslandt and Maskus (2001) offer proofs of these propositions.

[14]It is theoretically possible from the global standpoint that a segmented equilibrium with international price differentiation is more desirable than an integrated equilibrium with partial or complete price convergence.

In particular, such a constraint gives the manufacturing firm an incentive to accommodate parallel trade rather than to set an arbitrage-free price in the home market.

Starting in the final stage, the manufacturing firm maximizes profits according to

$$\mathrm{Max}_p(a - bp - X)p + X\bar{p_f} \tag{3}$$

which gives the following first order condition

$$a - 2bp - X = 0. \tag{4}$$

The profit-maximizing price as a function of the total PI quantity is consequently

$$p(X) = \frac{a - X}{2b} \tag{5}$$

which shows that the price falls in the volume of PIs.

Working backwards to the second stage, we solve for the non-cooperative quantities chosen by the PI firms. These firms would face a downward sloping demand curve, recognizing that they all possess some (limited) degree of market power. Each symmetric PI firm maximizes

$$\mathrm{Max}_x[p(X)x - (\bar{p_f} + t)x] \tag{6}$$

where $p(X)$ is given by equation (5). The n interior first order conditions follow immediately and are symmetric:

$$\frac{a - 2x - \sum_{-i} x_{-i}}{2b} - (\bar{p_f} + t) = 0 \tag{7}$$

where the subscript $-i$ refers to all PI firms other than i. The unique sub-game equilibrium is

$$x(n) = \frac{a - 2b(\bar{p_f} + t)}{n + 1} \tag{8}$$

which increases in the size of the home market (a) and decreases in the sum of foreign price and the unit trade cost. In the sub-game equilibrium with n PI firms, the total PI quantity is

$$X(n) = \frac{n}{n+1}[a - 2b(\bar{p_f} + t)] \tag{9}$$

which increases in the number of PI firms. Accordingly, the equilibrium price as a function of n is

$$p(n) = \frac{1}{2b}\left[a - \frac{n(a - 2b(\bar{p_f} + t))}{n+1}\right]. \tag{10}$$

Thus, the equilibrium price in h falls in the number of PI firms. It is also worth noting that the equilibrium price in the home market converges to the foreign market price plus variable trade cost as the number of PI firms goes to infinity.

Working backwards to the first stage we find the equilibrium number of PI firms. It is assumed that a PI firm will only enter if the expected profit in equilibrium is non-negative. Thus, the free-entry equilibrium condition for PI firms is

$$(p(n) - (\bar{p_f} + t))x(n) - T \leq 0 \tag{11}$$

where $p(n)$ is the sub-game equilibrium price given by equation (10) and $x(n)$ is the corresponding sub-game equilibrium quantity given by equation (8). Assuming for simplicity that the number of PI firms is a continuous variable, we solve equation (11) for the equilibrium number of PI firms:

$$n^* = \frac{a - 2b(\bar{p_f} + t)}{\sqrt{T}} - 1. \tag{12}$$

This expression is positive for sufficiently small fixed costs, i.e. $T < (a - 2b(\bar{p_f} + t))^2$. Accordingly, the equilibrium number of PI firms decreases in the foreign price ($\bar{p_f}$) and in the variable trade cost (t). It increases in the size of the market (a) and decreases in the fixed cost (T).

Finally, we can obtain the equilibrium price and the equilibrium volume of PIs by inserting the number of PI firms into equations (10) and (9), respectively. These various expressions constitute our accommodation equilibrium.

4. Empirical Analysis

We undertake an empirical analysis of the price effects of parallel imports. The analysis focuses on two issues. First, we quantify the average price impact of PIs in the import market (Sweden) by comparing products not subject to competition from that source with products subject to such trade. Second, we estimate an econometric model of the impact of arbitrage from parallel trade, accounting for the endogeneity of entry decisions. We consider both actual competition as PI firms enter and potential competition arising from the entry of Sweden into the European Union.[15]

For these purposes we collected detailed data about the pharmaceutical market in Sweden. The Swedish market provides a natural test for our theoretical hypotheses. When Sweden joined the European Union on the 1st of January 1995 the policy on PIs was drastically changed. Before membership in the EU, parallel imports of pharmaceutical products were prohibited, but when Sweden became a member they had to be allowed.[16] Sweden is, therefore, a natural experiment for considering an exogenous change between two completely different policies.

4.1. *Current Regulatory Structure*

Our analysis is motivated by the differences in regulation that should sustain international price gaps as analyzed in the model. There is a significant difference between the pricing of pharmaceuticals in Sweden, on the one hand, and the pricing of drugs in strictly regulated markets such as Italy, Spain, Portugal and Greece, on the other. The system in Sweden can best be characterized as a negotiation between the national health-care authority and the pharmaceutical companies. When a new drug is introduced, the pharmaceutical company suggests a price and justifies its proposal by listing the benefits of the product, the expected patient volume, prices of comparable treatments and prices in other countries. After some negotiation, upon approval of the price by the authority, the product is covered

[15] As noted below, the endogeneity of entry precludes a sharp distinction in this regard.
[16] Parallel imports are subject to a regulatory approval period, which effectively delayed the regime shift.

in the national health-insurance program. The pharmaceutical company is free to propose changes to the price at a later date, subject to an approval by the authority. Thus, this policy is sufficiently flexible that it constitutes a relatively moderate constraint when the manufacturing firm makes its pricing decisions.

The situation in Southern Europe is different. The prices of pharmaceuticals are strictly controlled in such countries as Italy, Portugal, Spain and Greece. In Italy the prices of reimbursed drugs are regulated with a price cap. The price of a drug eligible for reimbursement within the health system can be no higher than the average price in twelve EU countries. This maximum price cannot be exceeded else the product is removed from the reimbursement list. In Portugal the regulation is even broader in scope and the prices of all prescription drugs are regulated. The price must be equal to or lower than the lowest price in three reference countries (Spain, Italy and France). If the average price in the two remaining countries is more than 30 percent higher than the lowest price, 30 percent of the average in the two remaining countries is added to the lowest price. Spain also controls the prices of all prescription drugs. Prices are set based on cost, profit allowance and anticipated volume of sales. Finally, Greece regulates drug charges with a system of reference prices.

4.2. *Data*

We employ three data samples. The primary data were provided by LIF, the Swedish Association of the Pharmaceutical Industry. The first data set includes the 50 molecules with the highest sales values in the Swedish market in 1994, 1995, 1996, 1997 and 1998. Approximately 35–38 percent of the Swedish pharmaceutical market in value terms is included in this data set. An observation is a "product", defined as a molecule with a specific anatomical therapeutic classification (ATC) code, form and concentration. More specifically, the data include 164 different forms and concentrations distributed over 50 molecules.[17] For each product there are quantities and prices for

[17]Whether a particular molecule is on patent in Sweden is not public information. Our best assessment is that 44 of the 50 molecules were protected by patents during the sample period because the time since initial launch was well within the average patent duration. Moreover, 49 of the 50 molecules were sold under a protected trademark.

both original manufacturing firms and all PI firms on a yearly basis. Note that because we use data assembled on a detailed product level the price comparisons are not subject to the methodological problems with aggregated price comparisons discussed in Danzon and Chao (1995) and Danzon and Chao (2000).

Second, we have constructed a biweekly, unbalanced panel with prices for the 164 products in our sample from January 1, 1994 to September 1, 1999, for a total of 149 periods.[18] We also have collected data on approvals to parallel-import the products in our sample for 1995–1999 and added this information to our biweekly panel. An approval is a formal decision by the Swedish Medical Products Agency and it allows a specific parallel importer to import a unique patented or trademarked molecule in a specific form and dose from a specified export country. Thus, in the biweekly panel we have information about the date when PI firms are approved as well as the date when they enter as actual competitors (the first-pricing date).

In a third sample we have detailed prices for subsets of the 50 molecules in France, Portugal, Spain, and Italy for 1994, 1995 and 1998. Spain and Italy are the two main parallel-exporting countries in the sample and we make particular use of their prices for 26 available molecules. These prices were obtained from IMS Health, a private consulting firm. The narrower sample was restricted to drugs that are precisely comparable on an international basis. Thus, they have the same ATC code, form, and concentration and originate from the same manufacturers across all markets.

4.3. *The Pharmaceutical Market and Parallel Imports*

Figures summarizing the Swedish pharmaceutical market are presented in Table 1. Pharmaceutical sales were approximately 0.8–0.9 percent of GDP valued at wholesale prices during the period 1995–1998. They ranged from 13.4 billion Swedish Kronor (SEK) in 1995 to 16.6 billion SEK in 1998. In 1998 the 50 top drugs accounted

[18]The panel is unbalanced in the sense that new products and PI approvals enter the data set over the period, implying that the number of products differs over time. In particular, there were 196 products by the end of the period.

Table 1. The pharmaceutical market in Sweden, 1995–1998.

Sweden	1995	1996	1997	1998
Gross Domestic Product (MSEK)	1649922	1688200	1738859	1816042
Total pharmaceutical sales (MSEK)	13393	15808	14263	16567
Sales of top 50 molecules (MSEK)	4576	5977	5201	6203
Top 50 as share of total sales	34%	38%	36%	37%
Parallel imports (MSEK)	0	> 0	269	1007
Parallel imports of top 50 (MSEK)	0	> 0	269	920
Parallel imports/Total sales	0%	0%	2%	6%
Parallel imports/Top 50 sales	0%	0%	5%	16%
Total number of PI approvals	0	1	45	226
PI approvals for top 50 molecules	0	1	31	131
Total number of PI firms	0	1	4	10

Sources: SCB, LIF, MPA.

for 37 percent of total sales in the pharmaceutical market valued at wholesale prices.

Parallel imports increased substantially after Sweden joined the European Union, both in terms of actual sales and approvals to engage in the activity. In 1995 no PIs occurred and no applications to import were filed. By 1998 PIs had grown to 1.0 billion SEK, which corresponded to six percent of the total market, and 226 approvals to import pharmaceutical products had been granted by the Medical Products Agency. Initial PI entry was gradual, for by 1996 only a single firm had been granted permission to start parallel trade. However, by 1997 the number had grown to four firms and in 1998 the number was ten. The largest PI firm accounted for 100 percent of total parallel trade in 1996, 85 percent in 1997 and 59 percent in 1998. The four largest companies accounted for 96 percent of all parallel trade in 1998. Parallel imports from 13 different countries had been approved by 1998. The sources of PIs were, however, heavily concentrated in a few low-price countries in Southern Europe. Two countries, Spain and Italy, were the exporters in approximately two-thirds (63 percent) of all cases. The three most important export countries (Spain, Italy, and Greece) accounted for 74 percent of all approvals.

In 1998 parallel trade in Sweden was concentrated in the highest-volume products. Parallel imports accounted for 16 percent of the sales of the top 50 molecules, but the distribution of parallel trade as a share of total sales for these 50 molecules was not equal. Measured on an unweighted product basis the median share of parallel trade was zero percent and the maximum was 72.1 percent. Parallel imports existed in approximately 15 percent of all products. Ranking products from lowest to highest parallel-import shares, at the 95th percentile such imports accounted for 53.6 percent of total sales.

There is a similar picture for approvals on a product basis. Approvals to parallel-import drugs into Sweden in 1998 were concentrated in a few high-value products. For 68.3 percent of all products on the top 50 list, no entry of a PI firm had occurred. More than one approval had been granted for 21.9 percent of products.[19]

To summarize this overview, our sample covers approximately 35–38 percent of the pharmaceutical market in Sweden. The growth of parallel imports from 1995 to 1998 was considerable and such imports accounted for 16 percent of sales of the top 50 molecules in 1998. A large and rapidly growing number of PI firms entered the market, in some cases to sell the same products as other firms already present. Overall, sales were achieved largely by four major parallel importers. Parallel trade was concentrated in a minority of the products in the sample and the share of PI was considerable for approximately 15 percent of the top-value products. Italy and Spain were the main source countries and were either individually or jointly specified as the export country in 63 percent of all PI approvals. A final observation is that because growth in PI activity seemed to be accelerating at the end of our sample, the data we have available presumably do not reflect a long-run equilibrium.

4.4. *Average Price Effects in the Import Market*

Next, consider the effects of parallel imports in the Swedish market. We start with a comparison between products that are subject to

[19]Eighteen approvals had been granted for one specific product. That this number could be so large has two explanations. First, for some products a specific parallel-importing firm was approved to import the goods from several exporting countries. Second, more than one parallel-importing firm had an approval to import certain drugs, possibly from the same exporting country as other such firms.

competition from PIs and products that are not. Note that this is not the fully appropriate counterfactual comparison, which would compare prices of those products facing PI competition with fictional prices of the same goods had there been no possibility of parallel imports. Unfortunately, this comparison cannot be made. Neither is it possible to make a strict comparison between prices of patented drugs and those not on patent because exact information on patent status is not available.[20]

We calculate the relative price change for all specific products between a base year (1994 or 1997) and 1998. The relative price change is defined as the price in SEK in 1998, divided by the corresponding price in the base year, minus one. We calculate the change for the average price including PIs as well as the change for the manufacturing firms' prices.

Table 2 reports the unweighted and weighted average price changes for all products. The unweighted average is a simple arithmetic mean. In the weighted average, however, product weights are computed as the product's sales in 1998 divided by the sum of sales for all products included in the average.

Our discussion primarily focuses on the unweighted average. Over the whole period 1994–1998, prices increased on average 6.64 percent for all products and manufacturing firms' prices increased somewhat more at 7.34 percent. On average, prices for products subject to parallel imports increased 2.88 percent while manufacturers raised their prices 6.38 percent for these products. In contrast, prices for products not subject to PI competition rose 7.57 percent.

The difference was even more pronounced over the shorter period 1997–1998. Average prices increased 0.25 percent among all products. Manufacturing firms' prices declined 0.34 percent for products subject to PIs, but rose 0.96 percent for products not facing this competition. Prices of products that were parallel imported fell on average by 3.12 percent. Thus, this data overview seems to confirm that the prices of PI products, and of products facing such competition, fell

[20]However, note our assessment that most of the drugs were patented, including virtually all of the large-volume molecules. Thus, it is unlikely that these differential price impacts are associated with patent status.

Table 2. Relative price changes of pharmaceuticals in Sweden, 1994–1998.

		Unweighted Average		Weighted Average	
Sweden		1994–1998	1997–1998	1994–1998	1997–1998
All Products	Including PI	0.0664 (0.134)	0.0025 (0.035)	0.0080 (0.028)	−0.0137 (0.006)
	Manufacturer's price	0.0734 (0.133)	0.0073 (0.031)	0.0279 (0.022)	−0.0016 (0.003)
Products facing PI competition	Including PI	0.0288 (0.121)	−0.0312 (0.041)	−0.0438 (0.044)	−0.0385 (0.009)
	Manufacturer's price	0.0638 (0.120)	−0.0034 (0.035)	0.0031 (0.036)	−0.0067 (0.005)
Products not facing PI competition	Manufacturer's price	0.0757 (0.137)	0.0096 (0.030)	0.0365 (0.017)	0.0015 (0.004)
No. of Obs.		125	151	125	151

Source: Authors' calculations based on data from LIF. Figures in parentheses are standard deviations.

in the import market relative to the prices of products not subject to parallel trade, with much of the impact concentrated at the end of the period.

We undertake simple tests of the hypotheses generated by our basic theory. First, to test whether the differences between the change in the manufacturing firms' prices for products subject to PIs and those not subject to PIs are significant, we perform t-tests, assuming unequal variances, of the hypothesis that the mean change is the same. The hypothesis that the manufacturing firms' price changes for goods facing PIs and those not facing such imports is the same cannot be rejected at the ten-percent level of significance for the period 1994–1998 ($t = 0.43$). The hypothesis is, however, rejected at the five-percent level for 1997–1998 ($t = 1.77$), which confirms that the manufacturing firms' prices increased significantly less for products subject to PIs than did prices of other products in the latter part of the period.

Our preliminary conclusion is that the data are consistent with accommodation of parallel trade by manufacturers for two reasons. First, the very existence of PI firms indicates that manufacturers choose not to block arbitrage. Second, the change in the manufacturing firms' prices is not significantly different between goods facing PIs and other goods over the period 1994–1998. However, the mean price of goods facing such competition, including both the PI volume and the original manufacturers' volume, increased significantly less than product prices of the manufacturing firms not facing such competition. Over the short period 1997–1998, the change of the manufacturing firms' prices was significantly lower for products facing import competition than for other drugs. This result suggests that manufacturing firms react to the volume of arbitrage with a lag, rather than reacting to parallel imports immediately upon their potential entry into the market.

4.5. *The Price Impact of Competition from Parallel Imports*

To investigate the price effect of parallel imports further we examine econometrically how changes in manufacturers' prices are affected by competition from that channel. For this purpose we use our biweekly panel from January 1994 to September 1999 with original manufacturers' prices and PI approvals in Sweden. The manufacturing firm's price in Sweden for each product is normalized in each period of our biweekly panel compared with the price of the same product during the first period in which it was sold. That is, in the first period that a product is sold by the manufacturing firm in Sweden the relative price is defined as 1.0, while subsequent prices are computed relative to this basis. This variable is denoted PRICEMNF_{it} in period t, where subscript i refers to product i. It should be noted that these prices tend to remain constant for long periods and then change discretely over time in a piecewise-continuous process. In this regard, the estimated constant term provides the strongest explanation of prices in the panel.[21]

We adopt two approaches to estimating the impact of entry by PI firms on relative manufacturers' prices in the panel. The first is an ordinary least squares (OLS) specification in which prices are affected by the number of PI entrants, the potential for PI competition, and a time trend. Thus, we define a variable DENTRY to take the value one if there are one or more PI firms active in the product at time t and zero otherwise. We also define DENTRY1, DENTRY2, DENTRY3, and DENTRY3P as indicator variables for one, two, three, and more than three active PI firms in a period. These entry variables directly measure the presence of competition. In contrast, potential competition is measured by the variable D1995, which is zero for the periods in 1994 and one after that. The notion here is that original manufacturers in principle needed to think strategically about the possibility of PIs after Sweden's entry into the EU, but not prior to that. Thus, D1995 is designed to capture the effect of the discrete change in policy facing manufacturers.

An obvious difficulty is that entry decisions of PI firms are endogenous to changes in the manufacturer's price and OLS

[21]We also estimated specifications with an AR(1) process within each product in the panel, finding that the lagged price coefficient was very close to unity and highly significant in the same way as the constant terms in the static specifications. However, we do not think that AR processes adequately capture price-setting in the data because, as noted earlier, the markets are not in long-run equilibrium during the period. Consequently we focus on the static specification of price changes.

accordingly would generate biased and inconsistent estimates. Thus, our second approach is to estimate a series of instrumental variables (IV) equations, in which various instruments are used to explain DENTRY in the first stage, while predicted values for DENTRY are used in the second stage as a determinant of price changes, along with D1995 and the time trend. For this purpose, we selected combinations of the following instruments. First, we included the number of periods since initial launch of the manufacturer's product in Sweden (LAUNCHPER), which should capture the longevity of a product and the evident fact that PI competition acts with a lag. Second, we used the volume of drugs (by unit) sold by product in 1995 (QUANT1995) or the value of these sales (SALES1995) in Sweden. This variable reflects the idea that PI firms would focus on products with large markets. Third, we used also the one-year lagged volumes or sales (QUANTLAG and SALESLAG, respectively) to account for growth in these markets.

We also wish to incorporate measures of the potential price gaps between export sources and Sweden, for these gaps directly account for the potential returns to PIs. Thus, a fourth instrument was the minimum price of matched available products in a set of potential exporters (France, Italy, Greece, Portugal, and Spain), defined relative to the contemporaneous Swedish manufacturer's price in Kronor (MINRELPR). Because these prices existed in our data for 1994 (and 1998) we used the 1994 local prices adjusted for changes in bilateral exchange rates over the estimation period. This wide definition of supply sources and biweekly price (exchange rate) changes permitted us to retain a relatively large sample size. However, to avoid any bias arising from attributing all price changes to exchange rates and to focus on key exporters, the next instruments were relative prices in Italy and Spain of these products in 1994 (RELIT94 and RELSP94). This choice considerably reduced the available products for inclusion in the analysis. Finally, we included as instruments the bilateral exchange rates of the Swedish Kronor with the Spanish Peseta (ESP), the Italian Lire (ITL), and the Greek Drachma (GRD).

It is evident that foreign prices and exchange rates could be construed by original manufacturing firms as determinants of potential competition. Because we use these variables as instruments for entry decisions, it is difficult to differentiate between the impacts of actual and potential competition.[22] Thus, the IV approach can only test for the effect of total competition, which may be a mix of actual entry and potential competition.

Our approach is most consistent with that in Frank and Salkever (1997). They estimated first a fixed-effects OLS specification in which the number of generic entrants was considered to be exogenous. As noted above, they found that entry tended to raise the prices of brand-name products. Second, they developed a two-stage instrumental variables procedure with fixed effects, in which entry was predicted first on the basis of pre-patent volume, market age, years off patent, and a time trend. Their IV estimates discovered a larger impact of generic entry on brand-name prices.

Table 3 provides summary data on the key variables used. Table 4 lists correlation coefficients among relative manufacturer's price, entry, the 1995 dummy, and central instrumental variables. Note first that PRICEMNF and DENTRY are positively correlated, suggesting that entry is associated with increases in relative prices. Next, the correlations of DENTRY with the instrumental variables are reasonably high with the exception of QUANT1995.[23] Most importantly, we believe that each of the instruments may be considered essentially exogenous to the contemporaneous pricing decisions of manufacturers, making them asymptotically uncorrelated with the structural-equation residuals.

Turn next to the OLS results listed in Table 5. We estimate these equations using both fixed-effects and random-effects estimators to account for the product-specific influences. As may be seen, the choice of fixed effects or random effects makes virtually no difference, except for reducing slightly the magnitudes of the constant terms. In models (1) and (2) we estimate simply the effect of any entry by PI firms (that is, by one or more firms) on relative manufacturers' prices over the period. As may be seen, entry reduced the average relative price by around 1.6 percent by these estimates, which was highly significant. However, our proxy for potential competition due

---

[22]We are grateful to Thomas McGuire for this insight.

[23]This is true also of the correlations with the bilateral exchange rate variables.

Table 3. Summary statistics for variables in econometric model.

Variable	Observations	Mean	S. D.	Min	Max
PRICEMNF	24751	1.0432	0.0944	0.6500	1.5620
DENTRY	24751	0.0448	0.2069	0	1
DENTRY1	24751	0.0169	0.1290	0	1
DENTRY2	24751	0.0103	0.1011	0	1
DENTRY3	24751	0.0059	0.0764	0	1
DENTRY3P	24751	0.0089	0.0940	0	1
D1995	24751	0.8200	0.3842	0	1
LAUNCHPER	24751	57.2	46.8	0	150.0
QUANTLAG (m units)	24751	0.0534	0.3363	0	5.4847
QUANT1995 (m units)	24751	0.0639	0.3657	0	5.0986
SALESLAG (m SEK)	24751	12.7000	32.3000	0	350.0000
SALES1995 (m SEK)	24751	14.1321	32.9084	0	281.3443
MINRELPR	16350	0.7045	0.1397	0.3952	0.9449
RELIT94	9450	0.8450	0.1758	0.5605	1.1840
RELSP94	11700	0.8335	0.1329	0.5951	1.0313
ESP (Kronor per 100 Peseta)	24751	5.4281	0.2459	5.0801	5.9926
ITL (Kronor per 100 Lire)	24751	0.4493	0.0149	0.4159	0.4883
GRD (Kronor per 100 Drachma)	24751	2.8825	0.1946	2.4775	3.2689

Note: MINRELPR is from a sample of relative prices using data from France, Italy, Greece, Portugal, and Spain.

to Sweden's EU-membership in 1995, D1995, had no effect on pricing behavior.[24] A more nuanced picture emerges from distinguishing between the number of entrants. Specifically, the entry of two PI firms had little impact on manufacturers' prices.[25] However, a third entrant tended to reduce average price by 4.6 percent, and additional entrants by another 4.9 percent. This result stems from the fact that multiple entrants existed in the highest-volume products. Thus, the average price impact over all products was modest but in the drugs with large

[24]In other specifications, we used an indicator variable for the time between a PI approval and first-pricing date for that product in Sweden as a measure of potential competition, with similar results.

[25]DENTRY2 is significant at the 10-percent level but with this large a sample that inference is marginal at best.

Table 4. Correlations among key variables.

	PRICEMNF	DENTRY	D1995	LAUNCHPER	QUANT1995	QUANTLAG	SALESLAG	MINRELPR
PRICEMNF	1							
DENTRY	0.17	1						
D1995	0.14	0.10	1					
LAUNCHPER	0.32	0.46	0.49	1				
QUANT1995	−0.00	0.09	0.00	0.07	1			
QUANTLAG	0.06	0.14	0.07	0.11	0.81	1		
SALESLAG	−0.07	0.18	0.18	0.22	0.63	0.83	1	
MINRELPR	−0.26	−0.26	−0.00	−0.05	0.05	0.07	0.03	1

Table 5. Estimated impacts of PI entry on original manufacturers' prices using ordinary least squares.

Variable	OLS-FE (1)	OLS-RE (2)	OLS-FE (3)	OLS-RE (4)
DENTRY	−0.0160	−0.0158		
	(−9.05)***	(−8.95)***		
DENTRY1			−0.0039	−0.0037
			(−1.55)	(−1.47)
DENTRY2			−0.0056	−0.0055
			(−1.77)*	(−1.72)**
DENTRY3			−0.0457	−0.0455
			(−10.9)***	(−10.9)***
DENTRY3P			−0.0494	−0.0492
			(−13.6)***	(−13.5)***
D1995	−0.0007	−0.0006	−0.0013	−0.0012
	(−0.55)	(−0.49)	(−1.03)	(−0.97)
TREND	0.0006	0.0006	0.0007	0.0007
	(59.14)***	(58.94)***	(60.4)***	(60.2)***
CONSTANT	0.9919	0.9825	0.9918	0.9825
	(1089)***	(189.4)***	(1095)***	(188.2)***
Observations	24751	24751	24751	24751
Products	196	196	196	196

Notes: Figures in parentheses are t-statistics (for fixed effects) and z-statistics (for random effects). *Associated P-values are significant at the 10-percent level. ***Associated P-values are significant at the 1-percent level.

market volumes the price reducing effects of PIs were large and significant. These results are consistent with the simple calculations in Table 2. Note that the tendency for additional entrants to strengthen the reduction in prices provides additional support for the notion of accommodation as specified in our theoretical model.

As noted earlier, entry of PI firms is an endogenous choice and these OLS coefficients likely are misleading. Thus, Table 6 lists the results of various IV estimations, using different sets of instruments. In this regard, we could not estimate a system of first-stage equations for each of the differentiated entry variables (DENTRY1, DENTRY2, DENTRY3, and DENTRY3P). Thus, we were forced to estimate an IV equation only for DENTRY in each model, using its predicted

Table 6. Estimated impacts of PI entry on original manufacturers' prices using instrumental variables estimation.

Variable	IV-FE (5)	IV-RE (6)	IV-FE (7)	IV-RE (8)	IV-FE (9)	IV-RE (10)	IV-FE (11)	IV-RE (12)	IV-FE (13)	IV-RE (14)
DENTRYHAT	−0.1890	−0.1796	−0.1913	−0.1735	−0.1715	−0.1663	−0.1300	−0.1208	−0.1661	−0.1386
	(−11.3)***	(−11.0)***	(−11.4)***	(−10.8)***	(−10.3)***	(−10.1)***	(−10.2)***	(−9.84)***	(−9.03)***	(−8.33)***
D1995	−0.0172	−0.0163	−0.0175	−0.0157	−0.0157	−0.0152	0.0033	0.0047	−0.0047	−0.0001
	(−7.99)***	(−7.70)***	(−8.06)***	(−7.51)***	(−7.35)***	(−7.20)***	(1.32)	(1.93)*	(−1.15)	(−0.02)
TREND	0.0010	0.0010	0.0010	0.0010	0.0010	0.0010	0.0008	0.0007	0.0011	0.0010
	(26.5)***	(26.5)***	(26.6)***	(26.6)***	(25.7)***	(25.7)***	(16.0)***	(15.8)***	(14.2)***	(14.0)***
CONSTANT	0.9844	0.9726	0.9843	0.9733	0.9848	0.9730	0.9875	0.9745	0.9763	0.9628
	(763.0)***	(113.6)***	(760.4)***	(175.7)***	(775.2)***	(91.6)***	(631.4)***	(110.0)***	(356.1)***	(103.5)***
Observations	24751	24751	24751	24751	24601	24601	13138	13138	7234	7234
Products	196	196	196	196	196	196	196	196	196	196

Notes: In Models (5) and (6) the instruments are LAUNCHPER, QUANTLAG, and bilateral exchange rates with Italy, Spain, and Greece. In Models (7) and (8) the instruments are LAUNCHPER, SALESLAG, and the same exchange rates. In Models (9) and (10) the instruments are LAUNCHPER, QUANT1995, and the same exchange rates. In Models (11) and (12) the instruments are LAUNCHPER, MINRELPR, and the same exchange rates. In Models (13) and (14) the instruments are LAUNCHPER, SALESLAG, the relative prices in Italy and Spain, and the same exchange rates. *Associated P-values are significant at the 10-percent level. ***Associated P-values are significant at the 1-percent level.

values DENTRYHAT in the pricing equations shown.[26] Again, we use both fixed effects and random effects, the choice of which makes virtually no difference in the coefficients.

It is immediately obvious that the use of IV techniques generates considerably different results from OLS. First, the coefficients on DENTRYHAT are highly significant and negative, ranging from an impact of a 12–19 percent reduction in manufacturers' prices, depending on the instruments and sample. Thus, the earlier OLS estimates were flawed in that they failed to account for the endogenous entry of PI firms. Specifically, because such firms tend to enter in products with otherwise significant pricing power, the OLS equations confound entry decisions (a positive association with price) with the price-reducing impact of entry. When instrumented, we find that entry has a sharply negative and significant effect on original manufacturers' prices.[27]

Note also that with the IV approach the impact of Sweden's EU-membership, captured by D1995, becomes negative in models (5) to (10). However, at most the effect is 1.7 percent, far below the influence of entry by parallel importers later in the period. It is also interesting that the use of IV estimation reduces the coefficient on the time trend from 0.6 percent per year to 0.1 percent per year.

The results from our econometric analysis may be summarized in two points. First, both the OLS and IV equations demonstrate that competition from PIs had a significantly negative impact on the original manufacturers' prices in Sweden. Second, the OLS results are questionable in light of the evident endogeneity of entry by PI firms. Thus, we believe the IV estimates are more sensible and they show that the price-reducing impacts were substantial, ranging from 12 to 19 percent on average. Indeed, this price impact may be compared to the average price gap of 21 percent between Sweden and

[26]Similar estimation of a variable with the number of entrants was unsuccessful because our instruments could not well predict changes in entry decisions given such frequent data.

[27]In IV specifications using an AR(1) process in each product, we find even larger negative effects of PI entry when the coefficients are converted to "long run" impacts. As noted, however, we do not believe the data reflect a long-run equilibrium, making such estimates suspect.

either Italy or Spain in available products in 1994 (Ganslandt and Maskus, 2001).

5. Conclusion

In this paper we developed a model of manufacturer behavior in the presence of parallel imports. We showed that the price of products facing such arbitrage should drop relative to prices of products that are not subject to it in the home market and for the manufacturing firm's price to fall as the volume of PIs (or the number of PI firms) rises.

The Swedish market provides a natural laboratory in which to investigate these effects. The growth of parallel imports from 1995 to 1998 was considerable and accounted for 16 percent of sales of major molecules in our sample in 1998. A large number of PI firms entered the market, in some cases to sell the same products as other such firms already present. Parallel trade was concentrated on a minority of the products in the sample but its share was considerable for up to 15 percent of major drugs.

The initial empirical analysis confirmed that prices of goods subject to import competition, including parallel-traded products themselves, fell approximately 4 percent in the import market relative to the prices of products not subject to parallel trade. Simple tests for differences in price changes suggested that the price effect came with a lag.

The econometric investigation discovered that when the endogeneity of entry is taken into account, the estimated effects of PIs becomes powerful, suggesting that original producers cut prices by up to 19 percent, relative to other drug prices. Thus, we conclude that parallel imports represent a significant form of competition in markets such as Sweden.

References

Caves, Richard E., Michael D. Whinston, and Mark A. Hurwitz (1991), "Patent Expiration, Entry, and Competition in the U.S. Pharmaceutical Industry," *Brookings Papers on Economic Activity Microeconomics*, 1–66.

CEC (1998), "Commission Communication on the Single Market in Pharmaceuticals," COM(98)588 final, Brussels, November 1998.

Danzon, Patricia M. (1998), "The Economics of Parallel Trade," *PharmacoEconomics* 13: 293–304.

Danzon, P. and L. Chao (1995), "The Uses and Abuses of International Price Comparisons," in Helms *Competitive Strategies in the Pharmaceutical Industry*, AEI Press, Washington.

Danzon, P. and L. Chao (2000), "Cross-national Price Differences for Pharmaceuticals: How Large, and Why?," *Journal of Health Economics* 19: 159–195.

Frank, Richard G. and David S. Salkever (1992), "Pricing, Patent Loss and the Market for Pharmaceuticals," *Southern Economic Journal* 59: 165–179.

Frank, Richard G. and David S. Salkever (1997), "Generic Entry and the Pricing of Pharmaceuticals," *Journal of Economics and Management Strategy* 6: 75–90.

Ganslandt, Mattias and Keith E. Maskus (2000), "Parallel Imports of Pharmaceutical Products," mimeo, IUI, Stockholm, Sweden.

Ganslandt, Mattias and Keith E. Maskus (2001), "Parallel Imports of Pharmaceutical Products in the European Union," World Bank Policy Research Working Paper 2630, July, Washington DC.

Grabowski, Henry G. and John M. Vernon (1992), "Brand Loyalty, Entry and Price Competition in Pharmaceuticals after the 1984 Drug Act," *Journal of Law and Economics* 35: 331–350.

Holmberg, C., H. Kiellberg, and B. Andersson (2003), "Distribution of Pharmaceutical Products in the Nordic Countries," SSE/EFI Working Paper Series in Business Administration No. 2003:10, Stockholm, Sweden.

Klepper, Gernot (1992), "Pharmaceuticals: Who's Afraid of 1992?" CEPR discussion paper no. 675, London, UK.

Lu, Z. John and William S. Comanor (1998), "Strategic Pricing of New Pharmaceuticals," *Review of Economics and Statistics* 80: 108–118.

Maskus, Keith E. and Yongmin Chen (2002), "Parallel Imports in a Model of Vertical Distribution: Theory, Evidence and Policy," *Pacific Economic Review* 7: 319–334.

Pecorino, Paul (2002), "Should the US Allow Prescription Drug Reimports from Canada?" *Journal of Health Economics* 21: 699–708.

Wiggins, S. N. and R. Maness (1994), "Price Competition in Pharmaceutical Markets," unpublished manuscript, Texas A&M University, Department of Economics.

World Health Organization (2001), *Macroeconomics and Health: Investing in Health for Economic Development: Report of the Commission on Macroeconomics and Health* (Geneva: WHO).

Chapter 6

The Impact of Parallel Imports on Investments in Cost-Reducing Research and Development*

Changying Li and Keith E. Maskus[†]

1. Introduction

Parallel imports (PI), also called "gray-market imports", are genuine products sold under protection of a patent, copyright, or trademark in one country but shipped legally to another country without the original manufacturer's authorization. For example, in the European Union parallel trade among member countries is legal in all products under the principle of free movement of goods in the single market (Ganslandt and Maskus, 2003). Such imports account for a substantial proportion (up to 72 percent in some drugs and 16 percent of drugs overall) of domestic sales of blockbuster pharmaceuticals in higher-price markets in the EU, such as Sweden, and exert considerable downward pressure on prices (Ganslandt and Maskus, 2004). Recent industry estimates suggest that original drug manufacturers in the EU currently lose sales to PI competition in the amount of some $3 billion per year.[1] They account also for significant shares (often over 10 percent) of sales in a number of other products in the EU, including compact disks, soft drinks, cosmetics,

*Reprinted from *Journal of International Economics*, Vol. 68, No. 1, pp. 443–455. Copyright 2006, with permission from Elsevier.
†We thank Yongmin Chen, Jonathan Eaton, Frank S. T. Hsiao, Jack Robles and two anonymous referees for their insights and comments. Changying Li acknowledges the financial support of the Personnel Department of Nankai University.
[1]See "Drug Makers Try to Curtail Cheap Imports," *The Wall Street Journal*, 11 April 2002.

jewelry, clothing and automobile parts (National Economic Research Associates, 1999). In the United States, parallel imports are commonly found in cameras, electronics, clothing and cosmetics sold at such outlets as Wal-Mart and independent stores.

The legal status of parallel imports stems from the choice of territorial exhaustion of intellectual property rights (IPRs). Under the doctrine of national exhaustion, rights are exhausted upon first sale within a nation but IPRs owners may prevent parallel imports from abroad (Maskus, 2000). That is, national exhaustion amounts to a legal ban on parallel trade. The United States pursues this policy for patented and copyrighted products but PI are generally permitted for goods protected solely by trademarks. Under the doctrine of international exhaustion, rights are ended upon first sale anywhere and parallel imports are permitted. For example, Australia and New Zealand are open to PI in most copyrighted and trademarked goods, while international exhaustion is the typical policy in developing nations. An intermediate policy is to adopt regional exhaustion, in which parallel imports are permitted among member countries, but not from outside the region. This is the policy enforced by the European Union in an effort to encourage regional market integration. That the legal treatment varies across nations may also be seen from the fact that in 1998, 34 of 107 countries examined were closed to PI in copyrighted goods and the rest were partially or fully open (International Intellectual Property Association, 1998).

Parallel imports concern manufacturers, distributors and policymakers for several reasons (Maskus, 2000). One important issue is that by reducing expected profits, PI could reduce incentives of original manufacturers to invest in research and development (R&D). Thus, a central claim by the research-intensive pharmaceutical industry is that if the United States were to deregulate its restrictions on PI in brand-name drugs from Canada and Europe, pharmaceutical companies would be forced to cut R&D expenditures.[2]

[2]High prices of patented and brand-name drugs in the United States are an important policy issue and legislation was introduced in 2003 and 2004 to deregulate restrictions on parallel trade from Canada and Western Europe. In contrast, the wisdom of remaining open to parallel trade is at the center of policy discussions regarding means by which

The relationship between parallel trade and innovation has been the subject of much informal discussion. For example, Barfield and Groombridge (1998) state that the economic case for permitting copyright owners to preclude parallel imports rests on reductions in the associated returns to innovation.[3] Danzon (1997) argues that it is necessarily welfare-enhancing to ban parallel trade in pharmaceuticals for otherwise the inability to profit from segmented markets would significantly damage incentives for innovation. Indeed, the legal justification for national exhaustion policies in goods protected by copyrights and patents is precisely that PI interfere with the originator's ability to earn a return on investment (Maskus, 2000).

In this paper we provide the first formal linkage between PI and innovation. We develop a vertical-pricing model in which a manufacturer must compete in its own market with PI from a distributor in another market.[4] The manufacturer offers a contract to the distributor that simultaneously must attempt to achieve both efficient vertical pricing and to minimize the costs of facing PI. Prior to contracting, the manufacturer may conduct R&D to develop a cost-reducing technology. We show that the distortions associated with PI reduce the investment incentive, so that the informal intuition is correct. However, the amount by which R&D is diminished depends on both transport costs and the legal treatment of such trade.

We also analyze briefly the welfare implications of prohibiting PI in this model. We find that under circumstances in which PI occur in equilibrium, banning them has ambiguous effects on expected global welfare. However, when PI are deterred by the manufacturer's strategic decision to raise prices in the PI-source nation, banning them increases expected global welfare. The reason is that such deterrence relies on an inefficiently high wholesale price, which can be reduced without the threat of PI.

---

poor countries can gain access to medicines to combat HIV/AIDS and other diseases (Scherer and Watal, 2002).

[3]See also Chard and Mellor (1989), Duhan and Sheffet (1988), Maskus (2000), and Palia and Keown (1991).

[4]As noted by NERA (1999) and Maskus and Chen (2002), the bulk of PI occurs at the distributor level, not through retail arbitrage.

Table 1. Measures of trade costs for shipments to the United States, 1994.

Product (10-digit HS2)	Average CIF/FOB import value	Average (CIF + duty)/ FOB import value
Bourbon whiskies, <41	1.25	1.26
Dictionaries	1.09	1.09
Art and pictorial books	1.06	1.06
Rack-size paperbound books	1.09	1.09
Women's or girls' suits	1.09	1.34
Electric household refrigerators	1.04	1.06
Dishwashers, household	1.07	1.11
Digital computers	1.03	1.06
Video disks, recorded	1.05	1.10
Laser disk sound recordings	1.07	1.07
Passenger vehicles, 4–6 cylinder, $>3000\,cm^3$	1.04	1.06
Passenger vehicles, >6 cylinder, $>3000\,cm^3$	1.05	1.07
Film cameras, 1600 mm	1.04	1.09

Before turning to the analysis, it is of interest to consider the importance of trade costs, which play a large role in the theory. Table 1 offers some basic measures of trade costs in various highly detailed (10-digit Harmonized System) classifications for U.S. imports in 1994.[5] These products exemplify a number of goods that are likely the subject of parallel imports (NERA, 1999). The first data column shows the average, across several trading partners, of the ratio of the CIF value of imports to the FOB price at the export location. These figures suggest that transport costs to the United States are modest for the most part, ranging from 3 percent of FOB value in computers to 9 percent in dictionaries, although reaching 25 percent in whiskies. The second data column adds U.S. import duties, which makes little difference except in apparel. The fact that trade costs are generally low suggests that our model equilibria with positive PI may be relevant where PI are permitted legally. At the same time, these figures likely do not capture the entire costs of

[5]These data are used in the econometric analysis in Maskus and Chen (2004).

shipping PI. Firms engaged in such trade need to find sources of supply and convince potential buyers of their reliability (Ganslandt and Maskus, 2003).

The rest of the paper is organized as follows. In Section 2 we offer a brief literature review to place this paper into context. In Section 3 we present the basic model and contrast equilibrium when PI are prohibited with the case where they are allowed. In Section 4 we undertake the analysis, while in Section 5 we conclude.

2. Prior Literature

This paper marries two strands of literature: first, the determinants of parallel imports and second, incentives for investing in cost-reducing innovation in the presence of trade and imperfect competition.

Regarding the former issue, there are two basic models explaining the existence of PI. The first is retail arbitrage where price differences exceed transport costs. This concept was first analyzed in the mid-1980s, after a substantial real increase in the value of the U.S. dollar seemed to generate significant amounts of gray-market imports (Tarr, 1985; Hilke, 1988). A formal model of retail price discrimination and perfectly competitive arbitrage was introduced by Malueg and Schwartz (1994). In their model, countries differ in demand elasticities for homogeneous goods. Segmented markets permit discrimination, whereas PI establish a uniform international price. Global welfare effects of banning PI are ambiguous and depend on the balance of consumer surplus created and destroyed. Richardson (2002) showed that neither uniform pricing nor full segmentation could be supported globally as a Nash equilibrium.

The second model involves pricing decisions by an original manufacturer attempting to exert vertical control over its distributors, who may engage in PI.[6] This model was introduced by Maskus and Chen (2002, 2004), who found that the holder of an intellectual property right facing potential PI competition must balance three factors in

[6]There are other potential sources of PI, including free riding on market investments and the existence of varying price controls across nations. The former has not received formal attention in the literature, while the latter is the subject of Ganslandt and Maskus (2004).

maximizing profits: optimal vertical pricing, competition in recipient markets, and wasteful trade costs. In this world the wholesale price and trade costs interact in limiting or deterring parallel imports. Global welfare from permitting PI is U-shaped in the extent of transportation costs. The authors showed that banning PI is globally welfare-increasing if such trade is being deterred by the manufacturer but permitting them is optimal in the presence of low trade costs. Our model extends this result to the context of endogenous innovation.

Neither of these approaches integrated R&D decisions into models of parallel imports, which is the task of the current paper. Thus, the second strand of literature involves investment decisions in cost-reducing R&D in the presence of international trade competition and trade policy. A good example is Reitzes (1991), who compared tariffs and quotas in their effect on first-stage process R&D in a two-stage Cournot duopoly game. A quota could reduce R&D by both the domestic and foreign firms relative to a tariff because it would restrict their need to engage in strategic technology investments. This theory was extended to two-stage price competition by Costa Cabral *et al.* (1998), who showed that the domestic firm would increase its R&D and the foreign firm would reduce its expenditure as a quota becomes more restrictive.[7]

Finally, Bester and Petrakis (1993) found that the degree of substitutability between products and the nature of competition influence process R&D investments. Under Cournot competition, a high degree of substitution weakens investment incentives.

Our model bears similarities to these approaches but arises in a quite different institutional context. Here, a single firm decides in the first stage on its optimal investment in process R&D, knowing that its investment could affect the degree of quantity competition it faces in a later stage. The extent of this competition depends on the contract between the original firm and its distributor, the legality of parallel trade, and the severity of transportation costs. Whether the

[7]See Spencer and Brander (1983) and Leahy and Neary (2001) for broader analyses of strategic investment or export subsidies, which are more direct policies than trade restrictions.

importing government chooses to permit or to ban parallel imports essentially amounts to a decision either to permit free trade (subject to trade costs) or to impose a zero quota. This generates a substantial difference in incentives for investing in cost-reducing R&D. Our model also adds depth to the framework in Haaland and Kind (2004), in which duopoly investments in R&D depend positively on a reduction in trade costs by making the market bigger. Here, R&D has a non-monotonic relationship with trade costs because of the tradeoffs involved in vertical competition.

3. Model Framework

The paper extends the model set out by Maskus and Chen (2004) to a framework with endogenous process R&D. Assume a manufacturer, M, sells its product in two countries, A and B. The inverse demand in market A is $p_{\mathrm{A}} = 1 - q$, and that in B is $p_{\mathrm{B}} = a - q$. M sells the product directly to consumers in its home market, country A, but sells the product through an independent distributor, D, in country B. The distributor may sell the good back to A through PI if the activity is permitted by that country. When D sells the product back to A, it competes with M in Cournot fashion but incurs a marginal transportation cost of $t \geq 0$. However, M cannot engage in PI to market B.

The assumption of one-way parallel trade may be justified in at least two ways.[8] Most obviously, there are cases where one country is open to PI and the other is not. Examples of the former are Australia and New Zealand, which practice international exhaustion in copyright goods, while examples of the latter are the United States and Japan, which observe national exhaustion (Maskus, 2000). A second possibility is that there may be fixed costs in transporting PI and market B is small while market A is large. Distributors would not find it profitable to ship goods from A to B in this case, which may be exemplified by the United Kingdom and Portugal, respectively, within the European Union.

[8]Ganslandt and Maskus (2005) offer additional explanations.

The timing of the game is as follows. First, the manufacturer decides whether to invest in a cost-reducing process innovation with an uncertain outcome of success. Second, M makes the distributor a take-it-or-leave-it offer in the form (w, T), where w is the per-unit wholesale price and T is a transfer payment from D to M.[9] Third, the distributor decides to accept or reject this two-part tariff. If D rejects, the good is not sold in country B and M sets a monopoly output in country A. If it accepts, the distributor chooses its sales in A and B simultaneously and M determines its sales in A.

Regarding the R&D problem, suppose M engages in process innovation with an investment $k \geq 0$. Then the firm's marginal cost level becomes c_{L} with probability $\alpha(k)$ and remains c_{H} with probability $1 - \alpha(k)$, where this probability function is continuous, twice differentiable and strictly increasing with $\alpha(0) = 0$ and $\alpha''(k) < 0$. We assume reasonably that $0 \leq c_{\mathrm{L}} < c_{\mathrm{H}} < \min\{1, a\}$ or that cost levels are less than both market sizes. We call this our Basic Assumption and it rules out the case in which production is inactive in either country.

3.1. *The R&D Decision*

A question of interest is whether permitting parallel trade would decrease R&D in process innovation. To address this question, we compare the case in which parallel trade is banned by law and markets are segmented with that in which it is permitted and markets are integrated.

The quantities sold by M in country A and by D in country B are determined simultaneously when the distributor accepts the manufacturer's offer. Let w_i and c_i denote, respectively, the wholesale price and marginal cost in the cases $i = \mathrm{H}, \mathrm{L}$.

First, M must choose its optimal contract (w_i, T_i) in the absence or presence of PI, where the transfer payment is set to extract D's

[9]While we think this setup comes closer to the reality of R&D timing, a different model would have the R&D decision after the distributor contract is signed. If such contracts could set contingent wholesale prices there would be no qualitative impact on the model. However, if the contract could only be set at a fixed wholesale price before the R&D decision, the quantity of investment could be affected and some results of this model may not hold.

profit.[10] The wholesale price is D's marginal cost of acquiring a unit of the good. This price is set by M to influence D's sales and price decisions.

For convenience, let $R^j_{\mathrm{M}}(w_{\mathrm{L}}, c_{\mathrm{L}})$ and $R^j_{\mathrm{M}}(w_{\mathrm{H}}, c_{\mathrm{H}})$ denote M's operating profits (excluding R&D costs) when the innovation is successful and unsuccessful, respectively. Superscript $j = (n, p)$ denotes the case of banned PI and permitted PI, respectively. A central variable is the difference in operating profits. For example, in the case of banned PI this variable is

$$\Delta R^n_{\mathrm{M}} - R^n_{\mathrm{M}}(w_{\mathrm{L}}, c_{\mathrm{L}}) - R^n_{\mathrm{M}}(w_{\mathrm{H}}, c_{\mathrm{H}}) = \frac{(c_{\mathrm{H}} - c_{\mathrm{L}})[(1+a) - (c_{\mathrm{H}} + c_{\mathrm{L}})]}{2}.$$

Manufacturer M engages in process innovation and its expected profit is

$$E^j_{\mathrm{M}} = \alpha(k^j) R^j_{\mathrm{M}}(w_{\mathrm{L}}, c_{\mathrm{L}}) + [1 - \alpha(k^j)] R^j_{\mathrm{M}}(w_{\mathrm{H}}, c_{\mathrm{H}}) - k^j. \qquad (1)$$

We assume that $\alpha'(0)$ is positive and large enough to ensure that the manufacturer will invest in positive process innovation. The firm maximizes Eq. (1) in the first stage and the first-order condition yields the following implicit expression for the level of R&D undertaken:

$$\alpha'(k^j) = \frac{1}{\Delta R^j_{\mathrm{M}}}. \qquad (2)$$

Note that k rises as the difference in operating profits gets larger because $\alpha''(k) < 0$ and thus $\alpha'(k)$ is a decreasing function in k.

To determine whether PI reduce the manufacturer's incentives to invest in innovation, we need to compare Eq. (2) for $j = (n, p)$. Because $\alpha'(k)$ is decreasing in k, the condition $\Delta R^n_{\mathrm{M}} > \Delta R^p_{\mathrm{M}}$ implies that $k^n > k^p$, in which case parallel imports discourage innovation. If the condition were reversed PI would encourage more process innovation. Thus, what matters is the impact of PI on the difference in operating profits.

[10]The prices and transfer fees would differ between cases but for ease of notation we do not attach to these variables superscripts accounting for the legality of PI.

3.2. *Profit Maximization*

In this model, the manufacturing firm's objective is to maximize global profits less R&D expenditure. When PI are banned, M's profit-maximizing problem is equivalent to solving $\max_{wj\geq 0} \pi_{\mathrm{B}}^{n}(w_i, c_i) = \pi_{\mathrm{B}}^{n}(w_i) + (w_i - c_i)q_{\mathrm{B}}^{n}(w_i) - k^n$, where the first term is D's operating profit excluding the transfer fee. In this case M would offer $w_i = c_i$, which represents efficient vertical pricing, and obtain the monopoly profit in market B.[11]

When PI are allowed, M and D act as Cournot competitors in country A. In this case, the manufacturer faces a number of tradeoffs in setting its contract terms with distributor B. First, M wishes to minimize the vertical pricing inefficiency in market B by setting its wholesale price w_i as close to marginal cost c_i as possible. Next, a higher wholesale price (along with the exogenous trade cost) is the instrument available to M to limit competition in the A market, for this higher price raises D's marginal costs. Finally, higher volumes of PI, for any given positive t, raise the resources wasted in parallel trade and reduce manufacturer profits.

With this intuition, the first-order conditions from maximizing manufacturer profits and distributor gross profits with respect to sales in A yield the following results. First, in the presence of PI, the manufacturer's sales increase in w_i (because this reduces PI volume) and decrease in c_i. Second, combined sales in the A market decrease in w_i. Third, as c_i increases, both M sales and total sales fall because the price in market A goes up. Fourth, increases in t cause PI to fall and at some level to be eliminated. Finally, in country B, the distributor solves its gross profit function as a monopolist, taking w_i as its marginal cost.

4. Analysis of the Model

In this section we derive conclusions about the impacts of parallel imports on prices, sales, and R&D investment.

[11]This and other results in the paper are proved in a separate appendix available from the authors.

4.1. *Wholesale Price and Profits*

Analyzing the specific demand and profit functions to maximize global profits less R&D costs finds that the solutions depend on transport costs and the realization of marginal cost levels, as follows:

$$q_{\mathrm{AD}}(w_i, c_i) = \begin{cases} \dfrac{1 - 2w_i - 2t + c_i}{3} & \text{if } 2(w_i + t) < 1 + c_i \\[2ex] 0 & \text{if } 2(w_i + t) \geq 1 + c_i \end{cases} \tag{3}$$

$$w_i = \begin{cases} \dfrac{1 + 8t + 11c_i}{13} & \text{if } 0 \leq t < \dfrac{3(1 - c_i)}{14} \\[2ex] \dfrac{1 + c_i}{2} - t & \text{if } \dfrac{3(1 - c_i)}{14} \leq t < \dfrac{1 - c_i}{2} \\[2ex] c_i & \text{if } t \geq \dfrac{1 - c_i}{2} \end{cases} \tag{4}$$

$$T_i = \begin{cases} (V_i) & \text{if } 0 \leq t < \dfrac{3(1 - c_i)}{14} \\[2ex] \dfrac{(2a + 2t - 1 - c_i)^2}{16} & \text{if } \dfrac{3(1 - c_i)}{14} \leq t < \dfrac{(1 - c_i)}{2} \\[2ex] \dfrac{(a - c_i)^2}{4} & \text{if } t \geq \dfrac{(1 - c_i)}{2} \end{cases} \tag{5}$$

where V_i is discussed below. The first pair of expressions shows PI quantity, which is positive for wholesale prices and trade costs that are low in relation to production costs but becomes zero at higher costs. The second set of expressions sets out three solutions for wholesale price, depending on t and c_i. Consider the three cases in turn.

First is the situation where PI are positive. For the lowest range of t in Eq. (4), PI are positive from Eq. (3). Moreover, it is straightforward to show that the volume of PI falls linearly to zero as t rises in this range. Note also that successful innovation (lower marginal cost) would generate a lower wholesale price and imply a higher volume of

PI for any trade cost. The idea that innovation would increase PI is novel and central to the analysis to follow.[12]

It may seem surprising that the wholesale price increases in transport cost for positive PI, but there are two offsetting effects. First, the bigger is t, the less M has to worry about parallel imports. Thus, a larger transport cost allows M to reduce w_i and improve vertical pricing efficiency. Second, the bigger is t, the higher is the cross-hauling waste per unit of PI, which reduces M's profits and provides an incentive to limit trade through a higher wholesale price. The manufacturer needs to balance these effects, together with the impact of the higher wholesale price on profits in market B. In this case the second effect dominates and w_i goes up with t. Note also that w_i increases in c_i. Thus, a higher marginal production cost results in a higher wholesale price, thereby lowering the possibility of PI.[13]

It is also interesting to note that the relationship between the transfer payment T_i and marginal cost c_i is not monotonic in the range of positive PI.[14] In this range, the payment equals an expression V_i (in Eq. (5) above), which is a quadratic expression in a, t, and c_i. Intuitively, a higher c_i forces M to offer a higher wholesale price, raising the distortion in market B. But the lower volume of PI limits competition and raises the retail price in A. Thus, distributor profits, and therefore T_i, could rise or fall with c_i.

Finally, in this first case the manufacturer's profits are

$$\pi^p_{\mathrm{M1}} = \frac{1}{52}(12 + 13a^2 - 26ac_i - 8t + 36t^2 - 24c_i + 8tc_i + 25c_i^2) - k^p. \tag{6}$$

[12]It may seem that the reason for this finding is that we have only one firm innovating, rather than two firms engaging in strategic process R&D. In fact, in a framework with two competing manufacturers (M and N) in Market A and a single distributor competing Cournot with both of them by shipping M's good to A, successful process innovation by M (the vertical firm) still increases PI, while successful innovation by N reduces PI volume (Li, 2004). There is an important strategic difference in terms of the reverse causation. Specifically, because M can extract market share from N through profitable PI, it is possible in some circumstances for PI to increase firm M's R&D expenditure, a strategic incentive that is absent here.

[13]It is straightforward to see that PI decline as the wholesale price increases in this range.

[14]The payment declines with marginal cost at higher trade costs.

The second case involves the intermediate range of trade costs in Eq. (4). Here the trade cost offers a sufficiently strong natural barrier to trade that the manufacturer finds it profit-maximizing to complement that cost with a wholesale price that is high enough to deter PI. We call this the deterrence equilibrium and M need not offer w higher than $((1 + c_i)/2) - t$. In this case the manufacturer's profits are

$$\pi^p_{\mathrm{M2}} = \frac{1}{16}(4a^2 + 3 - 8ac_i + 4t - 4t^2 - 6c_i - 4tc_i + 7c_i^2) - k^p. \quad (7)$$

In the third case, high transport costs will block parallel trade. It is straightforward to show that transportation costs in the high range of Eq. (4) prevent parallel trade, operating like an outright ban. The manufacturer offers a wholesale price equal to its marginal cost and achieves vertical pricing efficiency. We refer to this equilibrium as blocked PI.

The price results may be summarized by noting that the profit-maximizing wholesale price is piecewise linear in transport cost. The price relationship takes on an inverted V-shape, going up with t when PI are positive (but declining), falling with t when PI are deterred, and then equaling constant marginal cost when PI are blocked. The price function for unsuccessful innovation is displaced upward and to the left from that for successful R&D, with a higher wholesale price in the former case for all trade costs.

From the analysis of profit functions we may state:

Proposition 1. *There exists a unique* $t^*, 0 \leq t^* < \frac{3(1-c_i)}{14}$ *such that* π^p_{M} *decreases in* t *when* $0 \leq t < t^*$, *increases in* t *when* $t^* \leq t < \frac{1-c_i}{2}$ *and is constant when* $t \geq \frac{1-c_i}{2}$.

Thus, the profit curve is U-shaped in terms of transportation cost.[15] When the trade cost is low, there are parallel imports. A rise in t generates higher resource wastes and also forces the manufacturer to raise the wholesale price, creating a distortion in vertical pricing. Thus, for low trade costs, the manufacturer's profit decreases as t rises. For intermediate trade costs, parallel trade is deterred and

[15]This result is akin to that in Maskus and Chen (2002), who find a similar U-shaped global welfare curve.

as t increases the vertical pricing inefficiency diminishes. However, the higher t has no impact on profits in market A and, therefore, M's total profit rises. Finally, for t high enough to block PI, the manufacturer achieves vertical pricing efficiency and earns monopoly profit in both markets. Any further increase in transport cost would not affect profits.

4.2. *Incentives to Innovate*

The manufacturer has an incentive to invest in R&D to the extent that its profits under successful process innovation are larger than under failed innovation. A first, and obvious, point is that if PI are permitted, profits under successful innovation are higher than under failed innovation for any trade cost. With success the manufacturer still has the option of choosing sales and revenues that would be the same as if innovation failed but has lower costs. The firm's optimal choices with successful innovation must, therefore, be still more profitable.

A second point, which we can derive under our Basic Assumption that marginal cost is small relative to the size of country A, is that successful cost-reducing innovation reduces the wholesale price. The implication is that the profit curve with successful R&D lies above that with failed R&D and the manufacturer's profit function is decreasing in marginal cost.

Combining these ideas, we can state that the manufacturing firm has a positive incentive to invest in cost-reducing innovation in the presence of legal PI. This statement is evidently true for low trade costs. For intermediate trade costs, where PI are deterred, a lower marginal cost would raise both the profits in market A due to higher sales and the profits in market B by reducing the pricing distortion. Finally, for high trade costs, parallel trade is blocked. This enables M to offer a wholesale price equal to marginal cost and achieve monopoly profits, which are increased with lower marginal costs, in both markets.

Our analysis finds that both profit functions are U-shaped in transport costs. However, we can show that the curve with lower marginal costs is both above and displaced to the right of the one with higher marginal costs. This means that the difference between the profit functions, which represents the manufacturer's incentives to engage in innovation, varies with transportation costs. We discuss this situation in the following subsection.

4.3. *Parallel Imports and Innovation*

Recalling that $\Delta R_{\mathrm{M}}^{p}$ is the difference in profits between successful and unsuccessful innovation, we can state:

Proposition 2. *There is a unique* $t^{**}, 3(1-c_H)/14 < t^{**} < 3(1-c_L)/14$ *such that* $\Delta R_{\mathrm{M}}^{p}$ *decreases in* t *when* $0 \le t < t^{**}$, *increases in* t *when* $t^{**} \le t < (1-c_L)/2$ *and is constant when* $t \ge (1-c_L)/2 > t^{**}$.

This proposition implies that $\Delta R_{\mathrm{M}}^{p}$ is continuous and takes a U-shaped curve as transport cost rises. The relationship between $\Delta R_{\mathrm{M}}^{p}$ and t is given in Fig. 1.

This result says that the difference between profits when process innovation is successful and not successful at first becomes smaller as t increases, achieves a minimum at t^{**}, then increases. If transportation cost is high enough to block PI regardless of innovation, the difference in these two profit levels is constant. Note

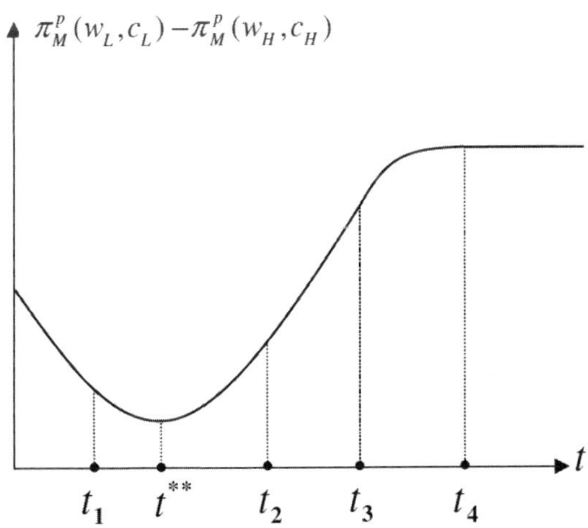

$$t_1 = \frac{3(1-c_H)}{14}, t^{**} = \frac{21 - 13c_H - 8c_L}{98}, t_2 = \frac{3(1-c_L)}{14}, t_3 = \frac{1-c_H}{2}, t_4 = \frac{1-c_L}{2}$$

Figure 1. Difference in profits between successful and failed innovation under legal PI.

from Eq. (2) that when this profit difference declines, so does investment in R&D, whereas R&D rises when the profit difference goes up. Thus, the level of R&D expenditures is also U-shaped in transport costs in the presence of PI.

Consider an alternative characterization of this result, in terms of Fig. 1. In the range of trade cost below t_1, PI exist regardless of the success of first-stage innovation. Here, the profit-reducing effects of competition outweigh any profit-enhancing effect on the post-innovation wholesale price in B. Interestingly, investment in R&D would rise as the natural trade-cost barrier goes to zero, despite the increasing competition from PI. In the range between t_1 and t_2, PI occur when M succeeds in reducing costs but are deterred by a higher wholesale price when M fails. The former possibility reduces the manufacturer's incentives to innovate. Note that the critical value of t, at which the difference in profits begins to rise, occurs in this range.

In the range between t_2 and t_3, PI are deterred under either innovation outcome but this sustains a pricing distortion that discourages M from making investments in process innovation. Clearly, this disincentive falls as transport cost goes up. Between t_3 and t_4 parallel trade is blocked by the transport cost if M fails to innovate, but the vertical pricing distortion in market B remains if M succeeds. Finally, above t_4, PI are blocked in all cases and there is no change in the difference in profits from innovation outcomes.

We turn now to assessing the impact of exhaustion policy on innovation expenditure. The usual intuition is that permission of PI reduces incentives to innovate because parallel traders free ride on innovation costs. However, as we have demonstrated, the difference in profits between successful and failed innovation rates depends on the legal treatment of PI. Thus, we wish to determine conditions under which profits with parallel trade are reduced by more than the amount by which profits without parallel trade are reduced. To address this question, we need to show that the distance between the profit functions where PI are banned is higher than in all other cases. Note that such a ban is equivalent to considering the range of transport cost that blocks PI. From Fig. 1 it is evident that this range generates the highest possible profit difference. It follows immediately from Eq. (2) that $k^p \leq k^n$ for all t. We therefore have:

Proposition 3. *Under our Basic Assumption, the permission of parallel imports reduces the manufacturer's incentives to invest in process innovation.*

To make this result more concrete, consider the manufacturer's expected profits with and without legal permission of parallel imports. We can show that for all k it must be that $E^n_{\mathrm{M}} > E^p_{\mathrm{M}}$. Thus, given any feasible investment the expected profit is higher in the case of no PI than that with PI. It follows immediately that the equilibrium expected profit is higher in the absence of PI than in their presence.

The central finding is that the distortions associated with PI reduce the manufacturer's incentive to innovate. Thus, despite the subtleties in how PI affect profits at various levels of trade cost, it remains the case that permitting parallel trade (as opposed to banning it) reduces the returns to innovation. Successful process innovation raises the manufacturer's profits under national exhaustion relative to all outcomes with international exhaustion.

4.4. *Impacts of Restricting Parallel Imports on Expected Welfare*

Because process innovation could change expected global welfare and expected welfare of both countries, the legal treatment of PI should influence these welfare levels. In this subsection, we discuss verbally the effects of restricting PI on changes in expected welfare.[16] Under our Basic Assumption we have the following results. First, restricting PI reduces expected consumer surplus in country A, raises expected consumer surplus in country B and has an ambiguous impact on expected global welfare when trade costs are in the low range (i.e., below t_2 in Fig. 1). Here, banning PI would raise price in A and reduce it in B. In terms of each market, country A would lose in the range of low trade costs because the loss in consumer benefits would outweigh the gain in manufacturer profits, while welfare would rise in country B.

[16]Again, these results are derived in the appendix available from the authors.

Second, when trade costs are in the intermediate range (i.e., between t_2 and t_4), such restriction does not affect expected consumer surplus in country A, raises expected consumer surplus in country B and increases expected global welfare. In that range, PI are deterred by the high wholesale price, regardless of the outcome of innovation. Banning parallel trade would not alter expected consumer surplus in country A but would increase profits and raise welfare in that market. The ban would also raise expected consumer surplus in country B, resulting in higher expected global welfare. Finally, a ban on PI affects neither expected consumer surplus in either country nor expected global welfare under high trade costs (i.e., above t_4).

To illustrate these results numerically, consider the difference in global welfare between permitting and banning PI at different trade costs. Taking parameters $a = 1$ and $c = 0.1$ (that is, we hold marginal cost constant and abstract from innovation), we can compute welfare impacts from the model. The static global welfare change of shifting from a ban to permission of PI would be $+0.012$ at $t = 0$, $+0.007$ at $t = 0.01, -0.023$ at $t = 0.036$ and -0.059 at $t = 0.18$. That is, at zero trade costs PI would raise welfare by 1.2 percent, while at 3.6 percent trade costs global welfare would decline by 2.3 percent. Given that our model shows that process innovation would decline in the presence of PI, the conclusion seems to be that permitting such trade under current levels of transport costs (see Table 1) would diminish economic well-being. The calculation is only illustrative, however, for market sizes in PI importing countries and marginal costs may be quite different, depending on circumstances.

5. Conclusion

In a model of vertical price control we formalized the common intuition that permitting PI reduces anticipated profits from process innovation. Our first result is that profits depend on the cost of transporting PI goods. When the transportation cost is small, the manufacturer raises its wholesale price in order to control PI. However, a higher wholesale price increases the distortion in the distributor's market and the net effect on the manufacturer's profit could be negative. As transportation cost rises, the manufacturer can offer a lower wholesale price to improve vertical pricing efficiency. Where the transportation cost is high enough to prevent PI, the manufacturer's profit achieves its maximum.

Our second result is that successful cost-reducing innovation is helpful in lowering the wholesale price, thereby diminishing the vertical pricing distortion. However, this in turn encourages more PI. The third outcome is that the difference in profits between successful and failed innovation takes a U-shape in terms of the cost of parallel trade. That is, the manufacturer's incentive to engage in R&D first decreases as the cost of parallel trade rises, then increases, and ultimately it remains constant. The final contribution of this paper is to show that openness to parallel imports inhibits R&D. In essence, by engaging in process innovation, the manufacturer achieves a higher expected profit without parallel trade than with parallel trade.

References

Barfield, C. and Groombridge, M., 1998. The economic case for copyright owner control over parallel imports. *The Journal of World Intellectual Property* 1, 903–939.

Bester, H. and Petrakis, E., 1993. The incentives for cost reduction in a differentiated industry. *International Journal of Industrial Organization* 11, 519–534.

Chard, J. and Mellor, C., 1989. Intellectual property rights and parallel imports. *The World Economy* 12, 69–83.

Costa Cabral, C., Kujal, P., and Petrakis, E., 1998. Incentives for cost-reducing innovations under quantitative import restraints. *Annales d'Economie et de Statistique* 49/50, 479–493.

Danzon, P. M., 1997. *Pharmaceutical Price Regulation.* American Enterprise Institute, Washington DC.

Duhan, D. and Sheffet, M., 1988. Gray markets and the legal status of parallel importation. *Journal of Marketing* 52, 75–83.

Ganslandt, M. and Maskus, K. E., 2003. Vertical restraints, distribution, and the price impact of parallel imports: implications for the European Union and Sweden. In: Lundvall, K. (Ed.), *High Prices in Sweden — A Result of Poor Competition?* Swedish Competition Authority, Stockholm, pp. 160–223.

Ganslandt, M. and Maskus, K. E., 2004. The price impact of parallel imports in pharmaceuticals: evidence from the European Union. *Journal of Health Economics* 23, 1035–1057.

Ganslandt, M. and Maskus, K. E., 2005. Vertical Distribution, Parallel Trade, and Price Divergence in Integrated Markets. Working Paper, University of Colorado, Boulder.

Haaland, J. and Kind, H., 2004. R&D Policies, Trade, and Process Innovation. Working Paper, Norwegian School of Economics and Business Administration, Oslo, Norway.

Hilke, J. C., 1988. Free trading or free riding: an examination of the theories and available evidence on gray market imports. *World Competition* 32, 75–92.

International Intellectual Property Association, 1998. *Parallel Import Protection in 107 Selected Countries*, report, Geneva.

Leahy, D. and Neary, J. P., 2001. Robust rules for industrial policy in open economies. *Journal of International Trade and Economic Development* 10, 393–409.

Li, C., 2004. The Impact of Parallel Imports on Cost-Reducing Innovation in the Presence of Competition. Working Paper, Nankai University.

Malueg, D. A. and Schwartz, M., 1994. Parallel imports, demand dispersion, and international price discrimination. *Journal of International Economics* 37, 187–196.

Maskus, K. E., 2000. Parallel imports. *The World Economy: Global Trade Policy* 23, 1269–1284.

Maskus, K. E. and Chen, Y., 2002. Parallel imports in a model of vertical distribution: theory, evidence, and policy. *Pacific Economic Review* 7, 319–334.

Maskus, K. E. and Chen, Y., 2004. Vertical price control and parallel imports: theory and evidence. *Review of International Economics* 12, 551–570.

National Economic Research Associates, 1999. *The Economic Consequences of the Choice of Regime in the Area of Trademarks.* National Economic Research Associates, London.

Palia, A. and Keown, C., 1991. Combating parallel importing: views of US exporters to the Asia-Pacific region. *International Marketing Review* 8, 47–56.

Reitzes, J. D., 1991. The impact of quotas and tariffs on strategic R&D behavior. *International Economic Review* 32, 985–1007.

Richardson, M., 2002. An elementary proposition concerning parallel imports. *Journal of International Economics* 56, 233–245.

Scherer, F. M. and Watal, J., 2002. Post-TRIPS options for access to patented medicines in developing countries. *Journal of International Economic Law* 5, 913–939.

Spencer, B. J. and Brander, J. A., 1983. International R&D rivalry and industrial strategy. *Review of Economic Studies* 50, 707–722.

Tarr, D. G., 1985. An Economic Analysis of Gray Market Imports. U.S. Federal Trade Commission, Washington DC.

PART II

INNOVATION AND INTERNATIONAL TECHNOLOGY TRANSFER

Chapter 7

Transfer of Technology to Developing Countries: Unilateral and Multilateral Policy Options*

Bernard M. Hoekman, Keith E. Maskus and Kamal Saggi[†]

1. Introduction

The importance of international technology transfer (ITT) for economic development can hardly be overstated. Both the acquisition of technology and its diffusion foster productivity growth. Developing countries have long sought to use both national policies and international agreements to stimulate ITT. *National* policies range from the general, such as education and intellectual property rights (IPR) protection, to the specific, such as tax incentives for purchase of certain types of capital equipment. A prominent episode of *international* efforts to encourage ITT came in the late 1970s, when many developing countries sought a Code of Conduct to regulate technology transfer under United Nations (UN) auspices.

Given the incentives of owners not to transfer technology without an adequate return and the problem of monitoring compliance with international regulations, it is not surprising that ITT is predominately mediated by national policies. Existing multilateral rules on, for example, subsidies, trade policy and IPR regimes, define limits on what is permissible. In contrast, agreements regarding actions

*Reprinted from *World Development*, Vol. 33, No. 10, pp. 1587–1602. Copyright 2005, with permission from Elsevier.
†We are grateful to three anonymous referees for their comments. The views expressed are personal and should not be attributed to the World Bank.

that governments should pursue to encourage ITT are largely of a best-endeavor nature.

Starting in the mid-1990s, multilateral disciplines on ITT-related policies began to deepen. The WTO Agreement on Trade-Related Aspects of Intellectual Property Rights (TRIPS), adopted in 1995, calls on countries to enforce comprehensive minimum standards of IPR protection on a nondiscriminatory basis. It also has provisions relating to ITT, as discussed later in this article. In 2001, WTO members established a Working Group on Trade and Technology Transfer to examine the relationship between trade and ITT and explore ways to increase technology flows to developing countries.

This paper discusses options that could be pursued by countries to promote ITT and the implications for international rule making and development assistance. Three dimensions are highlighted: safeguarding national "policy space" to address market failures; identifying actions by source countries to encourage ITT; and multilateral initiatives to address international externalities associated with technology markets or national policies. Section 2 reviews the major channels of technology transfer and the main policy instruments that could be used to enhance ITT. In Sections 3–5 our discussion turns to normative policy implications, focusing on policies of host (importing) countries, source (exporting) countries, and multilateral cooperation in the WTO, respectively. In Section 6 we summarize the discussion by means of suggested policy rules of thumb for countries at different stages of development. In Section 7 we conclude with a brief discussion of how trade agreements can help improve ITT policies.

2. Channels of Technology Transfer

We start with a brief discussion of the main avenues through which ITT occurs: trade in goods, direct investment and licensing, and movement of people.

2.1. *Trade in Goods*

Economic growth centrally relies on technological change through the creation of new products and processes (Grossman and

Helpman, 1991). New products embody novel ideas and international trade can transmit knowledge across borders. Indeed, barriers to technology adoption are a key determinant of international differences in per-capita income (Parente and Prescott, 1994) and greater trade openness can increase growth by lowering such barriers. Trade contributes to ITT by allowing local reverse engineering and access to new machinery and equipment. Coe *et al.* (1997) found that foreign R&D embodied in imported capital goods has a significantly positive impact on total factor productivity (TFP) of importing countries.[1] Later studies noted that this impact is greater the more open are countries, the more skilled is their labor force, and, in the case of developing countries, the more their trade is with developed countries (Schiff *et al.*, 2002). Further, variations in capital-goods trade explain cross-country differences in productivity better than does overall trade (Eaton and Kortum, 1999).

These results suggest that open trade policies are crucial for developing countries to be able to attract technology. But openness alone is not sufficient, for strong absorptive capacity and the ability to adapt foreign technology are important for ITT to effect local technical change. In developing countries, technology acquisition often amounts to adapting existing methods to local circumstances (Evenson and Westphal, 1995). Gradual adoption of new techniques or inputs may be optimal for risk-averse producers in the face of costly investment and uncertain returns. Producers need to learn how to apply the new technology and improve it gradually over time (Tybout, 2000). Countries tend to acquire international technology more readily if domestic firms have R&D programs, there are domestic research laboratories and universities, and there exists a sound basis of technical skills and human capital (Maskus, 2000). These factors reduce the costs of imitation, adaptation, and follow-on innovation.

---

[1]See Keller (1998) for a criticism of earlier work by Coe and Helpman (1995), the results of which may have been spurious. However, that trade embodies technical information is evident in numerous studies, including Eaton and Kortum (1996) and Keller (2002). Keller (2002) noted that information diffusion declines with distance but recognized that one channel must be trade.

2.2. *Foreign Direct Investment and Licensing*

Investment by multinational enterprises (MNEs) may provide developing countries with more efficient foreign technologies and result in technological spillovers. In addition to demonstration effects, spillovers may arise because of labor turnover and vertical linkages. Case studies suggest that substantial technology diffusion occurs due to FDI (Blomstrom and Kokko, 1997). However, econometric studies support a more diverse set of conclusions. Some found that firms in sectors with a relatively high MNE presence tend to be more productive (Kokko *et al.*, 1997), while others noted that competing domestic firms do worse as the foreign presence in their industry increases (Aitken and Harrison, 1999). Such negative horizontal spillover effects may occur if MNEs siphon off domestic demand or bid away high-quality labor. Similarly, there is a risk that imports of technology may reduce R&D efforts of local firms, which may have happened in some sectors in China in the 1990s (OECD, 2002).

Moran (2004) argued that there is a substantial difference in operating characteristics between subsidiaries that are integrated into the international sourcing networks of the parent MNEs and those that are prevented from such integration by policy barriers such as mandatory joint venture and domestic content requirements. These different characteristics include size of plant, proximity of technology and quality-control procedures to global best practices, speed with which production processes are brought to the frontier, and cost of output. Integrated subsidiaries have a more positive impact on the host country, often accompanied by vertical backward linkages and externalities. Isolated affiliates have a less positive, and sometimes negative, impact on the local economy.

Drawing upon case studies and econometric evidence, Moran claimed that this contrast in performance holds across different industries, countries, and time periods. He noted that failure to differentiate between export-oriented FDI and import-substitution FDI, or between foreign investors free to source from any location and those operating under domestic content requirements, or between foreign investors obliged to operate as minority shareholders and those with complete or majority ownership, accounts for the inability of earlier studies to isolate the influences of FDI on host-country welfare.

Studies focused on spillovers to local competitors of MNEs miss the fact that such firms typically transfer technology to local suppliers. Mexico's *maquiladora* sector is a recent example of vertical ITT. Most maquiladoras began as subsidiaries of U.S. firms that shifted labor-intensive assembly operations to Mexico. However, over time the maquiladoras adopted more sophisticated imported production techniques (Saggi, 2002).[2] A recent case study of the effects of Intel's investment in Costa Rica by Larrain B., *et al.* (2000) found that local suppliers benefited substantially from Intel's investment. Similar evidence exists for other sectors and countries (Moran, 1998 and 2001). For example, in the electronics sector, foreign investors helped their local subcontractors keep pace with modern technologies by assigning technicians to the suppliers' plants to help set up and supervise automated production and testing procedures.

Javorcik (2004) examined backward linkages and technology spillovers using data from Lithuanian manufacturing firms over 1996–2000 and found that productivity is positively affected by a sector's contacts with multinational customers but not by the presence of MNEs in the same industry. Thus, her results supported the existence of vertical spillovers from FDI. Blalock (2001) used panel data from Indonesian manufacturing establishments to check for similar effects. He found a strong positive impact of FDI on productivity growth of local suppliers, suggesting that effective technology transfer does occur. He also plausibly suggested that, since MNEs tend to source inputs that require relatively simple technologies, local intermediate suppliers are in a good position to learn from affiliates.

Licensing is an important source of ITT for developing countries (Correa, 2003). Contracts typically involve the purchase of production or distribution rights and the underlying technical information and know-how.[3] The general determinants of decisions to license are similar to those involving FDI, including market size and policy certainty and transparency. An important additional factor is the confidence of licensor firms that proprietary technologies will not leak into

[2]Similar findings apply to transition economies; see, for example, Javorcik (2004).
[3]With intra-firm ITT the MNE retains proprietary control of the know-how, while in the arm's-length case it must be provided the licensee.

the host economy through copying or defection of personnel. If this is likely, foreign firms may prefer FDI, may not engage in licensing at all or may transfer lagging technologies (Maskus, 2000).

Successful transfer typically requires capacity to learn and investments to apply technologies into local production processes. This factor explains why countries with substantial engineering skills and R&D programs for adaptation and learning are greater recipients of licensing flows than others (Yang and Maskus, 2001).

2.3. *Labor Turnover and Movement of People*

Little attention has been given to the role of labor flows as a channel for ITT. Some studies have found that domestic labor turnover from MNEs to local firms is limited (Gershenberg, 1987), while others claimed the opposite (Rhee, 1990). An explanation is that in countries where local firms are not far behind MNEs in technical terms, labor turnover is more likely. Thus, the ability of local firms to absorb new technologies is a determinant of whether labor turnover is a means of diffusion.

International movement of people, associated with nationals studying or working abroad for a limited period, or the inward movement of foreign citizens, is another potential channel for ITT. The recent experience of India in developing a software and related services industry illustrates that payoffs from such movements may take time to materialize but can be large. A policy challenge for developing countries is to facilitate temporary movement abroad and encourage returnees to undertake local research and business development.

2.4. *Market Failures and the Need for Policy*

Markets within which international technology transfers take place are subject to at least three major problems. The first is asymmetric information (Horstmann and Markusen, 1987; Arora *et al.*, 2001). Technology suppliers cannot fully reveal their knowledge without destroying the basis for trade, while buyers cannot fully determine the value of the information before buying it. This problem can lead to large transactions costs and stifle technology flows.

In the international context, information and contract enforcement problems may be severe. A second problem is that owners of new technologies typically have substantial market power resulting from lead-time and IPR (Maskus, 2000). Thus, the price of technology will exceed its marginal cost. While this exclusivity allows developers to profit from innovation, it reduces the static national welfare of those importing technologies. A third problem is that externalities arise if the costs and benefits of technology exchange are not internalized by participants (Saggi, 2004). A major share of benefits to recipient countries of ITT is likely to arise from uncompensated spillovers, wherein technological information is diffused into the wider economy and the technology provider cannot extract the associated economic value.

These market failures support the potential for policies to increase welfare by altering the incentives of private agents to engage in ITT. This task is complex and it is difficult for policymakers to identify optimal policies or even to rank-order them according to effectiveness.[4] However, the problematic characteristics of knowledge markets noted above suggest that policy should aim at three targets. First, increase access of local buyers to the international stock of knowledge and improve the ability of technology owners to signal the true value of their inventions to buyers. Second, reduce the costs of acquiring and absorbing existing technologies. Third, increase incentives for domestic innovation. We turn next to policy options for host countries, source countries and international cooperation.

3. National Policies

A basic challenge for host developing countries is to improve the local environment for ITT and its diffusion. Both FDI and licensing respond to such factors as an effective infrastructure, transparency and stability in government, and a reasonably open trade and investment regime (World Bank, 2004). Also important is an entrepreneurial environment that attracts skilled workers home

[4]In practice, the potential for beneficial policy may be frustrated by mistakes or rent seeking.

from developed countries. Given that vertical spillovers are strongest in countries where MNEs work with competitive suppliers, reducing entry barriers in upstream industries can also assist ITT.

An important determinant of the ability to absorb and adapt technology is the return to investing in at least simple R&D capacity. If existing technology policies, capital market regulations and taxes discourage such investments, they could be reformed to encourage innovation. Similarly, absorption and competitive deployment of ITT depend on an adequate supply of engineering and management skills, making domestic education and training policies important. Governments can reduce the technological distance between local and foreign firms by establishing innovation systems that encourage R&D, transfer knowledge from universities and public laboratories to domestic firms, and promote use of cost-saving technologies.

Host-country governments also need to recognize the risk-taking involved in adopting foreign technologies. Suppose domestic firms are considering the costly adoption of some foreign technology but significant uncertainty exists regarding the degree to which it can be used effectively. Firms adopting the technology first generate positive spillovers for others who can watch and learn. In the presence of such informational externalities, it makes sense to subsidize the adoption of foreign technologies. Hausmann and Rodrik (2003) have emphasized that in many developing countries the market undersupplies investment by firms in new activities because of appropriability problems. If so, a subsidy or similar incentive can expand innovation and risk-taking.[5]

3.1. *Trade and FDI Policies*

Available evidence suggests that diffusion of knowledge is facilitated by an open trade regime (Saggi, 2002). Firms need access to capital equipment and imported inputs that embody foreign knowledge. At the same time, the existence of market failures makes questionable any unconditional arguments against trade protection.

---

[5]In such cases IPRs may not be effective, for often the technology will already have been invented elsewhere. The objective is to encourage imitation of "revealed successes" through entry into what are new productive activities for the country concerned.

An important question is the scope of knowledge spillovers. International spillovers, for which there is considerable evidence (Eaton and Kortum, 1996), tilt the balance in favor of free trade. In contrast, intra-national spillovers create a role for policy intervention, though trade policies would be neither efficient nor effective. Instead, general policies encouraging agents to undertake activities that generate social benefits exceeding private returns, without simultaneously creating additional distortions, are more appropriate.

Historically, restrictive trade policies were complemented by restrictions on FDI. Thus, Japan, the Republic of Korea, and Taiwan imposed restrictions on FDI whereas policies were more welcoming toward other modes of ITT. Moran (1998) summarized the approach taken by Korea and Japan as a combination of import barriers coupled with export promotion; fiscal subsidies to create national champions in certain sectors; and restrictions on FDI coupled with insistence upon licensing and other arrangements. More recently, national FDI policies have become more liberal, but policies may differentiate between joint ventures and majority-owned FDI. For example, Chinese policy has encouraged joint ventures more than inward FDI. Whether this approach is beneficial depends on the incentives MNEs have to transfer technologies to recipient firms and there is evidence that joint ventures obtain less advanced technology (Mansfield and Romero, 1980; Moran, 1998).

While the magnitude of ITT undertaken by MNEs may not be socially optimal, evidence exists that they are keen to transfer technology to local suppliers (Moran, 1998). Policies that facilitate this process, rather than insisting that MNEs engage in ITT to local competitors, are more likely to succeed. In practice, many countries seek to attract foreign investors through special economic zones, subsidies, tax holidays, and other grants. Such investment incentives can be justified only if host countries enjoy positive learning and productivity externalities from inward FDI.

The prevalence of "follow the leader" behavior among MNEs provides a potential case for FDI incentives. One new entrant may attract investments by both other firms and upstream suppliers. If so, competition at multiple stages of production may increase, thereby improving both efficiency and overall output. An implication is that a host country may be able to unleash a sequence of investments by successfully inducing FDI from one or two major firms.

If the local economy lacks a well-developed network of potential suppliers, however, MNEs might not invest and suppliers may not develop. In such cases, growth is constrained by coordination problems that can partially be resolved by initiating investments from key firms. Such problems cannot be tackled solely through investment incentives, however. Policy efforts need to improve the investment climate and reduce the costs of absorbing technology. The latter task is complex and involves improving property rights, expanding innovation systems, and investing in efficient infrastructure. Thus, for incentives to be effective, important preconditions relating to the investment climate and absorptive capacity must be satisfied. Once upstream capacity exists, there may be a case for programs that help potential suppliers meet the needs of MNEs as customers. Such programs have been successfully implemented in a number of middle-income countries.

One set of policies often justified on the basis of regulating ITT are trade-related investment measures (TRIMs), such as local content rules and technology transfer requirements. However, TRIMs are implicit taxes on intermediate goods imports because manufacturers are forced to use higher-cost local inputs. They provide little incentive for the protected producers of intermediate goods to acquire knowledge and improve productivity. If the constraints imposed by TRIMS are too strict, MNEs may refrain from investing.

3.2. *General vs. Specific Technology-Related Policies*

Many studies indicate that absorptive capacity is crucial for obtaining significant spillover benefits from trade or FDI. For example, Borensztein *et al.* (1998) tested the effects of inward FDI on growth in 69 developing countries. They found that it contributes *more* to domestic growth than does domestic investment but only where the host country has a minimum threshold stock of human capital. Similarly, Keller (1996) argued that access to foreign technologies alone does not increase growth rates of developing countries.

Consider the role that subsidies can play in facilitating learning and technology acquisition when returns to such investments cannot be appropriated by private agents. Such commentators as Amsden (1989) argued that policy interventions, including implicit or explicit subsidies, lay behind the economic miracles in Korea and Taiwan. Their case is that carefully targeted subsidies allowed these governments to stimulate key sectors that became efficient and provided positive spillovers. It is important, however, to differentiate between sector-specific subsidies and general policies facilitating learning and enterprise development. In a recent retrospective on the East Asian development experience, Noland and Pack (2003) argued that sector-specific policies did not result in high rates of TFP growth for manufacturing. In Korea and Taiwan, TFP growth was not much higher than in OECD countries. In India, selective interventionist policies were associated with declining TFP growth rates, while the opening of the economy led to an increase in TFP growth (Krishna and Mitra, 1998).

The case for general policy supports for innovation, education, transport infrastructure and similar public goods is uncontroversial. Regarding more commercial activities, however, the efficient use of support policies requires that governments can both identify cases that justify intervention and implement it appropriately. In practice, governments may fail to avoid potential problems, including subsidies that support inefficiency and strategic subsidy seeking by firms. Because these problems become difficult to control, credible exit strategies are needed to weed out successes from failures.

3.3. *IPR and Technology Transfer*

Intellectual property rights can support markets in technology, including ITT (Arora *et al.*, 2001; Maskus, 2002). Patents and trade secrets provide a legal basis for revealing the proprietary characteristics of technologies to subsidiaries and licensees, supporting the formation of licensing contracts.[6] Patent protection both increases

[6]See Correa (2003) for the counterargument that IPR stifle ITT as firms exploit market power.

flows of ITT to countries with sufficient technological capacity and shifts incentives for investors between FDI and licensing.

The empirical literature supports several observations. First, patent applications from foreign firms are associated with productivity growth in recipient countries (Eaton and Kortum, 1996). Thus, "trade in ideas" is a significant factor in world economic growth. Second, information from patent citations suggests that the most significant patents are widely diffused, as is knowledge in highly technological sectors (Peri, 2003). Knowledge flows have a positive impact on international innovation.

Third, international trade flows, especially in patent-sensitive industries, respond positively to increases in patent rights among middle-income and large developing countries but not among poor countries (Smith, 2001). Next, the evidence on patent laws and inward FDI is mixed but recent studies find positive impacts among middle-income and large developing countries. However, in poor countries, patents do not expand FDI (Smith, 2001; Blyde and Acea, 2002). Fourth, strengthening IPR tends to shift ITT from exports and FDI toward licensing and also increases R&D expenditures undertaken on behalf of affiliates in recipient countries with strong imitative abilities (Smith, 2001). Fifth, the sophistication of technologies transferred rises with the strength of IPR protection and domestic capacities to absorb and improve technology.

A common finding is that the poorest countries are unlikely to benefit from strong IPR (McCalman, 2001). Stronger patent rights may be expected to raise monopoly rents earned by international firms as such rights become more valuable, obliging developing countries to pay more for protected technology. These are also countries where ITT-related spillovers are likely to be small. Thus, in poor countries policy should aim at lowering costs of imports of IPR-intensive goods and raising the capacity to absorb and adapt technologies.

3.4. *Summary: Theory and Experience*

Economic theory does not provide unambiguous guidance regarding the relative social payoff to alternative channels of ITT. Much depends on whether spillovers are international or intra-national, the capacities to absorb and improve technologies, and other factors. A "one size fits all" approach to policy is inappropriate. Consider, for example, the choice between licensing and FDI from a host-country's viewpoint. If the relevant market is imperfectly competitive the choice is not simple. The alternative to suffering market power at the hands of MNEs might be sustaining domestic incumbents with significant market power. While the profits of the latter add to national income, such incumbents will generally possess inferior technologies. If there exists a strong domestic incumbent, limiting FDI and encouraging technology licensing to that firm can increase its market power vis-à-vis weaker domestic rivals (Saggi, 1996). On the other hand, if FDI were restricted and open technology licensing encouraged, licensing would be preferable to FDI if the technology owner agreed to transfer the technology.

Given the limited guidance offered by theory, it is helpful to revisit briefly the history of successful efforts to move up the technology ladder. One reason for Japan's rapid growth and industrialization after World War II was that its patent system was designed for both small-scale innovation and diffusion. The system encouraged incremental innovation by Japanese firms and promoted the diffusion of knowledge into the economy (Maskus, 2000). Japan also encouraged foreign firms to license to Japanese firms, in part through restrictions on FDI.

Korea encouraged learning via "duplicative imitation" of mature technologies that were in the public domain or available cheaply (Kim, 2002). IPR protection was weak and encouraged imitation and adaptation. In the 1970s, Korea specialized in labor-intensive goods, with firms importing "off the shelf" technologies and adapting them to produce differentiated goods. Government promoted the development of technical skills through education and workplace training and significantly reduced anti-export bias. In the 1980s, Korean firms shifted to "creative imitation", involving more significant transformation of imported technologies. This shift required domestic R&D and in-house research capabilities. The government also became more welcoming to formal channels of ITT and strengthened the IPR regime.

Brazil, Mexico, Malaysia, and the export-intensive regions of China and India are other examples of movement from "pure" to "creative" imitation. In these cases, IPR protection was limited and firms took advantage of available foreign technologies. As the technological sophistication of production processes matured and the depth and complexity of knowledge for effective absorption grew, firms resorted more to formal means of ITT and governments strengthened the IPR regime.

These experiences suggest that the nature of ITT and appropriate policies follow a technology ladder. Many middle-income developing countries are at the duplicative imitation stage, hoping to absorb foreign technologies into labor-intensive export production and evolve into higher value-added strategies over time. The poorest countries have barely stepped onto this rung. Thus, a differentiated approach is needed for national policy and multilateral rulemaking. Specifically, the priority in poor countries with weak institutions and limited R&D capacity is to improve the business environment, with liberal trade policies to encourage imports of technology. Such countries should be exempt from strong IPR obligations and have access to mechanisms to reduce the cost of imports of IPR-protected goods. This could be achieved, for example, through either subsidies or differential pricing schemes.

Because absorptive capacity is weak in low-income countries, the emphasis should be on using trade to benefit from foreign knowledge and acquiring technology through FDI. Licensing is not a realistic option for least-developed countries (LDCs), given weak business environments and absorptive capacity. In poor countries there is a greater need for FDI because the incentives for licensing or joint ventures are weaker. Indeed, this situation may provide a case for outward FDI investment incentives by high-income countries as part of their development assistance.

As countries move up the income and technology ladder, they gain more from IPR, which becomes important for licensing and benefits home entrepreneurs and innovators. Based on the experience of Asian economies, developing countries should adopt standards for patentability, novelty, and utility that are stricter (raising a higher bar to patenting) than those found in the United States and the

EU. This approach is consistent with WTO rules, which do not specify substantive criteria on the basis of which IPR grants are awarded. Upper-middle-income countries would benefit less from subsidy schemes to lower the price of technology.

4. Source-Country Policies

Perhaps the most powerful indirect incentive for ITT that source countries could provide is to grant greater market access for products in which poor countries have a comparative advantage, including agricultural products and labor-intensive goods. The linkage is that better assurance of foreign market access would expand incentives to transfer new technologies to producers in developing countries.

Turning to specific measures aimed at ITT, fiscal incentives or subsidies are the most obvious candidate. Sub-central governments in OECD countries often offer tax incentives to induce firms to establish facilities or to remain in their own countries, especially in low-income areas. One option is for central governments to offer the same fiscal benefits to firms transferring technologies to developing countries as are available for domestic activities. Developed countries could also offer the same tax advantages for R&D performed abroad as for R&D done at home.

In designing such incentives, home countries could tailor interventions toward channels appropriate for countries at different stages of development. Given the forgoing arguments in favor of FDI over licensing in low-income countries, for example, policies that subsidize direct ITT through licensing may not be beneficial to poor countries. A better approach would be to ensure that incentives target (or minimally do not discriminate against) outward FDI. Source countries should also differentiate between nations in initiatives to lower the cost of technology-intensive imports through promotion of differential pricing schemes. While not a direct ITT policy, such price segmentation would avoid undesirable reverse transfers from South to North through arbitrage.[7]

[7]A recent example was the August 2003 decision by WTO members to permit poor countries to issue compulsory import licenses for essential medicines, in return for effective means to prevent the backflow of medicines to higher-price nations.

Several other options to increase ITT incentives exist. First, permit tax deductions for contributions of technology to non-profit entities engaged in ITT, taking the form of grants, technical assistance, or mature patent rights. Second, offer fiscal incentives to encourage enterprises temporarily to employ recent graduates from developing countries. Here there are potential synergies with efforts to expand the temporary movement of natural service suppliers under mode 4 of the General Agreement on Trade and Services (GATS). Third, public resources, such as those from the U.S. National Science Foundation, could be used to support research into the technology needs of developing countries. Technologies developed under such programs could be made publicly available.

Fourth, universities could be encouraged to recruit and train students from LDCs. Financial incentives that tap into development aid funding for setting up degree programs through distance learning or even foreign establishments may be particularly effective. Finally, additional initiatives could center on increasing information flows. For example, technical standards play a role in diffusing production and certification technologies. Thus, developed countries could finance participation by experts from developing countries in their standards-setting bodies.

Convincing OECD governments to offer incentives for transferring technology is a challenge, but could be part of their overall development assistance efforts.

5. International Cooperation

Many of the suggestions in the previous section will come at a cost to source countries. Similarly, a number of the developing country policy options discussed in Section 3 will require financing. Moreover, their effectiveness will depend importantly on design, discipline, and monitoring. These considerations provide rationales for using international agreements as commitment devices and enforcement mechanisms to increase the credibility and impacts of ITT policies. Safeguarding the ability of countries to pursue beneficial policies is also important. Certain avenues that were used in the past to achieve industrialization have been narrowed as a result of the WTO, suggesting a need to ensure that policy space exists to encourage ITT.

Helleiner (2000), Finger (2002), and Sabel and Reddy (2002), among others, emphasized that countries need the freedom to experiment with regulatory policies. Thus, detailed international policy harmonization is inappropriate. As argued by Hoekman (2005), multilateral monitoring and information exchange mechanisms can play a useful role in preventing capture and identifying effective policies. Such institutions as the OECD, UNCTAD, and UNIDO have provided a forum for exchanging experiences and discussing appropriate policies. The WTO Working Group on Trade and Technology Transfer provides a forum for dialogue that could be used more effectively.

There is a close connection between ITT discussions in the WTO and the concept of special and differential treatment (SDT) of developing countries. The argument that needs (market failures) differ depending on the type of country suggests that ITT policies should be differentiated. The same is arguably true of SDT more generally, although differentiation is resisted by many developing countries.[8] Similar questions arise in regional trade and investment integration agreements. Most such agreements do not address technology transfer explicitly in the form of binding commitments, instead imposing disciplines on the ability to use specific instruments such as subsidies or TRIMs. However, some North–South agreements put substantial emphasis on financial and technical assistance to be granted by the high-income partner(s). The best example is the EU, which has included such provisions in its partnership agreements with Mediterranean and other countries and has provided assistance for improving both absorptive and innovation capacity in its partners. To date, no such links have been made in the WTO and we now discuss areas where actions could be considered in that forum.

5.1. *Subsidies*

The WTO Agreement on Subsidies and Countervailing Measures (ASCM) divides subsidies into three categories: prohibited,

[8]Hoekman *et al.* (2004) offered an extensive discussion of the need for moving towards greater differentiation in SDT in the WTO.

actionable, and non-actionable. R&D and related technology subsidies are non-actionable if they are not specific, or, if specific, satisfy certain conditions. These conditions cover "assistance for research activities conducted by firms or by higher education or research establishments on a contract basis with firms." Fundamental research, defined as "an enlargement of general scientific and technical knowledge not linked to industrial or commercial objectives" is not subject to disciplines.[9] However, the provisions on non-actionability of R&D subsidies lapsed in 1999 and should be reinstated to permit developing countries to use them. In defining what is permitted, scope should exist to adopt measures that can be justified on the basis of externalities of the type identified by Hausmann and Rodrik (2003), discussed above. This does not mean complete freedom, for there is an important role for multilateral disciplines to help governments control subsidy policies and prevent capture. One approach would be to develop monitoring and surveillance mechanisms in the WTO aimed at increasing information on the effectiveness of policies to encourage innovation.

The current focus of the WTO is disciplines on national use of subsidies. A complementary approach would seek commitments by high-income countries to provide financing for risk-reducing entrepreneurial programs that promote entry into new activities. Such funding should be nondiscriminatory, made available to both foreign and domestic firms, because the nationality of innovating firms does not matter for realizing knowledge spillovers. The possibility of funding foreign entry through FDI or joint ventures might also enhance the support for such aid programs by creating a constituency in its favor in source nations.

5.2. *Temporary Movement of People and Labor Turnover*

Learning by doing, and subsequent labor turnover, are important channels of ITT. While most of the literature has focused on

[9] A distinction is made between industrial research and pre-competitive development activity. For the former the maximum amount of government participation is 75%; for the latter it is 50%.

within-country labor turnover associated with FDI, international movement of people has a potentially larger role to play in fostering ITT. In order to be most beneficial to developing countries, policies should encourage temporary movement of people. The classic problems with international migration are that it is often long-term and can give rise to a "brain drain" with potentially negative impacts on home-country welfare. Such problems would not arise if labor movements were temporary and returnees applied new skills and knowledge at home.

Negotiations over the temporary cross-border movement of people have been launched in the WTO. These discussions arise in the GATS, for one mode of supplying services is through the temporary movement of suppliers. While GATS is limited to people providing services, its approach could be extended to a category of personnel that relocate temporarily in order to increase their human capital and acquire new skills ("training services"). In effect, such movements could be regarded as a mechanism for host countries to export knowledge to developing countries. While it may not be feasible to incorporate this idea into the GATS, the mode 4 precedent might be used to negotiate a stand-alone arrangement under which developing countries would be granted additional temporary visa allocations for working in OECD countries, motivated by ITT objectives. This would also be a way to provide concrete SDT to developing countries.[10] The visa allocation mechanism could be similar to the "GATS visa" regime that has been suggested by a number of WTO members (Mattoo and Carzaniga, 2003).

Donor countries and organizations could also consider establishing special trust funds for training scientific and technical personnel, facilitating the transfer of technologies that are particularly sensitive for the provision of public goods, and encouraging research in developing countries (Roffe, 2002). To the extent that data and research results must be purchased, differential pricing schemes for governments and institutions in poor countries could be encouraged. At the margin, visa allocations could be aimed at students and

[10] A proposal by the LDC Group in the WTO, that they be given preferential access to a mode 4 'quota' as part of the GATS negotiations, goes in this direction.

researchers from poor countries. More generally, developed countries could help developing nations build capacity for improving education and science, including their ability to access international information and the Internet.

5.3. *IPR and TRIPS*

TRIPS Article 66.2 imposes an obligation on developed economies to find means of increasing ITT to the LDCs. One option would be for governments in developed countries to increase technical and financial assistance for improving the ability of poor countries to absorb technology and engage in trade. Among key issues here would be capacity building in IPR management and technical standards, establishing public and public-private research facilities, and facilitating trade in technology-related services. The terms of TRIPS Article 66.2 could also be expanded to include all developing countries without a significant domestic science and technology base. In addition, a special fee on international patent applications could be considered, through the Patent Cooperation Treaty, with revenues earmarked for improving IPR administrative systems in developing countries. Given different interests of developing countries regarding criteria for patentability, novelty, and utility, efforts towards harmonization of criteria or tests could be limited to the regional level, through, for example, cooperative examination offices that apply regional standards (Maskus, 2003).

Poor countries face major shortages in expertise for developing and enforcing anti-monopoly laws. One way for their governments to gain confidence in the system would be for authorities in developed countries to undertake enforcement actions against firms headquartered or located in their jurisdictions. A committed effort on the part of rich nations to prevent market-power abuses in developing-country markets by their sellers of technology could help achieve the goals of TRIPS Article 66.

5.4. *Protecting and Expanding the Global Commons for Knowledge*

Another proposal that has considerable potential to expand ITT to poor countries is to negotiate a WTO Agreement on Access to Basic

Science and Technology (ABST) (Barton and Maskus, 2003). This would place into the public domain the results of publicly funded research. The idea is to preserve and enhance the global commons in science and technology without unduly restricting private rights in commercial technologies. The agreement could encourage researchers from other countries to participate in, or compete with, local research teams for grants and subsidies, possibly combined with increased opportunities for temporary migration. It could also give researchers in other countries access to nationally generated science and data. It may be necessary to adopt a GATS-like approach to the ABST, permitting governments to reserve sensitive areas of technology and to designate different levels of commitment to open access. Safeguards for security-related regulation would be required as well.

5.5. *Information Exchange and Multilateral Monitoring*

To reduce problems of asymmetric information, international organizations such as the WTO could serve as an intermediary conduit for knowledge about successful technology-acquisition programs of national and sub-national governments. Firms in many developing countries may have little knowledge about the structure of international ITT contracts. What are reasonable royalty rates? What conditions have sellers of technology been willing to negotiate? Which contract clauses have proved helpful in encouraging local technological development? Answers to such questions exist but their dissemination requires efforts by both the private and public sectors of developed countries. Privacy concerns should not be a major obstacle, as summarizing expired licensing contracts would not reveal confidential data.

More emphasis on information exchange and multilateral monitoring would also be beneficial. This effort should focus on the success of various policies in achieving particular objectives. Rather than only seeking to regulate what countries may do to encourage ITT and innovation, a more productive approach could be to establish a broad framework that requires countries to engage in a regular exchange of information and policy assessments. Sabel and Reddy (2002) provided a conceptual sketch of such a "learning to learn" framework that could be applied to the ITT arena. The framework could also be devised to monitor the operation of subsidy programs supported by high-income countries.

6. Rules of Thumb

Determining optimal policies to maximize ITT is difficult. International technology transfers depend on many factors, including proximity to markets, size, growth, competition conditions, human capital basis, governance, and infrastructure. Significant uncertainty remains regarding the extent of market failures and spillovers, complicating the identification of good policies. Nonetheless, the forgoing analysis identified some rules of thumb for policy intervention as well as a number of specific proposals. We summarize the main policy implications in Table 1, distinguishing between low-income, lower-middle-income and upper-middle-income countries. These categories are illustrative only and are useful primarily in distinguishing between the types of general policies that are most appropriate for countries at different levels of development. We also suggest a rank ordering of policies on the basis of likely social benefits.

A main priority in all types of developing countries is effective general technology policies, including improving basic education, building appropriate infrastructure and reducing entry barriers for local firms that could be suppliers for MNEs. For local economies to gain productivity from ITT, such broader policy initiatives are important. This is a complex task that involves building human capital, expanding national innovation systems, and appropriately protecting IPR. Lower-income and middle-income economies could gain from investing in R&D support, especially as regards collaborations between public research entities and private enterprises, as part of their innovation systems. Source (donor) countries should support such investments.

Turning to specific policies, the evidence suggests that countries pursuing relatively closed trade policies fail to achieve the benefits from technology implicit in international trade. We therefore argue for liberal trade policies as a key priority, as important as sound general technology policies. As noted earlier, R&D-intensive

Table 1. A rule-of-thumb typology and examples of ITT policies.

	Trade in goods	FDI	Temporary movement of natural persons	Trade in knowledge (licensing)	IPR	General technology policies
Own policies:						
Low-income countries	(1) Liberal access	(2) Non-discriminatory investment promotion	(3) Incentives for education abroad and training-related movement	(4) Improve information flows about public domain and mature technologies	(5) Basic protection and minimum standards only	(1) Basic education; improve infrastructure; reduce entry barriers
Lower-middle-income countries	(1) Liberal access	(2) Non-discriminatory investment promotion	(4) Incentives for education abroad and training-related movement	(3) Improve information; limited incentives for licensing	(5) Wider scope of IPR protection; employ flexibilities	(1) R&D support policies; improve infrastructure; reduce entry barriers

(*Continued*)

Table 1. (*Continued*)

	Trade in goods	FDI	Temporary movement of natural persons	Trade in knowledge (licensing)	IPR	General technology policies
Upper-middle-income countries	(1) Liberal access	(3) Upstream supplier support programs	(2) Encourage two-way mobility	No active policy	(4) Apply full TRIPS	(1) R&D support policies; improve infrastructure; reduce entry barriers
OECD policies towards:						
Low-income countries	(1) Subsidize public-good-type imports; free trade	(2) Incentives for outward flows exceeding those for FDI to LMICs (see below)	(3) Preferential access; subsidies for education; incentives for universities to accept DC students in STI disciplines and temporary employment	(5) Subsidize transfer of public domain and mature technologies	(4) Forbearance in disputes; differential pricing for exports of IPR products; assistance in competition policy	(1) Support for general DC technology policies (see above); public and public–private research facilities

(*Continued*)

Table 1. (*Continued*)

	Trade in goods	FDI	Temporary movement of natural persons	Trade in knowledge (licensing)	IPR	General technology policies
Lower-middle-income countries	(1) Free trade; no controls	(2) Incentives equal to those granted for own disadvantaged regions	(3) Wider access for education and training; temporary employment of DC scientific personnel/engineers	(4) Assistance in establishment of joint venture partnerships; matching grants	(5) Differential pricing of public-good type IPR protected goods; assistance in competition policy	(1) Support for general DC technology policies (see above); fiscal incentives for R&D performed in DCs
Upper-middle-income countries	Free trade; no controls	No incentives	Encourage mode 4 type mobility	No active policy	No active policy	No active policy

Notes: Numbers in parentheses indicate authors' rank ordering of policy on basis of economic importance or anticipated payoff. DC: developing country. LMIC: lower-middle-income country.

capital-goods imports from high-income countries are associated with higher TFP in developing economies. Further, spillovers from technology-intensive imports can encourage exports because exporters need to deploy technologies that support international quality levels and standards.

Again, there is compelling evidence of vertical spillovers from FDI (Saggi, 2002; Javorcik, 2004). FDI is likely to be particularly important for low-income and lower-middle-income countries. The weak investment climates in many of these countries may justify a temporary case for encouraging FDI inflows, which we would rank second behind open trade policies. Such incentives should avoid discrimination across both sectors and foreign origins. Governments are not good at picking winners and should focus on encouragement of new activities. Given the possibility of coordination failures and learning externalities, any incentives for new "non-traditional" activities should apply to both domestic and foreign firms. Ideally, such support would be financed by development assistance.

A third specific priority for developing economies, especially poorer countries, is temporary international movement of technical workers abroad for education and training. Upper-middle-income countries could see growing two-way movements in skilled personnel. Such movements are an important source of learning, and will also lead to establishment of networks in foreign countries that are potential markets as well as sources of knowledge. China, India, and other emerging economies illustrate the importance of having a significant and sustained outflow of people who, upon their return, raise productivity. Even if they stay abroad for an extended period of time, they can be part of networks supporting trade and investment flows.

While licensing is an important source of technical transformation, successful transfer generally requires capacity to learn and adaptive investments by local firms. Poor countries are most likely to achieve these gains by taking advantage of mature technologies in the public domain or available cheaply. Thus, specific policies here are not as likely to be a high priority. They could aim at improving information flows about such technologies, as well as building skills and R&D capacity. Middle-income countries, in which firms have engineering skills and active R&D programs, are more likely to be the recipients of significant licensing flows. To enhance these flows, policy efforts could focus on reducing the costs of absorbing technology. The upper-middle-income economies require no active intervention in licensing, where technology markets may be expected to operate effectively. Note that our analysis in no case supports extensive government involvement in selecting technologies or placing restrictions on the use of technical information.

Despite the intense focus on IPR in the literature, we would argue that this is a relatively less important area for policy, so long as IPR regimes are tailored to levels of development and technological capacities. In Table 1 we hint at the issues and far more detail could be added.[11] Thus, low-income countries would find it advantageous to enforce basic protection of trademarks, trade secrets, utility models, and industrial designs in order to encourage both local small-scale innovation and inward FDI in labor-intensive technologies. However, it is inadvisable to move beyond minimum TRIPS standards, while requirements for patents, plant variety rights, and copyrights should be as pro-competitive as possible. The LDCs may be expected to do relatively little in terms of enforcing foreign patent rights in any case. There is a strong case for forbearance by OECD governments in pursuing enforcement-related dispute settlement at the WTO. For their part, lower-middle-income countries should take advantage of TRIPS flexibilities while offering somewhat wider scope of protection.

The bottom half of Table 1 encapsulates policy recommendations for high income source economies that desire to encourage ITT to poor countries. Here again, market access for goods and services produced in developing countries is a priority. Fiscal incentives may be an effective means for overcoming market failures in ITT. As mentioned, such financial support should be directed at the general technology policies that are a top priority in most developing countries.

In terms of specific support policies, FDI is most important for low-income countries as they have the most to gain from investments in "new" activities and from entry by efficient foreign firms. Such

[11] World Bank (2001) offers extensive discussion.

incentives should be at least equal to those offered for investments in OECD countries, and, as argued above, could extend to entry by local firms into new markets and activities, including as suppliers to MNEs.

Liberal policies towards, and explicit encouragement of, labor movement through improved access to educational establishments, scholarships and temporary employment of graduates and professionals is a third priority area for source-country action. This idea is politically sensitive and thus may be less feasible than the other areas. Finally, OECD governments could improve flows of public-domain technologies with appropriate subsidies and support the establishment of extensive price differentiation for exports of IPR products to low-income developing economies.

7. Leveraging Trade Agreements for ITT

Many of the suggestions made in this paper can be implemented unilaterally. However, some require action in the WTO and many can be made more credible by incorporating them as specific commitments. This idea could be achieved as part of a new approach to SDT in the WTO that provides greater scope for policy flexibility on the part of developing countries. Alternatively, various ITT-related policy initiatives could be embedded in a mix of existing and new WTO agreements.

Although binding disciplines help ensure policy certainty, there is a rationale for flexible government tools to encourage ITT and address externalities in information markets. Thus, a key option is to negotiate clear criteria that differentiate among beneficiary countries. Broad, transparent categories would minimize transactions costs and uncertainty, but would not do much to help countries achieve ITT objectives. An alternative suggested by Hoekman (2005) is to provide for greater flexibility in the enforcement of multilateral disciplines in specific areas, such as ITT-related policies, but to accompany this with multilateral monitoring and periodic discussions among WTO members about the incidence and effectiveness of the policies pursued.

A related issue is whether the proactive ITT measures we have suggested for source countries should take the form of binding and enforceable commitments in the WTO or regional trade agreements. In some cases, such as the suggested implicit exemption from enforcement of TRIPS for low-income countries, a change in the relevant WTO rules is needed. However, in areas that are not now covered by the WTO, a "soft law" approach that establishes broad guidelines and relies on ensuring transparency and accountability via regular multilateral monitoring of performance is likely to be more effective at increasing cooperation and compliance. What matters most is that "demandeur" developing countries elucidate the need for specific policies and engage in a process of analysis to identify the most appropriate instruments. Insofar as these are subsidies, proposals should be put forward in development assistance allocation discussions. The extent to which such requests are met should be part of the proposed multilateral monitoring mechanism. The same is true with regard to allowing developing countries to pursue general technology related policies and to encourage the use of specific channels of ITT.

References

Aitken, B. J. & Harrison, A. E. (1999). Do domestic firms benefit from direct foreign investment? Evidence from Venezuela. *American Economic Review* 89, (3), 605–618.

Amsden, A. (1989). *Asia's Next Giant: South Korea and Late Industrialization.* New York and Oxford: Oxford University Press.

Arora, A., Fosfuri, A., & Gambardella, A. (2001). *Markets for Technology: The Economics of Innovation and Corporate Strategy.* Cambridge: MIT Press.

Barton, J. H. & Maskus, K. E. (2004). Economic Perspectives on a Multilateral Agreement on Open Access to Basic Science and Technology. In S. J. Evenett and B. M. Hoekman, eds., *Economic Development and Multilateral Trade Cooperation.* New York: Palgrave/MacMillan, 349–368.

Blalock, G. (2001). Technology from Foreign Direct Investment: Strategic Transfer through Supply Chains. Mimeo, Cornell University.

Blomstrom, M. & Kokko, A. (1997). How Foreign Investment Affects Host Countries. Washington, DC: World Bank PRD Working Paper No. 1745.

Blyde, J. S. & Acea, C. (2002). The Effects of Intellectual Property Rights on Trade and FDI in Latin America. Washington, DC: Inter-American Development Bank, manuscript.

Borensztein, E., De Gregorio, J., & Lee, J-W. (1998). How Does Foreign Direct Investment Affect Economic Growth? *Journal of International Economics* 45, (1), 115–135.

Coe, D. T. & Helpman, E. (1995). International R&D Spillovers. *European Economic Review* 86, (1), 859–887.

Coe, D. T., Helpman, E., & Hoffmaister, A. W. (1997). North-South R&D Spillovers. *The Economic Journal* 107, (440), 131–149.

Correa, C. M. (2005). Can the TRIPS Agreement Foster Technology Transfer to Developing Countries? in K. E. Maskus and J. H. Reichman (Eds.), *International Public Goods and Transfer of Technology under a Globalized Intellectual Property Regime*. Cambridge: Cambridge University Press.

Eaton, J. & Kortum, S. J. (1996). Trade in Ideas: Patenting and Productivity in the OECD. *Journal of International Economics* 40, (3–4), 251–278.

Eaton, J. & Kortum, S. J. (1999). International Technology Diffusion: Theory and Measurement. *International Economic Review* 40, (3), 537–570.

Evenson, R. E. & Westphal, L. E. (1995). Technological Change and Technology Strategy. In J. Behrman and T.N. Srinivasan (Eds.), *Handbook of Development Economics: Vol. 3A*. Amsterdam: North-Holland, 2209–2300.

Finger, J. M. (2002). The Doha Agenda and Development: A View from the Uruguay Round. Manila: Asian Development Bank.

Grossman, G. M. & Helpman, E. (1991). *Innovation and Growth in the World Economy*. Cambridge: MIT Press.

Hausmann, R. & Rodrik, D. (2003). Economic Development as Self-Discovery. *Journal of Development Economics* 72, (2), 603–633.

Helleiner, G. K. (2000). *Markets, Politics, and Globalization: Can the Global Economy be Civilized?* Geneva: 10th Prebisch lecture, United Nations Conference on Trade and Development, manuscript at www.unctad.org.

Hoekman, B. M. (2005). Making the WTO More Supportive of Development. *Finance & Development* 42, (1), 14–18.

Hoekman, B. M., Michalopoulos, C., & Winters, L. A. (2004). Special and Differential Treatment in the WTO after Cancun. *The World Economy* 27, (3), 481–506.

Horstmann, I. & Markusen, J. R. (1987). Licensing versus Direct Investment: A Model of Internalization by the Multinational Enterprise. *Canadian Journal of Economics* 20, (3), 464–481.

Javorcik, B. S. (2004). Does Foreign Direct Investment Increase the Productivity of Domestic Firms? In Search of Spillovers through Backward Linkages. *American Economic Review* 94, (3), 605–627.

Keller, W. (1996). Absorptive Capacity: On the Creation and Acquisition of Technology in Development. *Journal of Development Economics* 49, (2), 199–227.

Keller, W. (1998). Are International R&D Spillovers Trade-related? Analyzing Spillovers among Randomly Matched Trade Partners. *European Economic Review* 42, (8), 1469–1481.

Keller, W. (2002). Geographic Localization of International Technology Diffusion. *American Economic Review* 92, (1), 120–142.

Kim, L. (2002). Technology Transfer and Intellectual Property rights: Lessons from Korea's Experience. Geneva: UNCTAD/ICTSD working paper.

Kokko, A., Tansini, R., & Zejan, M. (1997). Trade Regimes and Spillover Effects of FDI: Evidence from Uruguay. Stockholm: Stockholm School of Economics, manuscript.

Krishna, P. & Mitra, D. (1998). Trade Liberalization, Market Discipline, and Productivity Growth: India. *Journal of Development Economics* 56, (2), 447–462.

Lall, S., Weiss, J., & Zhang, J. (2005). The 'Sophistication' of Exports: A New Measure of Product Characteristics. Manila: Asian Development Bank Institute.

Larrain B., F., Lopez-Calva, L. F., & Rodriguez-Clare, A. (2000). Intel: A Case Study of FDI in Central America. CID Working Paper No. 58, Harvard University.

Mansfield, E. & Romeo, A. (1980). Technology Transfer to Overseas Subsidiaries by U.S.-Based Firms. *Quarterly Journal of Economics* 95, (4), 737–749.

Maskus, K. E. (2000). *Intellectual Property Rights in the Global Economy.* Washington, DC: Institute for International Economics.

Maskus, K. E. (2003). Encouraging International Technology Transfer. Geneva: UNCTAD/ICTSD working paper.

Mattoo, A. & Carzaniga, A. (Eds.) (2003). *Moving People to Deliver Services.* Washington, DC: World Bank.

McCalman, P. (2001). Reaping What You Sow: An Empirical Analysis of International Patent Harmonization. *Journal of International Economics* 55, (1), 161–186.

Moran, T. (1998). *Foreign Direct Investment and Development.* Washington, DC: Institute for International Economics.

Moran, T. (2001). *Parental Supervision: The New Paradigm for Foreign Direct Investment and Development.* Washington DC, Institute for International Economics.

Moran, T. (2004). How Does Foreign Direct Investment Affect Host Country Development: Do We Already Know the Answer? Using Industry Case Studies to Make Reliable Generalizations. In M. Blomstrom, E. M. Graham, & T. Moran (Eds.), *The Impact of Foreign Direct Investment on Development.* Institute for International Economics, Washington DC.

Nicholson, M. W. (2002). Intellectual Property Rights and International Technology Transfer: The Impact of Industry Characteristics. Washington DC: U.S. Federal Trade Commission.

Noland, M. & Pack, H. (2003). *Industrial Policy in an Era of Globalization. Lessons from Asia.* Washington DC: Institute of International Economics.

Organization for Economic Cooperation and Development (2001). *Growth, Technology Transfer and Foreign Direct Investment.* Paris: OECD.

Organization for Economic Cooperation and Development (2002). *Science and Technology Industry Outlook.* Paris: OECD.

Parente, S. L. & Prescott, E. C. (1994). Barriers to Technology Adoption and Development. *Journal of Political Economy* 102, (2), 298–321.

Peri, G. (2003). Knowledge Flows, R&D Spillovers, and Innovation. Davis, CA: University of California at Davis, manuscript.

Rhee, Y. W. (1990). The Catalyst Model of Development: Lessons from Bangladesh's Success with Garment Exports. *World Development* 18, (2), 333–346.

Roffe, P. (2002). Preliminary Note on the WTO Working Group on Trade and Transfer of Technology. Geneva: UNCTAD, manuscript.

Sabel, C. & Reddy, S. (2002). Learning to Learn: Undoing the Gordian Knot of Development Today. New York: Columbia University, manuscript at www.sopde.org/discussion.htm.

Saggi, K. (1996). Entry into a Foreign Market: Foreign Direct Investment versus Licensing. *Review of International Economics* 4, (1), 99–104.

Saggi, K. (2002). Trade, Foreign Direct Investment, and International Technology Transfer: A Survey. *World Bank Research Observer* 17, (Fall), 191–235.

Saggi, K. (2004). *International Technology Transfer to Developing Countries.* Economic Paper 64, Commonwealth Secretariat, London, UK.

Schiff, M., Wang, Y., & Olarreaga, M. (2002). North-South and South-South Trade-Related Technology Diffusion: An Industry-Level Analysis. Washington DC: World Bank, manuscript.

Smith, P. J. (2001). How Do Foreign Patent Rights Affect U.S. Exports, Affiliate Sales, and Licenses? *Journal of International Economics* 55, (2), 411–440.

Tybout, J. R. (2000). Manufacturing Firms in Developing Countries: How Well Do They Do and Why? *Journal of Economic Literature* 38, (1), 11–44.

United Nations Conference on Trade and Development (2002). *World Investment Report 2002.* Geneva: UNCTAD.

World Bank (2001). *Global Economic Prospects and the Developing Countries 2002: Making Trade Work for the World's Poor.* Washington DC: The World Bank.

World Bank (2004). *World Development Report 2005: A Better Investment Climate for Everyone.* Washington DC: The World Bank.

Yang, G. & Maskus, K. E. (2001). Intellectual Property Rights and Licensing: An Econometric Investigation. *Weltwirtschaftliches Archiv* 137, (1), 58–79.

Chapter 8

Intellectual Property Rights, Technology Transfer and Exports in Developing Countries*

Lei Yang and Keith E. Maskus[†]

1. Introduction

Since 1995 many developing countries have reformed their laws governing intellectual property rights (IPR). Reforms in IPR are commonly presumed by trade economists to raise imitation costs, reduce access to global information and place firms in developing countries at a competitive disadvantage in global markets (Helpman, 1993; Lai and Qiu, 2004). However, one essential purpose of IPR is to reduce the costs of technology transfer (Maskus, 2004). Indeed, empirical evidence supports the view that multinational firms expand technology flows through greater foreign direct investment (FDI) and licensing as local patent rights are improved (Smith, 2001; Branstetter *et al.*, 2005).

By expanding access to international technologies, strengthened IPR could improve the export performance of recipient firms, a possibility that has been little studied to date. In this paper we provide a model of contracting and technology transfer that illuminates one such mechanism. Specifically, we analyze a model of two-country competition between a Northern firm and an unaffiliated Southern firm, where the former may choose to provide cost-reducing technical

*Reprinted from *Journal of Development Economics*, Vol. 90, No. 2, pp. 231–236.

†We thank Yongmin Chen, Gordon Hanson, James Markusen, Walter Park and two referees for their helpful comments and suggestions.

information to the latter through licensing or FDI. We find conditions under which greater transfers are made in equilibrium under stronger patents and the consequent effect on exports of the Southern firm. Welfare in the Southern country increases if its firm has high absorptive ability, but could fall if it has a weak capacity to implement new technology.

In contrast to our strategic approach, the theoretical literature generally has set out general-equilibrium, North–South product-cycle models among atomistic firms competing dynamically. Helpman (1993) and Glass and Saggi (1999) assumed stronger IPR would raise imitation costs, tending to diminish technology flows and global innovation. Lai (1998) noted that innovation could be enhanced if FDI is the form of technology transfer. Yang and Maskus (2001) found that patent reforms would both raise imitation costs and reduce the costs of technology licensing, with the latter encouraging greater information transfer and innovation in equilibrium.

These insights are valuable. However, to make these dynamic models tractable the authors forego analysis of strategic interactions among firms. The primary advantage of our approach is to permit detailed analysis of the microeconomic tradeoffs involved in contracting in response to IPR changes. Our bargaining framework explicitly considers strategic choices among imitation, licensing, and FDI at various ranges of patent strength, generating a rich menu of tradeoffs and welfare calculations that cannot readily be analyzed in the more general context. Of course, the partial-equilibrium Cournot context within which we operate also makes strong assumptions for tractability.[1] Thus, we comment later on how the results would vary with alternative assumptions.

Empirical evidence in several developing countries suggests that investing in technology is important for entering export markets (Hasan and Raturi, 2001). Key sources of such investment are imports of technology and linkages to multinational firms (Kumar and Siddharthan, 1993). However, this literature has paid little attention to the combination of IPR, technology transfer and trade in

[1]Vishwasrao (1994) is an early example of a strategic model of IPR and technology transfer.

order to establish a linkage between IPR and exports. A recent exception is Branstetter *et al.* (2007), whose empirical analysis found an increase in export intensity of local affiliates of multinational firms after IPR policy changes. Again, none of these papers considered the detailed mechanisms under which IPR reforms could expand technology transfer and exports.

As empirical motivation for our framework, within which patent reforms may encourage more technology transfer through unaffiliated licensing, with a subsequent boost to exports, consider two recent histories from East Asia. South Korea engineered a major strengthening of its patent laws from 1988 to 1995 (La Croix and Kawaura, 1996), increasing its measured patent index by 47 percent, from 2.65 to 3.89, between 1985 and 1995.[2] Taiwan (Chinese Taipei) made substantial reforms in 1986 and 1994 (Diallo, 2003), raising its index by 152 percent, from 1.26 to 3.17, over the same period. South Korea's licensing payments to unaffiliated U.S. firms rose from \$38 million in 1987 to \$717 million in 1995, reaching \$1.686 billion in 2005. Taiwan's licensing payments to unaffiliated U.S. firms rose from \$17 million in 1986 to \$267 million in 1996, reaching \$1.165 billion in 2006. Finally, South Korea's merchandise exports rose from \$28.5 billion in 1985 to \$99.5 billion in 1995, reaching \$284.4 billion in 2005. Taiwan's exports similarly rose from \$33.4 billion in 1985 to \$128.4 billion in 1995, reaching \$223.7 billion in 2006.[3]

2. The Model

We study the effects of IPR on export development in an oligopolistic setting. Consider a world economy of two regions, North and South. Assume that at most one firm in each country can profitably produce the good. We denote these firms by N and S. Both firms produce a single homogenous good and compete in Cournot fashion. Assume also that the two markets are segmented, in the sense that firms can charge a different price in each market.

[2]This measure is the well-known Ginarte–Park index, explained in Ginarte and Park (1997).

[3]Sources for these data include the on-line WTO statistics database, World Bank *World Development Indicators* (CD-ROM), and U.S. Department of Commerce (2007).

2.1. *Consumption*

Let the utility functions in both regions be quadratic in the good we study, with an additive term for a second composite good. If A and B represent the market size of North and South, respectively, the inverse-demand functions for our good are

$$p_N = A - q_N; \quad p_S = B - q_S. \tag{1}$$

We assume that market sizes are sufficiently greater than marginal costs to ensure positive production.

2.2. *Decision on Mode of Supply*

Only the N firm engages in prior R&D, which achieves proprietary technological knowledge embedded in the production process for its good. It can retain production at home and export the good to market S, risking loss of its knowledge through imitation, or transfer the technology through licensing.[4] The choice depends on the absorptive capacity of the licensee, market size, the threat of imitation, and the legal protection of technology.

We assume that codified knowledge (e.g., blueprints and formulas) can be imitated by S, but tacit knowledge (e.g., know-how and information gained from experience) cannot. Imitation of codified knowledge is costly and can be achieved under the export mode through product inspection, reverse engineering, or trial and error. Imitation permits S to avoid paying license fees but the reduction in its production costs is less than it would be with licensing because the firm cannot acquire know-how this way.

N may instead offer to license production rights to S. In this event the licensing contract specifies a lump-sum fee and S is able to produce the good at reduced marginal cost with partial access to know-how. If S accepts the licensing contract it would have no incentive to imitate. Thus, its problem is a tradeoff between the license fee and imitation costs, with different impacts on marginal production costs.

---

[4]We extend the model to technology transfer through FDI in the next section.

Our specification of a lump-sum license fee without per-unit royalties captures the empirical reality that a large portion of technology contracts in developing countries have this feature. For example, Vishwasrao (2007) assembled data on all foreign technology licensing agreements entered into by manufacturing firms, unaffiliated with the licensors, in India between 1989 and 1993. Over the period 1991–1993, there were 968 contracts with only lump-sum fees, amounting to 45 percent of all licensing deals.

2.3. *Costs and Production*

We assume that labor is the only factor of production and that N's marginal production cost is c_N. Before any imitation or licensing, let S's marginal production cost be c_S, which is greater than c_N because the firm has no knowledge of N's improved technology. A key parameter, the absorptive ability of S, is denoted by $a \in [0, 1]$, where an increase in a indicates higher learning capacity. This capacity is exogenous and given by such characteristics of the South market as education level and infrastructure. Because a stronger learning capacity would permit more efficient production, we assume that imitation reduces marginal cost by more, the greater is a. The reduced cost is $c_S - m(a), m' > 0$.

Let $k \in [0, 1]$ be the strength of IPR in the South. Parameter k is 1 when patent protection is highest and 0 if patents are absent. Denote by $I(k, a)$ the S firm's imitation cost. Stronger IPR make it harder for S to imitate N's product. Indeed, as IPR protection approaches its maximum the costs of legally imitating around a patent become quite high. Thus, we suppose that imitation will not occur beyond some less-than-full level of protection.[5] At the same time, a higher absorptive capacity makes it easier for S to imitate.

There are costs of transferring technology through licensing. Two components of these costs involve setting enforceable contract terms and shifting codified knowledge. These costs typically fall as Southern IPR are tightened because enforceable patents and trade secrets

[5] One proof below relies on a convexity assumption that $\frac{\partial^2 I}{\partial k^2} < 0$, though this cost can get high enough to deter imitation.

reduce contracting problems under asymmetric information and limit the need for N to mask its proprietary knowledge (Taylor, 1994; Yang and Maskus, 2001). The third component of transfer cost is ensuring that local partners gain the know-how needed to produce efficiently. We assume that these costs increase with the proportion of know-how transferred, which we capture by parameter $x \in [0,1]$. Thus, let licensing incur a transfer cost $F(x,k) = \varphi + G(x,k)$, where φ is a fixed transfer cost and variable cost G decreases with the strength of IPR and increases with the proportion of know-how transferred.[6] This transfer cost F is borne by both partners and enters only their payoff functions through a reduction in joint surplus. It does not directly affect production profits.

If S accepts the licensing contract, the technology is transferred and its marginal production cost becomes $c_S - r(x,a)$. Here $r(x,a)$ is the advantage to S of accepting the license and learning know-how. We posit that this cost reduction is positively related to both S's absorptive ability and the proportion of know-how that N transfers, and is concave in both x and a. Finally, assume that $c_S > c_N > c_S - m(a) \geq c_S - r(x,a)$. Thus, either imitation or licensing reduces S's marginal cost below N's but the decline is greater in the latter case.[7]

Nash bargaining determines the lump-sum license fee L, which must be paid to gain production rights and know-how. Let τ be the bargaining power of N and $1-\tau$ that of S. This parameter determines the shares of joint surplus from sales under licensing. S's bargaining power reflects its lower labor cost and knowledge of local market conditions and N's power reflects the uniqueness of its technology.

2.4. *Decision Structure under Varying North Import Policies*

In this game there are two players, firms S and N. Initially, N is the global (two-country) monopoly and S may imitate the technology

---

[6] We assume that $\frac{\partial^2 G(x,k)}{\partial x \partial k} < 0$.

[7] That is, we assume that $m(a) = r(0,a)$, so that imitation is identical to the absence of know-how transfer.

through reverse engineering. Northern import policy related to IPR is important. The North may choose to permit imports from the S firm, even if that firm imitates the N technology without a license. However, such imitation could violate the patent issued in the North and N could direct its government to bar imitative imports.[8] We initially consider the case without import blockage and then analyze competition subject to this constraint.

The timing of the game without blockage is as follows. In the first stage, given the Southern IPR policy, N chooses either to export to the South, risking imitation, or to offer a license. If N exports, in the second stage S can choose to imitate, incurring a cost, or do nothing. If S imitates, competition emerges in the third stage and the firms simultaneously choose quantities in both markets. If S does not imitate, N remains the global monopoly.

If N offers a license to S, the firms bargain over the fee. If the negotiation succeeds, technology is transferred and both firms simultaneously choose optimal quantities produced. Otherwise, S can choose to imitate or not enter the market.

2.5. *Equilibrium Analysis*

We begin by analyzing the equilibrium where N exports. Initially suppose that S chooses to imitate. Then the respective final-stage maximization problems under export are

$$\begin{aligned}\text{Max}\Pi_{NE} &= (A - q^E_{NN} - q^E_{SN})q^E_{NN} + (B - q^E_{NS} - q^E_{SS})q^E_{NS} \\ &\quad - c_N(q^E_{NN} + q^E_{NS}) \qquad (2)\\ \text{Max}\Pi_{SE} &= (A - q^E_{NN} - q^E_{SN})q^E_{SN} + (B - q^E_{NS} - q^E_{SS})q^E_{SS} \\ &\quad - c_{SE}(q^E_{SN} + q^E_{SS}) - I(k,a). \qquad (3)\end{aligned}$$

Here the E superscript refers to the exports equilibrium and $c_{SE} = c_S - m(a)$. Solving for the optimal quantities, equilibrium profits

[8] Such is the policy in the United States under Section 337 of the 1930 Tariff Act, as amended.

under exporting are given by

$$\Pi_{NE*} = \frac{A^2 - 4Ac_N + 2Ac_{SE} + 8c_N^2 - 8c_Nc_{SE} + 2c_{SE}^2 + B^2 - 4Bc_N + 2Bc_{SE}}{9} \tag{4}$$

$$\Pi_{SE*} = \frac{A^2 + 2Ac_N - 4Ac_{SE} + 2c_N^2 - 8c_Nc_{SE} + 8c_{SE}^2 + B^2 - 4Bc_{SE} + 2Bc_N}{9} - I(k,a). \tag{5}$$

If S chooses not to imitate and produce, its profit is 0. Let superscript NB refer to no Northern import blockage and define $k = \bar{k}^{NB}$ as the IPR level at which imitation cost just offsets production profit and $\Pi_{SE*} = 0$.[9] S will imitate when $k < \bar{k}^{NB}$, with profit declining in k, and will not enter when $k > \bar{k}^{NB}$.

Next, if N offers a license and the negotiation fails, the equilibrium is the same as that under export. If licensing is offered and the negotiation succeeds, N could have two choices in principle after transferring its process technology. It could continue producing and compete with S in both markets. Alternatively, it could commit not to produce, leaving S as a monopolist. However, in our game structure this situation cannot survive as the equilibrium.[10] The reason is that the license fee is a one-time transaction paid prior to the production stage, so that a commitment not to compete is not credible. We therefore analyze the case where a license payment is made but both firms produce.

Maximizing the relevant profit functions in this case shows that the N firm's equilibrium production profits increase with the reduction in S marginal costs, sustaining the incentive to license technology.[11] In the negotiation stage, N decides how much know-how will be transferred to S under the contract. The N profit equals the sum of the license fee and production profit under successful licensing. It chooses x to maximize[12]

$$\text{Max}\Pi_{NL} = \Pi_{NE*} + \tau[S_1(x,a) + I(k,a) - F(x,k)] \tag{6}$$

[9]By assumption this level is below full IPR protection. Later we analyze a functional form where this assumption holds.

[10]We demonstrate this fact in a mathematical appendix available on request.

[11]From this point forward, profit functions are suppressed to save space.

[12]Determination of the license fee is in the mathematical appendix.

where $S_1 > 0$ is an expression involving market sizes, marginal costs, and reductions in costs from imitation and licensing. The first term is N's profit under exports and the second term represents its share of joint surplus (in brackets), which N maximizes through its choice of x. Because S_1 is positive, this joint surplus rises in k. We denote the first-order condition of Eq. (6) as $f(x^*, k, a)$. The equilibrium know-how transferred satisfies

$$f(x^*, k, a) = \left[\frac{2}{9}A + \frac{2}{9}B + \frac{16}{9}c_N - \frac{20}{9}(c_S - r(x^*, a))\right]\frac{\partial r(x^*, a)}{\partial x} - \frac{\partial F(x^*, k)}{\partial x} = 0. \quad (7)$$

Consider briefly the situation in which the Northern government blocks imitative imports but freely permits imports of goods produced by S under license. By restricting one competitive outlet ($q^E_{SN} = 0$), the S firm's tradeoffs are altered. Performing the relevant substitutions and computing profit functions, we find a threshold IPR level $\bar{k}^B$ above which S does not imitate and N remains a monopoly in both markets with blocking. Below that level, N monopolizes its own market but the firms compete in South.

In the bargaining game under licensing, N's profit expression is the sum of its exports under blocking-aided monopoly (its reservation value) and its share of the joint surplus, where S can now export to North. We can show that the licensing equilibrium becomes more likely the greater the reduction in S's marginal cost. Further, the larger is North market size, the less likely is licensing (import blockage sustains the monopoly in a large market) but the larger is the South market the more likely is licensing. Thus, under some circumstances, duopoly profit under licensing can dominate monopoly profit under export with imitation risk, even where the North government blocks imitative imports.

3. The Impacts of IPR Reform

In this section we study the effect of a stronger patent regime on the extent of know-how licensed, exports, and the choice between export and licensing.

3.1. *Extent of Know-How Transferred*

Suppose that licensing is chosen. Differentiating expression (7) shows that

$$\frac{dx^*}{dk} > 0 \quad \text{if } A + B + 8c_N - 10c_{SL} > \phi(x^*, a), \quad \text{where}$$

$$\phi(x^*, a) = \frac{10\left[\frac{\partial r(x^*,a)}{\partial x}\right]^2}{-\frac{\partial^2 r(x^*,a)}{\partial x^2}}. \tag{8}$$

This result demonstrates that stronger patent rights increase the amount of know-how transferred under large joint market size, high marginal cost in N, and low marginal cost in S post-transfer. We can similarly show that the amount of know-how transferred increases with the absorptive ability of the S firm under the same conditions. These results hold under any licensing equilibrium, regardless of North's import-blockage policy.

3.2. *Southern Exports*

Suppose first that exports with imitation pertain at low k range. Our assumption that a rise in k affects S imitation cost but not marginal cost implies that imitative exports are insensitive to policy changes in this region. Consider next the role of IPR on exports under successful licensing. We saw that know-how transferred rises with IPR under large joint market size and large difference in marginal costs. Under these circumstances, tighter patents reduce production cost in S, shifting outward that firm's reaction function and expanding exports to the N market. Similar analysis finds that stronger IPR would increase S's domestic production and imports from N would fall. However, in the case of small joint market size and small difference in marginal costs, S activities would decline.

The intuition is that stronger patents in the South increase (decrease) the marginal benefit (marginal cost) of transferring know-how, raising the amount transferred. Also, there are fixed costs of technology transfer and it requires large market sizes to guarantee that the total surplus is positive. Under such circumstances, stronger technology protection would make the S firm more competitive in the international market.

3.3. *N Firm's Export or Licensing Decision*

Assume imitative imports are not banned. Working back to the first stage of the game, N compares export profit with licensing profit in deciding the mode of entry. We determine k_1, the threshold value of IPR that induces licensing as the equilibrium outcome. This occurs where the profit from licensing just equals that from export, or the joint surplus in brackets in Eq. (6) is zero. Under our assumptions this joint surplus at $k = 0$ is negative because the technology transfer costs are high and S's productivity gain is small, despite the low imitation cost. Further, the joint surplus at $k = 1$ is positive. Finally, because this surplus is non-decreasing in k, the threshold value lies between zero and one.

There is no solution for this value under our general functional forms for changes in costs of technology transfer, production and imitation. We therefore adopt specific functional forms that conform to our assumptions and calculate the threshold value of k.[13] With this specification, N licensing profits follow an inverse-U shape in the amount of know-how transferred. These profits are maximized at some $x^* < 1$, implying incomplete know-how transfer. Further, licensing profit rises throughout as patents are strengthened as long as licensing is the equilibrium. However, this relationship is concave so that the profit gain under licensing diminishes as k rises.

When patent protection gets sufficiently high, S's imitation cost would be prohibitive. In that case, N exports to South and achieves a monopoly profit that exceeds its profit under duopoly with licensing. We find that monopoly is the equilibrium in this unblocked case if IPR exceeds a threshold value $\bar{k}^{NB}$, which depends on imitation costs and absorptive ability. In the case where North blocks imitative imports, however, the corresponding IPR level at which S chooses not to imitate is $\bar{k}^B$. It is straightforward to show that $\bar{k}^B < \bar{k}^{NB} < 1$. In short, if the North blocks imitative imports, the altered payoffs deter licensing and sustain a full N monopoly at a lower level of Southern patent rights. As noted above, we define k_1 as the critical IPR value that induces a shift from exporting with imitation risk to licensing, assuming no import blockage. Similarly, we define k_2

[13]These forms are listed in the mathematical appendix.

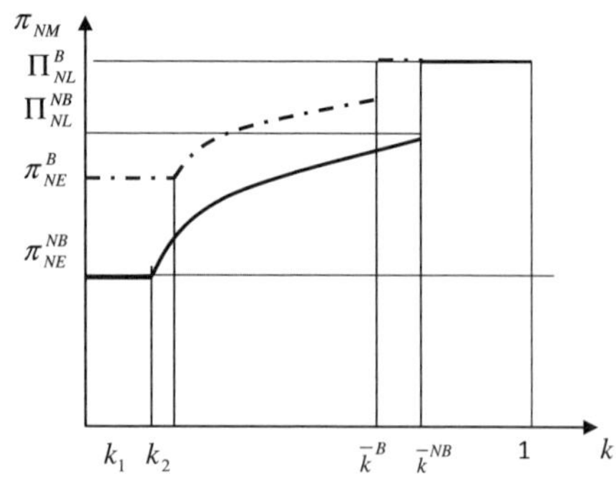

Figure 1. Northern profits from export and licensing under North free imports and blocked imports.

as this critical IPR value when there is import blockage. It can be shown that $k_1 < k_2$, so that the North policy of banning imports deters imitation up to a higher level of Southern patent protection.

In Figure 1 the horizontal lines represent N's export profits and the other curve represents profits from licensing.[14] Thus, without blockage the equilibrium outcome is export with S imitation if $0 < k < k_1$, licensing if $k_1 < k < \bar{k}^{NB}$, and the export monopoly if $\bar{k}^{NB} < k < 1$. Note that the North import ban raises the lower critical level and reduces the higher critical level of Southern patent rights, shrinking the licensing range. Further, the N firm's profits are the same under either import policy for an IPR level higher than $\bar{k}^{NB}$. Otherwise the import ban favors N profits at any level of Southern patent strength.

3.4. *A Note on Wage Impacts*

This is a partial-equilibrium model in which it is relatively straightforward to describe wage effects of IPR policy.[15] Assume that labor is the only factor of production and is in fixed supply. Let α be the unit labor requirement for producing the good in the South, where α

[14]The N profit level is constant under the imitation threat (low k) because a change in IPR does not affect relative marginal costs, though it does reduce S profit through higher imitation costs.

[15]The mathematical appendix demonstrates these results.

declines with x. As the entry mode changes from exports to licensing, the unit labor requirement decreases and total quantity produced by S goes up. The first effect reduces labor demand but the second increases it and the latter impact dominates under a sufficiently large increase in S market share. In the North the unit labor requirement does not change while total quantity produced falls in the shift from exporting to licensing, thereby reducing labor demand and the wage rate.

Once licensing is the equilibrium, further strengthening of patent rights increases the S firm's market share. The increase in quantity dominates the reduction in labor requirement under large market sizes, raising the Southern wage rate. Correspondingly, the North wage rate falls as more know-how is transferred. Finally, when IPR gets sufficiently strong that the S firm exits and N asserts a monopoly, the Southern (Northern) wage rate declines (rises).

3.5. *Alternative Market Assumptions*

We offer brief comments on the robustness of our results to alternative market structures.[16] First, suppose there are two N firms but one has a superior process technology. In this case the critical Southern patent value at which licensing is chosen becomes higher because the S firm has two technologies to imitate. Beyond this value the lower-cost N firm would undertake licensing and the extent of know-how transferred would rise with k. The marginal gain from licensing rises faster with k because it steals profits from the higher-cost N firm, making it possible for a greater amount of know-how to be transferred.

Second, suppose there are two S firms to which the single N firm can license, giving it more bargaining power. Here the critical k inducing licensing is smaller than in the benchmark case and more know-how is transferred at every level of IPR in the licensing range. These results hold also when there is free entry into imitation in the South but N can license to a single firm. In this case the number of S imitators falls with k in the licensing range. Overall, our findings

[16]Each of these claims can be readily proven analytically.

remain qualitatively intact with different numbers of firms engaging in quantity competition.

Finally, return to two firms and suppose N and S compete in prices. Here either successful imitation or licensing permits S to capture the full global market at a price just below N marginal cost. Again, a critical k exists above which N will sacrifice its production to gain a share of higher licensing profits. Within the licensing range, increases in patent strength do not affect S exports or price. While this case is perhaps less interesting analytically, it is consistent with firms fully outsourcing their production.

4. Welfare Impacts of Strengthened IPR

Welfare calculations in this context are complex and, to save space, we simply overview the results here.[17] We consider the range of k in which there may be imitation or licensing because once IPR protection becomes sufficiently high that N is a monopolist, further increases are redundant.

For $0 < k < k_1$, S competes through imitation. Here an increase in patent rights reduces Southern well-being, defined as local consumer surplus plus the S firm's production profit, by diminishing the latter. Upon reaching k_1 there is a discrete jump in Southern welfare because the decision to license generates a positive know-how transfer, reducing S marginal cost and expanding competition and consumer gains. Beyond this point, stronger patent rights have offsetting effects. First, this policy shift reduces technology-transfer costs and increases the proportion of know-how transferred under licensing. These impacts increase S's profit and expand consumer gains from greater competition, with these effects being larger under higher local absorptive capacity. Second, the higher imitation cost reduces the value of the S firm's outside option, decreasing its licensing profits. Southern welfare increases if the former effects dominate. In fact, this outcome pertains for a sufficiently small N bargaining share. In brief, for IPR reform to raise local welfare under licensing,

[17] Again, these computations are available in the mathematical appendix.

S must have a relatively strong bargaining position. Similar analysis shows that Southern welfare increases unambiguously with a rise in local absorption capacity. Once the equilibrium regime becomes an N monopoly at the higher critical value of k, Southern welfare falls discretely to its lowest level and remains constant.

We can show further that Northern welfare rises with k in the low range because that policy decreases the threat of imitation, raising N profits. Upon reaching the licensing equilibrium range, stronger patents both raise N profits and expand competition, increasing welfare. In the high range of k, where a monopoly is established, the impact depends on the net effect of higher N profits and lower consumer welfare. A specific result is that Northern welfare in the monopoly range exceeds that in the imitation range if market sizes are large.

Finally, global welfare rises with a stronger IPR regime in the South under licensing, so long as the condition in Eq. (8) holds. Recall that this condition requires the joint market size to be large and the difference in post-transfer N and S marginal costs to be high. The policy change decreases the technology-transfer cost and expands know-how transfers, thereby increasing competition and consumer welfare. These effects outweigh any reduction in global profits. An expansion in absorptive capacity also increases global welfare by shifting more production to the South where marginal cost is lower. When the full N monopoly is reached, however, global welfare is diminished in comparison with licensing. Thus, in this model full protection would not be globally optimal.[18]

A final note is that the welfare impacts of a Northern ban on imitative S imports vary by country. It reduces S profits under imitation and also diminishes the S firm's share of licensing profits. These factors dominate and Southern welfare is diminished by blocking throughout the range of k before the full monopoly is established. In contrast, this ban tends to raise Northern welfare by raising profits in the imitation range and achieving a higher profit share in the licensing range.

[18]See Grossman and Lai (2004) for additional insights.

5. Adding Foreign Direct Investment

We now incorporate a simple specification of FDI into the model to permit a third important form of technology transfer. Suppose that the N firm has the option of creating an S subsidiary at a cost of $g = \gamma + z(x)$. Here, γ is the fixed cost of setting up a plant and transferring technology, while $z(x)$ is the variable transfer cost, which increases with the proportion of know-how transferred. Since the S subsidiary is fully owned by the N multinational firm, it has no incentive to imitate. However, the subsidiary faces imitation risk from a local firm at the same cost as above.

In this setup, assuming sufficient market sizes, the N firm would transfer the full amount of know-how to its subsidiary.[19] The essential reason is that under FDI all profits are returned to N, rather than shared under licensing. Accordingly, N transfers its full know-how to achieve the maximum reduction in production cost. Thus, in any FDI equilibrium the transfer cost $g = \gamma + z(1)$ is constant and the marginal production cost of the S subsidiary firm becomes $c_{DD} = c_S - r(1, a)$. Indeed, under FDI the N firm exits production altogether and serves both markets from its subsidiary, another case of full offshoring.

Consider next the entry mode of the N firm. Compared to licensing it now has two possible reservation values in the bargaining game. If the FDI transfer cost is less than the difference between production profits under FDI and production profits under exports, N chooses total profits with the subsidiary (Π_{DD}) as its reservation value. Otherwise it chooses profits under production at home and export. To summarize, FDI involves a fixed setup cost but transfers know-how fully and achieves lowest marginal cost, while facing imitation risk. Licensing incurs a technology-transfer cost and generates partial knowledge transfer but does not face imitation risk.

Working through the profit functions we find the following results. For sufficiently high investment costs, the reservation option for N is exports with imitation. That is the equilibrium for $0 < k < k_1$, with the remaining equilibria as patent rights are strengthened the same as those described earlier. For lower setup costs the

[19]This claim is proven in the mathematical appendix.

equilibrium is FDI with imitation for $0 < k < k_D$, where $k_D > k_1$, with licensing occurring at yet-stronger patent rights before the monopoly ensues.

In essence the results of this simple model may be characterized as follows. In countries where subsidiary setup costs are large and patent protection is weak, the N firm competes through exports and accepts the risk of local imitation. In this case a tighter patent regime diminishes S profits and Southern welfare. However, if these countries strengthen IPR they can induce a licensing equilibrium, which favors S licensee exports and likely expands welfare. Alternatively, countries with weak patents but lower subsidiary setup costs are more likely to attract FDI than N exports. Here, as IPR are strengthened the imitative S firm becomes a licensee and increases in k again expand local exports and well-being. The shifts from trade to FDI to licensing at various configurations of setup costs and patent rights are reminiscent of the literature on internalization and IPR (Markusen, 2001).

To complete the analysis, note that if the North government blocked imitative imports under FDI with imitation, the impact would be to raise even further the minimum Southern IPR level at which licensing occurs and a diminution in the latter country's welfare.

6. Summary and Conclusions

In this paper we developed a model of strategic competition to study the effects of stronger IPR protection in developing countries on their export performance, the mode of technology transfer and welfare. By reducing the Southern firm's marginal cost related to the degree of know-how absorbed, and also the cost of technology transfer, patent reforms could expand export opportunities. We found also that Southern welfare could rise with both IPR and absorptive capacity if S has relatively high bargaining power.

This model is stylized but offers insights into means by which policymakers in developing countries might fashion their reforms in intellectual property. Specifically, stronger patent rights and trade secrets are likely to expand exports and improve welfare if domestic firms gain a more advanced ability to absorb and implement available international technologies. A complementary policy is to reduce the costs of establishing a subsidiary.

References

Branstetter, Lee, Fisman, Raymond, and Foley, C. Fritz, 2005. Do stronger intellectual property rights increase technology transfer? Empirical evidence from U.S. firm-level panel data. *Quarterly Journal of Economics* 121, 321–349.

Branstetter, Lee, Fisman, Raymond, Foley, C. Fritz, and Saggi, Kamal, 2007. Intellectual property rights, imitation and foreign direct investment: Theory and Evidence. NBER Working Paper, vol. 13033.

Diallo, Barrou, 2003. Historical perspectives on IP protection for software in selected countries worldwide. *World Patent Information* 25, 19–25.

Ginarte, Juan C. and Park, Walter G., 1997. Determinants of patent rights: a cross-national study. *Research Policy* 26, 283–301.

Glass, Amy J. and Saggi, Kamal, 1999. Intellectual property rights and foreign direct investment. *Journal of International Economics* 56, 387–410.

Grossman, Gene M. and Lai, Edwin L.C., 2004. International protection of intellectual property. *American Economic Review* 94, 1635–1653.

Hasan, Rana and Raturi, Mayank, 2001. Does investing in technology affect exports? Evidence from Indian firms. Honolulu: East-West Center Working Paper, vol. 21.

Helpman, Elhanan, 1993. Innovation, imitation, and intellectual property rights. *Econometrica* 61, 1247–1280.

Kumar, Nagesh and Siddharthan, N.S., 1993. Technology, firm size and export behavior in developing countries: the case of Indian enterprises. UNU/INTECH Working Paper No. 9.

La Croix, Sumner J. and Kawaura, Akihiko, 1996. Product patent reform and its impact on Korea's pharmaceutical industry. *International Economic Journal* 10, 109–124.

Lai, Edwin L.C., 1998. International intellectual property rights protection and the rate of product innovation. *Journal of Development Economics* 55, 115–130.

Lai, Edwin L.C. and Qiu, Larry D., 2004. The North's intellectual property rights standard for the South? *Journal of International Economics* 59, 183–209.

Markusen, James R., 2001. Contracts, intellectual property rights and multinational investment in developing countries. *Journal of International Economics* 53, 189–204.

Maskus, Keith E., 2004. Encouraging Technology Transfer. Report for UNCTAD/ICTSD Project on Intellectual Property Rights and Sustainable Development, Geneva, Issue Paper no. 7.

Smith, Pamela J., 2001. How do foreign patent rights affect U.S. exports, affiliate sales, and licenses? *Journal of International Economics* 55, 411–440.

Taylor, Scott M., 1994. TRIPs, trade, and growth. *International Economic Review* 35, 361–381.

U.S. Department of Commerce, Bureau of Economic Analysis, 2007. *U.S. International Services: Cross-Border Trade 1986–2006 and Sales through Affiliates 1986–2005*. At http://www.bea.gov/international/intlserv.htm.

Vishwasrao, Sharmila, 1994. Intellectual property rights and the mode of technology transfer. *Journal of Development Economics* 44, 381–402.

Vishwasrao, Sharmila, 2007. Royalties versus fees: how do firms pay for foreign technology? *International Journal of Industrial Organization* 25, 741–759.

Yang, Guifang and Maskus, Keith E., 2001. Intellectual property rights, licensing, and innovation in an endogenous product-cycle model. *Journal of International Economics* 53, 169–187.

Chapter 9

Southern Innovation and Reverse Knowledge Spillovers: A Dynamic FDI Model*

Yin He and Keith E. Maskus†

1. Introduction

The massive expansion of foreign direct investment (FDI) in recent years to such emerging countries as China, India and Brazil has increasingly been accompanied by growth in innovation by local firms in those markets (OECD, 2005; WIPO, 2007). It seems likely that these trends are jointly related in that more inward FDI permits additional domestic imitation and learning, from which innovation springs. Moreover, the expanded innovation in developing economies sets up the possibility of *reverse spillovers*: firms in developed countries may learn from inspecting new products and technologies imported from the developing world.

The evolution of optical storage media in audio and video systems illustrates this process. The earliest technology was the audio compact disc (CD), available in Western countries by the early 1980s. In 1993, Wanyan Electronics, a Chinese firm, invented the more affordable video compact disc (VCD) and VCD players, which quickly became popular in mainland China but were rarely known in the Western world (Xu, 2006). This invention was made possible

*Reprinted from *International Economic Review*, Vol. 53, No. 1, pp. 281–304. Copyright 2012, with permission from John Wiley and Sons.
†We thank Martin Boileau, Alan Deardorff, Murat Iyigun, Wolfgang Keller, Robert McNown, Raymond Riezman, Changcai Zhu and two anonymous referees for valuable comments.

through inspecting the technology sold by C-Cube Microsystems (an American company, which combined the CD with the MPEG-1 standard) and learning the production process of firms producing compact disks in China's special economic zones. However, Wanyan Electronics failed to patent their products, which were duplicated by domestic and international firms (Xu, 2006). Among these were such Japanese and European companies as Sony and Philips, which have international patents covering VCD standards (Linden, 2003). These companies, along with the US innovator of the MPEG-2 standard, later jointly invented the Digital Video Disc (DVD) technology.

This history illustrates the possibility we analyze here: a global technology is imported into an emerging economy, where it is imitated and then improved through local innovation. The new product, in turn, is learned by companies in the developed world and serves as the basis for further innovation. In essence, we develop the first product-cycle model in which there are two-way innovation and technology transfers in equilibrium.

For this purpose, we set out a dynamic, general-equilibrium model in which FDI (or imports), imitation, learning-by-doing (LBD) and innovation are jointly determined and ultimately affect economic growth. The model combines elements of those in Baldwin, Braconier and Forslid (BBF, 2005) and Lai (1998) but extends both. In BBF, firms in two symmetric developed countries perform R&D to generate new knowledge, subject to spillovers from the general knowledge pool and local LBD. We extend their analysis by permitting different learning productivities between a developed nation (the North) and a developing economy (the South) and higher-cost innovation in the latter. Lai introduced FDI into an endogenous North–South model but did not permit innovation in the developing country, nor did he consider LBD, both of which exist here.

In our benchmark model, after new knowledge is developed Northern firms choose the physical location of their production between the North and South. Firms in the South learn from observing the operations of multinational firms, imitate some of the new varieties, and compete in these varieties. In our extension, Southern firms are able to invent new products based on the local technology pool and LBD spillovers. Finally and most importantly, Southern new knowledge also spills over back to the North to reduce the unit cost of innovation there. To our knowledge this is the first model to study Southern innovation and reverse knowledge spillovers in a general-equilibrium, endogenous product-cycle model.

Within this framework we analyze how changes in intellectual property protection (IPP) in the South, indexed by the imitation rate, affect these processes. In the benchmark model, without the possibility of innovation in the South, a rise in imitation reduces inward FDI in equilibrium. This tends to reduce global growth in new products because less labor is allocated to innovation in the North. When the South is able to learn from FDI and develop its own products, innovation without reverse spillovers restricts FDI flows compared to the benchmark. Since Southern innovation efficiency is lower than its Northern counterpart, the equilibrium growth rate is lower than the benchmark rate at each given imitation rate. However, the relationship between imitation and multinationalization exhibits a U-shape because increases in the former first reduce FDI directly but ultimately increase it through a competition effect arising from the expansion of Southern new varieties. Finally, with a significant reverse spillover the enhanced innovation possibilities in the North generate more rapid growth than in the case without such learning. As a consequence, the growth rate is also U-shaped in imitation and may exceed its benchmark value without Southern innovation.

To motivate our work we offer stylized facts and briefly survey relevant literature in Section 2. We develop our benchmark model, where the South imitates multinational varieties without engaging in R&D or generating new knowledge, in Section 3. In Section 4 we introduce Southern innovation and reverse knowledge spillovers into the model. We perform extensive simulation analysis in Section 5 to compare models with and without Southern innovation and determine how the extent of reverse spillovers affects other key variables, including steady-state growth. We conclude in the final section.

2. Motivation and Relevant Literature

There are two novelties in this paper: permitting Southern innovation in response to inward knowledge flows, and the possibility of reverse flows of this innovation to the North. To motivate this work we first discuss basic evidence of these phenomena and then place our analysis into the broader literature.

2.1. *Stylized Facts*

Our story, in essence, is that countries in the South receive inward technology transfer (TT), learn to use incoming technologies through imitation and learning, and then build on this knowledge to become innovators themselves, with the potential for reverse TT to the North. As will become evident in the model, this dynamic is most likely to occur in developing nations that are both large and have a significant capacity for absorbing and improving technical information. Thus, we should expect to see stylized evidence of this evolution in data for such countries as Argentina, Brazil, Mexico, China, India and the Republic of Korea.

The three columns in the top panel of Table 1 list basic figures in 1995 for these countries in the three canonical forms of inward TT: high-technology imports (measured as a share of total manufacturing imports), foreign direct investment (measured as the position of U.S. multinational enterprises), and licensing (measured by royalties and license fees paid for patents, trademarks, brands and other forms of intellectual property).[1] In 1995 each of these nations had a significant share of manufactured imports arrive in high-technology items, which include pharmaceuticals, electrical machinery, scientific and controlling instruments, and aerospace goods. These shares ranged from around 10 percent in Mexico and Korea to over 20 percent in Brazil and China. Further, by 1995 U.S. multinationals had established significant FDI positions in Mexico and Brazil but those positions exceeded one billion dollars in China and Korea also. Finally, each nation paid far more in royalties and license fees than it earned (compare the payments column to the 1995 receipts in the fifth column of the bottom panel), with those payments ranging from $90 million

[1]Keller (2010) reviews evidence regarding trade and FDI as means of technology transfer and their resulting spillovers into higher local productivity. On licensing see Maskus (2004).

Table 1. Indicators of technology transfer.

Country	High-Tech Imports (%)[a]	U.S. FDI Position Abroad ($m)[b]	Royalties & Fees Payments ($m)
	1995	1995	1995
Argentina	13.8	3,289	392
Brazil	22.3	17,385	529
Mexico	9.7	10,580	484
China	20.7	1,263	1,281[c]
India	13.9	399	90
Rep. Korea	10.3	2,083	2,385

Country	High-Tech Exports (%)[a]		FDI Position in U.S. ($m)[b]		Royalties & Fees Receipts ($m)		Patent Application Shares in USPTO & EPO	
	1995	2005	2000	2005	1995	2005	1995	2005
Argentina	3.5	6.5	neg	neg	12	51	0.03	0.04
Brazil	4.8	12.8	429	1,588	32	102	0.06	0.11
Mexico	15.1	19.6	3,448	5,674	114	171	0.04	0.05
China	10.5	30.6	neg	175	80[c]	157	0.07	1.01
India	4.3	4.7	neg	94	1	206	0.05	0.56
Rep. Korea	25.9	32.3	1,141	525	299	1,908	1.26	5.61

Notes: [a]Percentage of manufacturing imports or exports. [b]FDI position in total manufacturing at historical cost. [c]Figures for 2000.
Sources: Computed by authors from UN Comtrade; World Bank World Development Indicators; US Bureau of Economic Analysis; World Intellectual Property Organization.

in India to $2.4 billion in Korea. Chinese payments in the year 2000 approached $1.3 billion.

The figures in the eight columns in the bottom panel strongly suggest that these inward TT flows were followed by significant increases in technological knowledge generated in this set of countries. Thus, each country experienced a rise in the share of high-technology products in its manufacturing exports from 1995 to 2005. There were particularly large increases in Argentina, Brazil, China and Korea. Next, by 2000 firms headquartered in Brazil, Mexico and Korea had established FDI positions in manufacturing in the United States, though positions from the other countries were negligible.

By 2005, however, all except Argentina had become significant investors in U.S. manufacturing plants, raising prospects for FDI-mediated reverse spillovers. There were also substantial increases in current-dollar royalties and license fees, suggesting that Korean, Chinese, Indian and Mexican brands and technologies were penetrating global markets. Finally, each nation saw growth in its share of patent applications at the United States Patent and Trademark Office and the European Patent Office. These increases were registered in a period of massive expansion of applications in both of these offices, implying major growth in the numbers of applications from the developing world. The growth in China's share from 0.07 percent to 1.01 percent was remarkable, as were the expansions from India and Korea. The latter nation now accounts for over five percent of applications at these two major authorities.

These stylized facts cannot demonstrate causality. However, they do suggest that the story we tell here is consistent with actual data. Thus, larger developing countries with emerging technical capacities to imitate, absorb, and improve upon international technologies ultimately become sources of new information that finds its way to the developed world. Our model is aimed at capturing this process.

2.2. *Literature*

The notion of a product cycle with one-way TT arising from international trade or FDI stems from Vernon (1966). It received initial analytical treatment by Krugman (1979) in a model with exogenous innovation and imitation. Early general-equilibrium models (Grossman and Helpman, 1991a; Helpman, 1993; Segerstrom *et al.*, 1990) of product-cycle dynamics with North–South TT retained exogenous imitation activity. More significantly for our model, they did not include knowledge spillovers, so that multinational corporations (MNCs) played no direct role in determining the growth rate. Later models, such as Lai (1998) and Glass and Saggi (2002), began to fill the latter gap by assuming that the efficiency of Northern innovation depends on the stock of existing knowledge. However, these models mostly assumed that knowledge contributes the same to further innovation regardless of its location and characteristics, which is not consistent with empirical studies. For example, Jaffe *et al.* (1993), Sjöholm (1996), and Keller (2002) showed that the scale of spillover effects from knowledge transfers is geographically limited and the scope of technology diffusion is severely limited by distance.

BBF (2005) is the first study that distinguished among channels of cross-border knowledge spillovers in a theoretical FDI model. In their model, innovators fully understand domestic knowledge but can only partially make use of foreign knowledge. The spillover from LBD to innovation is introduced into a symmetric North–North framework, where innovators become more efficient as they observe more local production processes. FDI activities promote innovation and growth in both regions, since innovators everywhere become more efficient by learning the production processes of increasing varieties.

Whether FDI exists in equilibrium in BBF is determined by comparing two exogenous parameters: the trade cost and the fixed-cost premium when setting up a firm abroad. The rate of multinationalization is undetermined and the authors "take it as determined by factors outside of the model". Therefore, although the share of domestic varieties transferred abroad is crucial to the efficiency of innovation and growth in both regions, the equilibrium level of this share is left unexplained.

An exogenous rate of multinationalization as in BBF is questionable in a North–South model. Bringing knowledge to the South to take advantage of the low wage rate is the incentive for many firms in the North to engage in FDI. At the same time, transferring production to the South would alter the wage differential between the host and home countries, thereby changing firms' incentives for FDI. Thus, the multinationalization rate, investment in Northern innovation, the growth rate, and the wage gap should be made endogenous in a fuller model.

These variables were endogenized in various specifications by dynamic North–South FDI models, such as Helpman (1993), Lai (1998) and Glass and Saggi (2002), but Northern innovation in these models depends only on local R&D investments. An interesting additional question posed in those articles is the effect of more rigorous regulation in IPP in the South on the rate of Northern innovation and global growth. Helpman showed that in the presence of FDI, a fall in the imitation rate would cause the South to suffer from both a deterioration of its terms of trade and higher prices paid for a larger fraction of products. Lai found that a lower rate of imitation encourages the North to increase the rate of both innovation and multinationalization, which raises the growth rate of the South. However, Glass and Saggi observed that a lower imitation rate exposes both MNCs and Northern firms to less risk, producing larger monopoly profits in production and thereby reducing both FDI and innovation. These papers generated numerous insights but did not consider the possibility of induced Southern innovation and what we call reverse spillovers.

There are models of induced Southern innovation that help inform our results. Chen and Puttitanun (2005) set out a static oligopoly model in which firms in developing countries in two sectors could imitate a foreign variety, imitate a domestic variety, or invest in a new variety. These decisions depended on the rigor of local IPP, providing a policy tradeoff for the authorities. They showed that there is a U-shaped relationship between optimal patent strength and a country's level of economic development, consistent with the empirical fact first highlighted by Maskus and Penubarti (1995).

A closer antecedent to our work is the paper by Chui *et al.* (2001). They set out a model in which Northern firms always innovate high-technology goods, while the Southern location passes through four stages as potential steady-state equilibria: (1) specializing in a traditional good; (2) copying the high-technology goods; (3) copying and innovating jointly; (4) specializing in innovation. These outcomes arise in various configurations of three exogenous asymmetries. The South is less efficient at adopting technologies, the diffusion rate from North to South is slow while it is instantaneous in the other direction, and the South has a lower endowment of skilled labor. Stronger values of Southern patent protection move firms in that location through these stages. Their full model has no closed-form solution and they rely on simulation, as we do here. Our model is more general in that it considers the crucial role of FDI as the channel of gradual technology transfer, solves for endogenous multinationalization decisions, and permits within-region and across-region learning at different rates. In particular, the degree of South-to-North technology spillovers, which Chui *et al.* take to be infinite and instantaneous, is a critical determinant of impacts of imitation on multinationalization and growth.

Finally, Glass and Wu (2007) presented a dynamic model in which innovative firms in the North choose between vertical quality gains and horizontal variety increases. Followers can imitate or innovate against either type, which are transferred via costless FDI. Southern imitation raises Northern innovation and FDI in quality gains but reduces them in variety expansion. There is no learning-by-doing or Southern innovation.

Thus, in this paper we develop an endogenous product-cycle model that extends the existing literature significantly. The multinationalization rate is endogenous to imitation and other factors but variations in this rate subsequently affect the growth rate. We distinguish among channels of knowledge spillovers by assuming that the extent of spillovers depends on the geographic location and ownership of general knowledge and the location of production processes. Finally, we analyze three scenarios regarding Southern innovation, where the imitation rate has different impacts on growth.

3. Benchmark Model with Costly Imitation in the South

We begin with a North–South model of MNCs with endogenous knowledge innovation in the North based on different knowledge spillover effects. The theory combines and builds on those of Grossman and Helpman (1991b), Lai (1998) and BBF (2005).

There are two final goods, with Y the homogeneous good and X the manufacturing good with differentiated varieties. The market for the homogeneous product is perfectly competitive, while that for the manufacturing good is monopolistically competitive. There is only one factor, labor. That is, no physical capital is necessary for setting up a firm on producing either X or Y. However, some labor must be allocated to innovating a new technology before any variety of X can be produced. For now we assume that Southern labor has sufficiently lower skill than its Northern counterpart that only Northern labor has the capacity to innovate. Without inward

FDI, firms in the South can only produce the homogeneous product. Let the Southern wage be normalized to unity and define the endogenous Northern wage to be w, which is greater than one. This implies that the North would produce only the differentiated product since it has a cost disadvantage in the homogeneous sector. Thus, the two regions would engage in complete specialization if no MNCs existed.

However, MNCs emerge in this model due to the wage gap between North and South. All innovation activities take place in the North by investing part of the Northern labor in R&D. When an element of new knowledge is generated and ready for industrial application, one unit of labor is needed for producing each unit of the new variety. Thus, firms in the North can choose to keep all production at home and serve the Southern market by exports. For simplicity we normalize constant unit trade costs to be zero. However, because of the lower wage rate, firms in the North might want to transfer the technology to the South, establish and produce in a subsidiary there, and export back to the North. The disadvantage of being a multinational lies in both the one-time cost of setting up a plant and the risk of being imitated, reducing its monopoly power and profits.

It is possible, as suggested by Lai (1998), that the physical appearance of production in MNCs in the host country permits Southern imitators to learn production processes more easily by inspection than by reverse engineering imported goods. There are also empirical studies showing that technology spillovers from inward FDI are larger than those from imports (Keller and Yeaple, 2009). Therefore, for simplicity in the benchmark case, we assume that exports from the North to the South are free from the risk of imitation, but local production within MNCs is not. However, in an extension described later we consider the impact of permitting also imitation of imported goods.

Imitation by Southern firms takes time. To capture this possibility we model the duration between the time of an MNC setting up its plant in the South and the time of successful imitation as a random variable with a Poisson arrival rate. We assume that if the technology is imitated the South sets up a firm with no physical cost and produces for both markets the same variety it imitated from the targeted MNC, which exits production.

We acknowledge that our assumption of exogenous imitation risk, while common in the literature (e.g., Helpman, 1993; Lai, 1998), limits the potential richness of the outcomes we may explore. In particular, while imitation affects innovation in our model, the reverse feedback is absent. If imitation were endogenous, faster innovation could reduce the gains from imitation as new varieties are introduced more rapidly. Unfortunately, introducing this channel into the model would make the analysis considerably more complex.

3.1. *Consumption*

Consumers have identical and homothetic preferences over the two final goods. The intertemporal utility function at time τ is assumed to be:

$$\begin{aligned} U_\tau &= \int e^{-\rho(t-\tau)} \ln[(C_X)^\phi (C_Y)^{1-\phi}]dt \\ C_X &\equiv \ln\left\{\int_0^{n+m+b} (c_h)^{1-1/\varepsilon} dh\right\}^{\frac{1}{1-1/\varepsilon}} \end{aligned} \tag{1}$$

where $\rho > 0$ measures the time preference of consumers, $\varepsilon > 1$ is the elasticity of substitution between varieties of X, and c_h is the consumption of variety h. Letters n, m, and b represent the number of Northern, multinational, and Southern imitated varieties, respectively. There is one firm per variety.

The utility maximization problem yields a standard instantaneous CES demand function of variety h:

$$c_h = \frac{p_h^{-\varepsilon}}{\int_0^{n+m+b} p_j^{1-\varepsilon} dj} \cdot \phi \cdot E = \eta_h \cdot \frac{\phi \cdot E}{p_h}. \tag{2}$$

The consumption share of variety h is $\eta_h \equiv p_h c_h / \phi \cdot E = (p_h/P_X)^{1-\varepsilon}$, where $P_X = (\int_0^{n+m+b} p_h^{1-\varepsilon} dh)^{1/(1-\varepsilon)}$ is the price index for manufacturing good X, E is global expenditure, and p_h is the price for variety h. Because varieties are assumed to be imperfect substitutes, goods with higher prices would have smaller market shares.

3.2. Production, Innovation and FDI

The market for good Y is perfectly competitive while that for good X is monopolistically competitive with product differentiation. The production of either type requires only one unit of labor (L), regardless of the location and ownership of the firm. However, knowledge (K) is required before the manufacturing goods are produced and the knowledge is generated by investing labor in R&D.

The Northern innovation sector performs all R&D. Each unit of new knowledge is produced with a_I^N units of Northern labor. Following BBF (2005), we assume that

$$a_I^N = \frac{1}{K^N + K^M + \mu n} \quad (\mu \geq 0). \tag{3}$$

Here, K^N and K^M are the cumulated knowledge stock owned by Northern firms and MNCs, respectively, with their sum constituting the world stock. This innovation function exhibits two types of spillovers from existing knowledge.[2] The first comes from the global knowledge pool. Northern innovators have complete access to knowledge held by Northern and multinational firms, since all knowledge is originally invented in the North. The second type of spillover is LBD in the North. We assume that LBD is governed by parameter μ, which determines the ability to learn from existing local varieties in developing new goods. Thus, the efficiency of innovation is positively proportional to the number of varieties produced in the North. Note that the cost of innovation is then $\Omega = wa_I^N$.

Once a blueprint (i.e., a new variety of X) is invented in the North, firms have the choice of either producing domestically or doing so abroad. If the firm chooses to transfer the technology to the South, it faces fixed costs to set up the plant. That is, if MNCs want to use one unit of knowledge for production in the South, they need to take $(1 + \Gamma)$ units of knowledge since the proportion Γ would need to be

[2]In this model, as in BBF, Northern knowledge is the same as the number of Northern varieties but we permit two means by which this knowledge spills over into reduced innovation costs. The stock of multinationals' knowledge is more than the number of their varieties due to the cost of engaging in FDI.

expended during the transfer process to cover contracting problems, language differences and the like.

Finally, Southern firms imitate varieties introduced by MNCs with some time lag. Following Lai (1998), define the Poisson arrival rate as the change in varieties produced by Southern firms as a proportion of the available multinational ones: $i = \Delta b/m$ $(1 \geq i \geq 0)$. The rate i is also the probability that any variety produced by MNCs is imitated by the South in each instant, which depends on the strength of IPP and the imitation ability of workers in the South. The higher is the imitation capacity and the weaker the strength of IPP, the sooner knowledge will diffuse from MNCs to Southern imitators. The number of imitated varieties per time period by the South is then $\Delta b = im$. Once the variety is copied, only the Southern firm produces it, which drives the price down to its marginal cost.

The profit-maximization price for Northern varieties is $p = w/\alpha$ and that for multinational varieties is $p = 1/\alpha$, where $0 < \alpha = 1 - 1/\varepsilon < 1$. The price index of the differentiated products is then $P_X = (\int_0^{n+m+b} p_j^{1-\varepsilon} dj)^{\frac{1}{(1-\varepsilon)}} = (\frac{w}{\alpha})^{1-\varepsilon} n + (\frac{1}{\alpha})^{1-\varepsilon}(m - im) + im$. The instantaneous profit functions for (unimitated) multinational firms and Northern firms are, respectively,

$$\begin{aligned} \pi^M &= \left(\frac{1}{\alpha} - 1\right) c^M = \frac{(1-\alpha)\phi E}{w^{1-\varepsilon} n + m - im + im\alpha^{1-\varepsilon}} \\ \pi^N &= \left(\frac{w}{\alpha} - w\right) c^N = \frac{(1-\alpha) w^{1-\varepsilon}\phi E}{w^{1-\varepsilon} n + m - im + im\alpha^{1-\varepsilon}}. \end{aligned} \tag{4}$$

As noted above, Southern profits are zero.

Assume that the North invests L_I^N units of Northern labor in R&D. The amount of new knowledge generated in each instant is

$$\Delta K = \frac{L_I^N}{a_I^N} = L_I^N (K^N + K^M + \mu n). \tag{5}$$

In this model we study the steady state equilibrium. In any equilibrium there is only one knowledge share of multinationals. Thus, the growth rates of knowledge held by Northern firms and multinationals are the same and both equal the growth rate of world knowledge

$$g = g^N = g^M = \frac{\Delta K}{K^N + K^M} = L_I^N[1 + \mu(1 - \Sigma^M)] \tag{6}$$

where $\Sigma^M = K^M/(K^N + K^M)$ is the share of knowledge owned by MNCs in all technology innovated in the North, which is between zero and one. Note that, although the growth rate in equation (6) is written in terms of endogenous variable L_I^N and Σ^M, these two variables can be expressed by exogenous parameters at the steady state so that the growth rate is also time-invariant in equilibrium. Also define $\Sigma^N = K^N/(K^N + K^M) = 1 - \Sigma^M$ to be the share of knowledge held by Northern firms. Since the South imitates existing multinational varieties at rate i at each instant, it grows at the same rate g as the North.

The discounted operating profit of a Northern firm at time τ is $\Pi^N \equiv \int_{t=\tau}^{\infty} e^{-r(t-\tau)}\pi^N dt$. Note that a higher interest rate reduces discounted profits. Further, the faster new varieties are innovated (i.e. the faster the growth rate), the lower the expected future profit for each variety. Therefore, the expected profit of a Northern firm is

$$\prod^N = \frac{\pi^N}{\rho + g}. \tag{7}$$

A multinational firm can only enjoy its monopolistic profit before its variety is imitated. Therefore, the expected profit for MNCs is $\Pi^M = \int_{t=\tau}^{\infty} [(\int_{\delta=0}^{\tau} \pi^M e^{-r\delta} d\delta) \cdot \text{prob}(t = \tau)]dt$, where $\text{prob}(t = \tau)$ is the probability that a variety produced by any MNC has been copied at time τ. With the standard Poisson arrival rate, the duration τ between the time of an MNC setting up in the South and the time of imitation follows the probability density function $\text{prob}(t = \tau) = e^{-i\tau}$ where i is the Poisson arrival rate at which a variety would be imitated in the next instant under the condition that it has not yet happened. Thus, the expected profit function for an MNC is $\Pi^M = \int_{t=\tau}^{\infty} (\int_{\delta=0}^{\tau} \pi^M e^{-r\delta} d\delta) \cdot e^{-i\tau} dt$. Again, the interest rate and the growth rate of new varieties need to be accounted for, as does imitation risk. The multinational's expected profits become

$$\prod^M = \frac{\pi^M}{\rho + i + g}. \tag{8}$$

3.3. *Equilibrium Choice of FDI*

Northern firms choose their production location after the new technology is innovated. In equilibrium, the choice between being an MNC or a Northern firm is based on the following equation with complementary slackness:

$$\Sigma^M(1-\Sigma^M)\left[\frac{\Pi^N}{\Omega}-\frac{\Pi^M}{\Omega(1+\Gamma)}\right]=0. \tag{9}$$

Equation (9) shows that, after successfully inventing the new variety, firms in the North compare the ratio of the expected operating profits and setup costs in order to choose location. Plug equations (7) and (8) into equation (9) and rewrite the equation with complementary slackness as follows:

$$\begin{aligned}
\Sigma^M &= 1 \text{ and } \frac{\pi^M}{(\rho+i+g)(1+\Gamma)\Omega} \geq \frac{\pi^N}{(\rho+g)\Omega}\\
\Sigma^M &= 0 \text{ and } \frac{\pi^M}{(\rho+i+g)(1+\Gamma)\Omega} \leq \frac{\pi^N}{(\rho+g)\Omega}\\
0 < \Sigma^M &< 1 \text{ and } \frac{\pi^M}{(\rho+i+g)(1+\Gamma)\Omega} = \frac{\pi^N}{(\rho+g)\Omega}.
\end{aligned} \tag{10}$$

The first equation states that all firms in the North would move their production to the South ($\Sigma^M - 1$) if the expected profits of being a multinational are higher than those of being a Northern firm. The second indicates that all firms would remain in the North ($\Sigma^M = 0$) if the opposite happens. The final condition states that firms are indifferent among locations ($1 > \Sigma^M > 0$) if the discounted returns of the two types of firms are the same.

3.4. *Market-Clearing Conditions*

The instantaneous expenditure of the South simply equals the labor income of the whole economy, which also equals the total population of the South since the Southern wage is one. Northern workers get labor income and we assume that all profits are earned by the owners of active firms. Thus, the North enjoys the profits from both Northern firms and MNCs remaining in operation. However, part of this income stream needs to be allocated to generating new knowledge. The extra cost that MNCs pay to set up the plant in the South does not show up explicitly in the expenditure function but is embodied in their profit. Therefore, the instantaneous expenditure function for each region is the following[3]:

$$E^S = L^S \tag{11}$$

$$E^N = wL^N + \pi^N n + \pi^M (m - im) - wL_I^N \tag{12}$$

$$E = E^S + E^N \tag{13}$$

where L^S and L^N are the population in the South and North, respectively.

The labor endowment in the North is used in R&D and the production of Northern varieties and the Southern labor endowment is utilized in producing multinational varieties not yet imitated, imitated varieties (B), and the homogeneous product. Thus, the labor market clearing conditions are as follows:

$$L^S = L_Y + (m - im) \cdot x^M + im \cdot x^B \tag{14}$$

$$L^N = L_I^N + n \cdot x^N. \tag{15}$$

In these equations, x refers to production of a differentiated variety, which also equals consumption in equilibrium.

3.5. *Dynamic Equilibrium*

In our model, the monopolistic-competition framework implies free entry in the differentiated-products sector. Thus, the expected operating profits of Northern firms and multinationals, if they exist, should exactly cover their costs of acquiring the technology and other costs before production takes place in dynamic equilibrium. We therefore have

$$\frac{\Pi^N}{\Omega} = 1 = \frac{\Pi^M}{(1+\Gamma)\Omega}. \tag{16}$$

[3]As in BBF (2005), we assume there is no population growth. Thus, when the number of varieties increases in the presence of a fixed labor force, the quantity of consumption for each variety would decrease to balance the labor market.

Comparing this to the inequality conditions in (10) on FDI choices we find that if all firms choose to be Northern firms ($\Sigma^M = 0$) or all decide to be multinationals ($\Sigma^M = 1$), either Northern firms or multinationals are making long-term positive profits and dynamic equilibrium is not possible. Therefore, we exclude those two inequality conditions and analyze only the equation stating that Northern and multinational firms should coexist.[4] It is illuminating to rewrite that condition as

$$\frac{1}{\{\rho + L_I^N[1 + \mu(1 - \Sigma^M)] + i\}(1 + \Gamma)} = \frac{w^{1-\varepsilon}}{\rho + L_I^N[1 + \mu(1 - \Sigma^M)]} \quad (\text{for } 0 < \Sigma^M < 1). \tag{17}$$

This condition shows that an increase in the imitation rate (i) or Southern relative cost premium (Γ) makes MNCs less profitable, *ceteris paribus*; while an increase in the Northern LBD spillover (μ) has the opposite effect.

As described in Baldwin and Forslid (2000), analyzing dynamic models of this kind requires choice of the state variable, numeraire and solution methodology. In a product-innovation model like ours, investment determines the rate of knowledge accumulation, which in turn determines the rate of growth. Thus, the natural state variable is the amount of resources devoted to investment. Since our model has only one primary factor, labor, it naturally becomes the numeraire and investment in Northern innovation (L_I^N) is the natural state variable.

Use equations (11) to (15) to express Σ^M, w, E, E^N and E^S in terms of the state variable L_I^N and parameters (Γ, i, μ, ϕ, ε, ρ, L^N and L^S) and substitute these variables into equation (16). Then the equilibrium level of investment in Northern innovation can be solved with equation (17) satisfied for $0 < \Sigma^M < 1$. The equilibrium growth rate g can also be solved by substituting Northern R&D investment (L_I^N) and the multinationalization rate (Σ^M) into equation (6).

---

[4]It is interesting to note that in the "corner solutions" in which there is either full Northern production or full technology transfer of varieties to the South, the rate of multinationalization has no impact on innovation or growth.

With appropriate substitutions, all other endogenous variables may be determined as a function of the wage differential, exogenous parameters and the state variable. However, the fundamental equation has the form $aw + bw^{-\varepsilon} + cw^{1-\varepsilon} = 0$, which has no general solution.[5] In turn, the primary solution cannot be used to compute comparative-static effects of parameter changes on central endogenous variables. Thus, after next presenting our extended model, we engage in simulation analysis in Section 5 to examine these relationships.

4. Southern Innovation and Reverse Knowledge Spillovers

Next we extend the model to permit investments in R&D by Southern firms. The introduction of Southern innovation and reverse knowledge spillovers alters the innovation sector directly. In addition to its own knowledge pool and LBD from the number of varieties produced locally, the North also absorbs knowledge from the South. The more knowledge the South creates and the more easily it flows to the North, the lower would be the unit labor cost for Northern innovation.

Like their Northern counterparts, Southern innovators learn from two sources, general knowledge and LBD. However, since the South is less developed, we assume its innovators are less productive in using knowledge, so that innovation functions are asymmetric. Specifically, we assume that Southern innovators cannot access Northern general knowledge, either because it is too difficult to reverse engineer or the IP protection in the North is too strong. Put differently, Southern innovators only use knowledge from the local pool, which consists of existing Southern inventions and knowledge that leaked out from MNCs to domestic imitators. As for LBD, we assume that Southern firms learn from production methods of Southern imitators and innovators only. That is, MNCs are capable of disguising their processes sufficiently to preclude that form of spillover directly to Southern rivals. Thus, LBD comes from the number of varieties the South

[5] In a technical appendix, which is available on request, we present the primary solution, which is a complex expression in the wage. We also discuss why there is no analytical solution for the wage gap for general values of the elasticity of substitution.

previously invented and is producing (s) and the number of varieties it imitates from MNCs (im).

Collecting ideas, the unit labor cost of Northern innovation (a_I^N) and Southern innovation (a_I^S) are as follows:

$$a_I^N = \frac{1}{K^N + K^M + \lambda K^S + \mu n} \quad (0 \leq \lambda \leq 1 \;\&\; \mu \geq 0) \tag{18}$$

$$a_I^S = \frac{1}{K^S + \frac{K^M}{1+\Gamma} + \theta(im + s)} \quad (\mu \geq \theta \geq 0). \tag{19}$$

Southern knowledge is defined to be only Southern-innovated varieties and not the ones imitated from MNCs. Parameter λ is the reverse knowledge spillover parameter that measures how easy it is for Northern innovators to observe and absorb Southern new knowledge. The Southern LBD spillover parameter (θ) is similar to the Northern LBD spillover parameter (μ) and measures the extent that innovators learn from watching local production processes.

The rates of knowledge accumulation for the North (g^N) and South (g^S) depend on R&D investment levels in each region, the unit labor costs of innovation, and the initial knowledge stocks. Define the South–North knowledge stock ratio to be $\Sigma^S = \frac{K^S}{K^N+K^M}$. The regional growth rates become

$$g^N = g^M = \frac{\Delta(K^N + K^M)}{K^N + K^M} = L_I^N[1 + \lambda\Sigma^S + \mu(1 - \Sigma^M)] \tag{20}$$

$$\begin{aligned} g^S = \frac{\Delta K^S}{K^S} &= \frac{L_I^S\left[\frac{K^M}{1+\Gamma} + K^S + \theta(im + s)\right]}{K^S} \\ &= \frac{L_I^S\left[\Sigma^M(1 + i\theta) + (1 + \Gamma)\Sigma^S(1 + \theta)\right]}{(1 + \Gamma)\Sigma^S}. \end{aligned} \tag{21}$$

Again, although the growth rates are written in endogenous variables, these variables can be expressed by exogenous parameters at the steady-state equilibrium so that the growth rates are time-invariant.

Consumption behavior is the same as in the benchmark model except that consumers also access unimitated Southern-innovated varieties and imitated Southern varieties. Accordingly, the global price index of the differentiated-goods sector now takes into account Southern goods:

$$P_X = \left(\int_0^{n+m+b+s} p_j^{1-\varepsilon} dj\right)^{\frac{1}{(1-\varepsilon)}} = \left(\frac{w}{\alpha}\right)^{1-\varepsilon} n + \left(\frac{1}{\alpha}\right)^{1-\varepsilon} (m - im) + im + \left(\frac{1}{\alpha}\right)^{1-\varepsilon} (s - is) + is. \quad (22)$$

The only difference in production lies in the South, which, in addition to multinational and imitated varieties, also produces its own innovated varieties. These enjoy the same instantaneous monopolistic profit as unimitated multinational varieties. Instantaneous profit for firms producing Northern (N), unimitated multinational (M), and unimitated Southern-innovated (S) varieties are as follows:

$$\pi^N = \left(\frac{w}{\alpha} - w\right) x^N = \left(\frac{1-\alpha}{\alpha}\right) \frac{w^{1-\varepsilon}\phi(E^N + E^S)}{\alpha^{-\varepsilon} \cdot P_X} \quad (23)$$

$$\pi^S = \pi^M = \left(\frac{1}{\alpha} - 1\right) x^M = \left(\frac{1-\alpha}{\alpha}\right) \frac{\phi(E^N + E^S)}{\alpha^{-\varepsilon} \cdot P_X}. \quad (24)$$

The FDI choice of firms in the North is based on the same condition as in the benchmark model (equation (9)).

The market-clearing conditions include two equations in each of the goods markets and labor markets. The amount of income in the South that can be spent in consumption is the sum of Southern workers' labor income and monopolistic profits of unimitated Southern innovative firms, excluding R&D investment. Spendable income in the North is the sum of Northern workers' labor income and monopolistic profits of Northern and unimitated multinational firms, excluding R&D expenditure. Northern labor is used to invest in R&D and produce Northern varieties, while Southern labor is now allocated to R&D as well as production of homogeneous products, unimitated multinational varieties, imitated varieties, and Southern varieties:

$$E^S = L^S + \pi^S(s - is) - L_I^S \quad (25)$$

$$E^N = wL^N + \pi^N n + \pi^M(m - im) - wL_I^N \quad (26)$$

$$L^N = L_I^N + n \cdot x^N \tag{27}$$

$$\begin{aligned} L^S &= L_Y + (m - im) \cdot x^M + (s - is) \cdot x^S \\ &\quad + (im + is) \cdot x^B + L_I^S. \end{aligned} \tag{28}$$

Finally, applying the free-entry condition, the expected operating profits of firms need to be the same as the cost of innovating a new variety:

$$\frac{\Pi^N}{wa_I^N} = \frac{\Pi^M}{(1+\Gamma)wa_I^N} = 1 = \frac{\Pi^S}{a_I^S}.$$

The question remaining is how to discount the expected profits (Π^M, Π^N and Π^S). All variables in the instantaneous operating profit functions (23) and (24) are time-invariant in steady-state equilibrium except for Σ^S and $(K^N + K^M)$. The Northern knowledge stock grows at a constant rate g^N at the dynamic equilibrium as before. Therefore, the system is solvable only when Σ^S is also time-invariant at the steady-state equilibrium. This implies that the growth rates of Southern and Northern knowledge are equal to each other in the steady-state equilibrium:

$$\begin{aligned} \frac{g^S}{g^N} = 1 \Rightarrow \frac{L_I^S\left[\Sigma^M(1+i\theta) + (1+\Gamma)\Sigma^S(1+\theta)\right]}{(1+\Gamma)\Sigma^S} \\ = L_I^N[1 + \lambda\Sigma^S + \mu(1 - \Sigma^M)]. \end{aligned} \tag{29}$$

In addition, since both multinational firms and innovative firms producing in the South suffer the risk of imitation in the local market, the free-entry conditions for multinational and Southern firms become

$$\frac{\frac{\pi^M}{\rho+g+i}}{(1+\Gamma)wa_I^N} = 1 \tag{30}$$

$$\frac{\frac{\pi^S}{\rho+g+i}}{a_I^S} = 1. \tag{31}$$

Note that the instantaneous profits for multinational and Southern firms are the same and they face the same risk of imitation.

It immediately follows that

$$\frac{1}{a_I^S} = \frac{1}{(1+\Gamma)wa_I^N}.$$

Because setting up a firm in a foreign (developing) country is more costly than doing so in the domestic (developed) country, the setup cost premium (Γ) is positive. Thus, the unit cost of innovation in the South (a_I^S) is always larger than that in the North (wa_I^N) in an equilibrium with FDI. This inefficiency means that expanded innovation in the South bears the potential to reduce growth.

Finally, expected profits must be discounted and the FDI choice of Northern firms is again characterized by the system (10). Now the following equation must hold for an interior solution of the multinationalization rate:

$$\frac{1}{\{\rho + L_I^N[1+\lambda\Sigma^S + \mu(1-\Sigma^M)] + i\}(1+\Gamma)}$$

$$= \frac{w^{1-\varepsilon}}{\rho + L_I^N[1+\lambda\Sigma^S + \mu(1-\Sigma^M)]} \quad (\text{for } 0 < \Sigma^M < 1). \quad (32)$$

This system of equations from (25) to (32) describes the steady-state equilibrium.

As with the benchmark case, the analytical solution of this extended model cannot be used to describe how parameter changes affect key variables. We therefore turn to simulation.

5. Simulation Analysis

The commodity-markets and labor-markets clearing conditions (11) to (15), free-entry condition (16), global growth rate in (6) and the FDI-choice condition (17) are simulated as a system for the benchmark model. For the extended model with Southern innovation, simulations of equations (25) to (32) describe the long-run, steady-state equilibrium.

We consider three cases for each simulation: the benchmark model, the situation in which Southern firms innovate but there is no reverse spillover ($\lambda = 0$), and the case with partial reverse spillover ($\lambda < 1$). We are interested in the impacts of Southern innovation

Table 2. Parameter values for simulation.

	Benchmark	Low	High
Reverse spillover (λ)	—	0.33	0.67
Southern imitation (i)	0.2	0	0.4
Relative investment cost (Γ)	0.65	0	1.3
Northern LBD (μ)	0.2	0	0.4
Southern LBD (θ)	0.13	0	0.2
Relative imitation risk (τ)	0.33	0.2	0.5
Time preference (ρ)	0.02	—	—
Elasticity of substitution (ε)	6.5	—	—
Consumption share of differentiated goods (Φ)	0.3	—	—
Northern labor	1	—	—
Southern labor	12	—	—

on multinationalization choices and the global growth rate as other policy-related parameters change. These parameters include the imitation rate (i), which can be interpreted as a measure of the strength of Southern patents, relative investment barriers (Γ), the Northern LBD spillover (μ), and the Southern LBD spillover (θ). The benchmark values of these parameters and the range over which they vary are taken from the literature or calculated from available statistics. We describe these calculations in the Appendix and the selected parameters are shown in Table 2.

The fundamental difference between models with and without Southern innovation lies in the role of the South. In the benchmark model, Northern firms select innovation and the multinationalization rate in light of the imitation threat and cost parameters. In the extended model, both Northern and Southern firms choose R&D investment levels in response to available knowledge and production processes from which they learn. The relative labor demands and the rate of multinationalization are codetermined by the two regions. Specifically, Southern innovation increases domestic labor demand for both innovation and production of Southern varieties. Thus, *ceteris paribus*, the rate of multinationalization in the extended model will be lower than that in the benchmark model because there is less labor available for international firms.

5.1. *Changes in the Imitation Rate*

In the benchmark model, the North is the only innovator but MNCs are subject to imitation risk. When the Southern imitation rate decreases, either because local firms are less productive at imitation or there is stronger IPP, this risk diminishes, encouraging more firms in the North to move their production abroad. This rise in multinationalization is shown by the dotted line in Figure 1, reading from right to left.

The impact on growth is determined by two offsetting forces. On the one hand, diminished imitation risk raises the profits of Northern innovative firms, inducing a rise in R&D investment. On the other hand, the rise in the proportion of firms moving production abroad decreases innovation efficiency by having fewer local production processes in the North for LBD. The net impact on the knowledge growth rate is governed by equation (6): $g = L_I^N[1 + \mu(1 - \Sigma^M)]$. This states that the overall effect depends on the relative strength of the increase in R&D investment (L_I^N) and the decrease in innovation efficiency

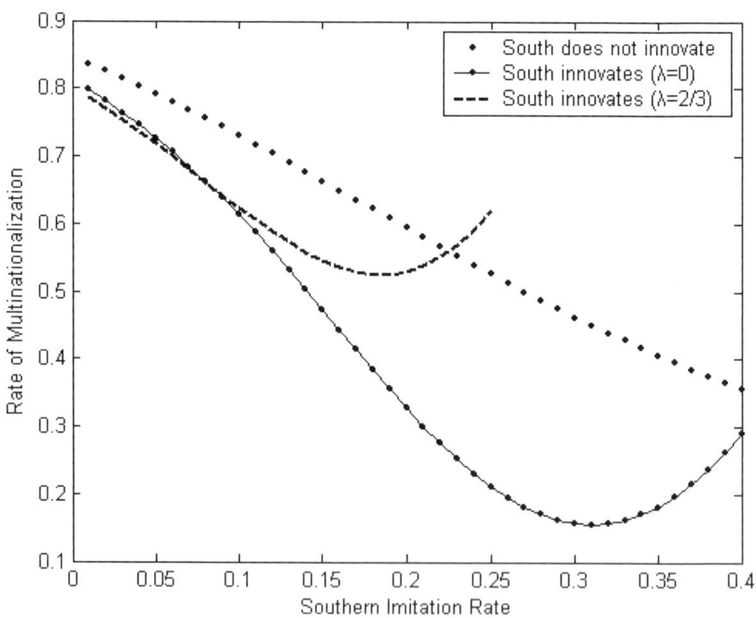

Figure 1. Southern imitation rate and the rate of multinationaliza-tion.

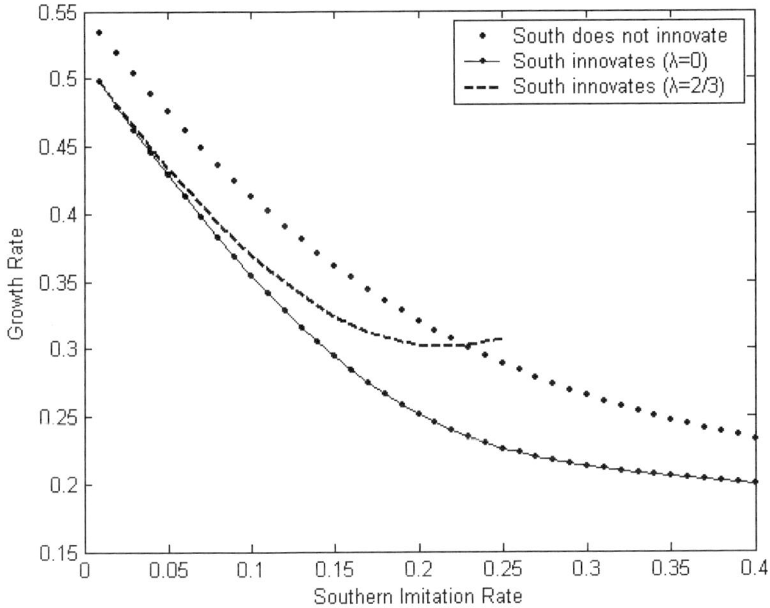

Figure 2. Southern imitation and growth.

from having diminished LBD $(1 - \Sigma^M)$. Our simulation shows that, as depicted by the dotted line in Figure 2, the effect of the change in R&D dominates so that the growth rate increases as the Southern imitation rate goes down.[6] Put differently, higher rates of FDI activity are associated with higher global variety growth. This finding is consistent with Lai's (1998) result that, in the presence of FDI, stronger Southern IPP tends to attract more firms to become multinationals and also boosts innovation and growth.

Finally, the higher FDI activity raises Southern labor demand, resulting in a narrowing wage gap as imitation falls, as indicated by Figure 3. Thus, in this benchmark model, stronger IPP increases the relative Southern wage through the induced impact on FDI.

Next, we introduce Southern innovation into the model without reverse spillover $(\lambda = 0)$. Focus initially on the impact of this change

[6]In fact, simulations across many parameter values show that this result always holds for an interior multinationalization rate.

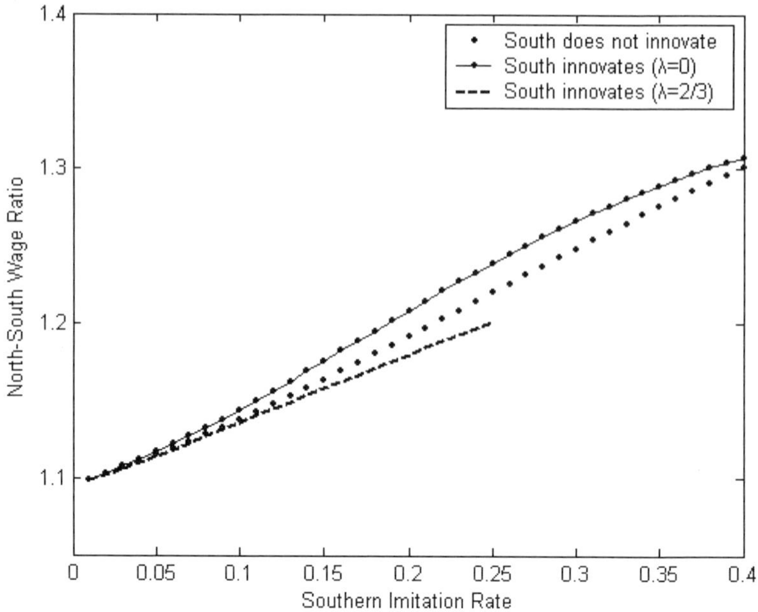

Figure 3. Southern imitation rate and North–South wage ratio.

on multinationalization, holding Southern imitation constant. Southern innovation will have two offsetting impacts. First, compared to the benchmark case, some Southern labor is allocated to innovation, squeezing out some multinationals and keeping more production in the North. This direct effect is offset by the fact that the appearance of new Southern varieties takes market share and profits from existing Northern varieties. This competition effect induces more Northern firms to become MNCs in order to exploit lower Southern production costs, raising the multinationalization rate.

Simulation shows that when the reverse spillover possibility is excluded the direct negative effect on FDI dominates so that the rate of multinationalization, shown by the connected-dot line in Figure 1, is lower than in the benchmark model for any imitation rate within the range depicted. It is possible, however, that at yet-higher imitation rates the competition effect would overwhelm the direct effect, generating a higher degree of multinational activity even with Southern innovation.

Note that there is a U-shaped relationship in Figure 1, where multinationalization first falls then rises with imitation. Higher imitation risk through, say, weaker IPP diminishes MNC profits and reduces the multinationalization rate, which dominates at lower imitation rates. On the other hand, an increase in the imitation rate also expands the Southern innovation ability through LBD, which makes Southern varieties expand faster. This effect would erode the profits of existing varieties and, since Northern varieties have a higher price, they would suffer relatively larger reductions in market share and profits. In this sense, more firms in the North would want to be MNCs as the imitation rate goes up. This impact dominates at higher imitation rates and MNC activity expands. To our knowledge, this is the first model that generates this non-monotonic effect of imitation on MNCs through the operation of Southern innovation.

Turning to Figure 2, we find that the impact of Southern innovation is always to reduce global variety growth at any level of imitation risk. The fact that knowledge growth rates are lower than in the benchmark reflects the inefficient specialization permitted by Southern innovation. Specifically, where there is reduced MNC activity, more labor is used in production in the North, decreasing R&D there. In this sense, some Southern labor is allocated to innovation and additional Northern labor to production. This substitution of labor in both regions expands activities in which each has a comparative disadvantage. As a consequence, global knowledge accumulates at a lower rate. Interestingly, even where there is increased multinationalization at higher imitation rates, the net effect is still to reduce global growth.[7]

Next, the introduction of Southern innovation increases local labor demand but this is more than offset by squeezing out MNCs, which raises labor demand in the North. Thus, the North–South wage gap becomes larger than that in the benchmark model, as noted by the connected-dot line in Figure 3.

We are also interested in how changes in the imitation rate affect Southern R&D investment, measured here by labor allocated to innovation. As shown by the connected-dot curve in Figure 4,

[7]Note that the global innovation curves converge at higher degrees of imitation.

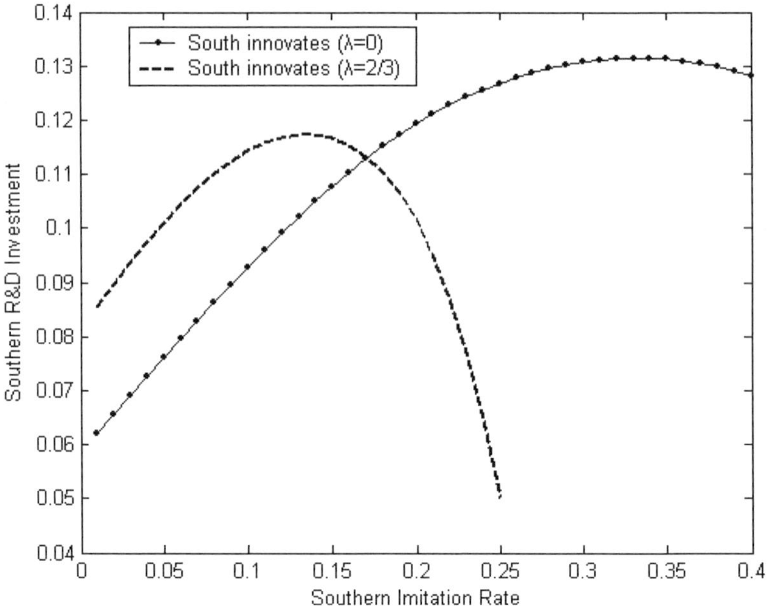

Figure 4. Southern imitation and Southern R&D investment.

the impact of higher imitation risk (weaker IPP) is non-monotonic: investment in Southern R&D first rises and then falls. This reflects the balance between changes in multinational activity and the growing numbers of imitated varieties. In fact, there are multiple processes in operation. First, a rise in imitation risk reduces (raises) MNC activity in the lower (higher) range of imitation rates. In turn, there is a fall (rise) in knowledge available to spill over to Southern innovation, reducing (raising) Southern R&D. Second, increases in i expand the efficiency of imitation, generating more imitated varieties per unit of both new MNC and Southern varieties. One impact is to generate more LBD, which expands Southern R&D investment. At the same time, however, greater imitation efficiency pushes more Southern labor into production of imitated varieties, which reduces labor available for R&D. Third, the higher imitation risk, which reflects weaker patent protection, would reduce the rents paid to new Southern-owned varieties. This would also limit investments in innovation in the South. This inverted-U relationship between patent rights and local innovation has been noted in other theoretical contexts (Park, 2008) but only in single-country models. It arises in our North–South model from a novel mix of imitation and innovation in the current framework.

Consider next a high degree of reverse spillover ($\lambda = 2/3$). In general, the same processes described above still hold. However, compared to the zero-spillover case, the new feature is that the South–North spillover increases innovation efficiency in the latter region. The immediate impact is that more new varieties would be generated, which establishes a higher global innovation growth rate, as noted by the dashed curve in Figure 2. As noted by the corresponding curve in Figure 1, the result of more Northern innovation is a higher degree of multinationalization as firms migrate to take advantage of cheaper production costs. And in Figure 3 we find that the North–South wage ratio is lowest in this case since Southern labor is in high demand from MNC production in addition to its allocation to R&D.

To examine the impacts of changing imitation risk in this scenario, first consider Figure 4, in which the dashed curve depicts Southern labor invested in R&D. We again get an inverted-U shape: R&D first rises, then falls as patents get weaker and imitation rates rise. We now find that Southern R&D investment bends downward at a lower imitation rate and it drops faster than in the no-spillover case.[8] The intuition is that the reverse spillover is sufficiently significant that a greater proportion of new Southern varieties face more rapid competition from Northern innovation per unit of time. Further, the expansion of Northern innovation induces a greater FDI flow, reducing Southern labor available for innovation. In combination with the growing risk from higher imitation, Southern firms find it less profitable to invest in R&D.

We note one potential policy implication in Figure 4. If the Southern government is interested in expanding local investments in innovation, our analysis suggests that it can be fairly lax in its IPP

[8]In fact, in the simulation, Southern innovation reaches zero at an imitation rate of approximately 0.25 and would become increasingly negative beyond that point. Negative innovation cannot be an equilibrium. Indeed, at this same point, the multinationalization rate would no longer be computed to lie within the interior solution range we consider in equation system (10). Beyond this rate of imitation, the remaining variables in Figures 1–3 are indeterminate with the high spillover. Note that the same cutoff would happen at a higher imitation rate (not shown) in the case of zero reverse spillover.

in the absence of any reverse spillover. However, if new Southern varieties easily leak back to the North, the need for stronger patent rights becomes paramount in encouraging local R&D.

Note in Figure 1 that the U-shaped relationship remains and that multinationalization starts to rise at a lower imitation risk compared to the case with no spillover. This indicates that the competition effect we described above is stronger with the reverse spillover. The intuition is that more new Southern varieties leak back into the North, generating a higher innovation rate and raising the share of firms becoming multinationals to take advantage of lower production costs.

A new feature is that, as shown by the dashed line in Figure 2, there is now a U-shaped relationship between the rate of imitation and variety growth. The reason is that when the multinationalization rate increases with a rising imitation rate (Figure 1), Southern innovation efficiency expands (even as overall investment in R&D falls in Figure 4) because there are both a higher level of general knowledge brought to the South and a larger LBD spillover. Moreover, the Northern innovation efficiency benefits through the reverse spillover. Finally, these high rates of multinationalization and imitation imply that the North transfers a larger share of its varieties to the South. In consequence, both regions engage more in activities in which they have greater relative advantage. Thus, the global growth rate can rise with weaker Southern IPP and may even surpass the level in the benchmark model.

To summarize this section, if the South only imitates, a low imitation rate (strong IPP) attracts more FDI, expands global innovation and promotes Southern imitation, so that both regions grow at a higher speed. The novelty here is to consider the effects of endogenous Southern innovation, with and without the reverse spillover. Our model demonstrates that the relationship between the rate of imitation and multinationalization becomes U-shaped. The intuition is that a high imitation rate implies rapid learning in the South, which expands Southern varieties and raises FDI incentives through a competition effect.

Concerning innovation growth, we find the following. With no reverse spillover, Southern innovation pushes labor in both regions to engage more in activities in which they have a comparative disadvantage. Thus, the growth rate is always lower than in the benchmark model. However, this impact can be reversed by a significant South–North spillover. In this case, the higher rate of multinationalization increases Southern innovation efficiency and also benefits Northern innovation. The result is to restore the efficiency loss from Southern innovation and generate higher knowledge growth.

5.2. *Changes in Other Parameters*

We briefly describe the results of other policy simulations.[9] As may be expected, a reduction in the Southern FDI investment cost (Γ), which in this model is a pure resource cost, raises the rate of multinationalization for any fixed imitation rate. The proportion of firms that become MNCs is always lowest in the case where Southern firms innovate but there is no South–North spillover. It becomes highest when there is such a spillover and the setup cost falls to low levels. A reduction in the investment cost also raises global innovation in all cases and raises the Southern wage relative to the Northern wage.

Next, a rise in the Northern learning spillover (μ) directly increases innovation efficiency in the benchmark model and expands both multinationalization and growth. Where there is a significant South–North spillover these impacts are magnified as Northern LBD improves. In essence, this situation expands the learning gain from the reverse spillover. Finally, increases in the Southern learning parameter (θ) diminish multinationalization and growth in both cases of Southern innovation. However, the FDI proportion and global innovation growth remain higher in the situation with a significant reverse spillover.

5.3. *Adding Imitation of Northern Exports*

Our assumption that Southern imitation targets only products made in multinational firms is stark. We relax it here by permitting also imitation of goods imported from the North, which generates an

[9]Precise results are available on request.

additional channel of North–South technology transfer.[10] Prior literature (Lai, 1998; Glass and Saggi, 2002) suggests that it is easier for Southern imitators to learn production techniques from local MNCs than from Northern exporters due to geographic proximity, demonstration effects and the potential for technical personnel to migrate to imitative firms. Thus, following Lai (1998), we assume that it takes longer to imitate Northern exports, implying that the Poisson arrival rate in imitating such varieties (i^e) is smaller than that of multinational goods (i^m).

In this setup, note that once the Northern variety is imitated, the exporter would abandon production since the price is driven down to the Southern imitator's marginal cost. An important implication is that imitated export varieties no longer contribute to Northern innovation through LBD. Further, in our extended model these varieties contribute to Southern innovation in the forms of both general knowledge and LBD.

We simulate a new case in this situation with a benchmark value of the export imitation parameter equal to either 20 percent or 33 percent of the FDI imitation rate, holding other parameters unchanged. Rather than presenting all outcomes we show differences between our earlier and new results when there is both Southern innovation and the South–North technology transfer.[11]

In Figure 5 the monotonically declining line (A1) is the same as our benchmark solution, with no Southern innovation and no export imitation. The lowest U-shaped curve (A2) repeats our scenario with Southern innovation and no reverse spillover, while the dashed curve (A3) repeats the situation with Southern innovation and a high rate of reverse spillover. The light-solid curve (A4) is the simulated result when we permit imitation to target both FDI and Northern exported goods with the 20 percent relative risk, while there is a significant reverse spillover ($\lambda = 2/3$). As may be seen,

[10] Equations are available in a technical appendix on request.

[11] There are no qualitative differences to report in the first two cases (no Southern innovation and Southern innovation with no reverse spillover) except to note that both the multinationalization and growth rates diminish marginally more rapidly as the imitation rate increases in the case of export-variety imitation threat.

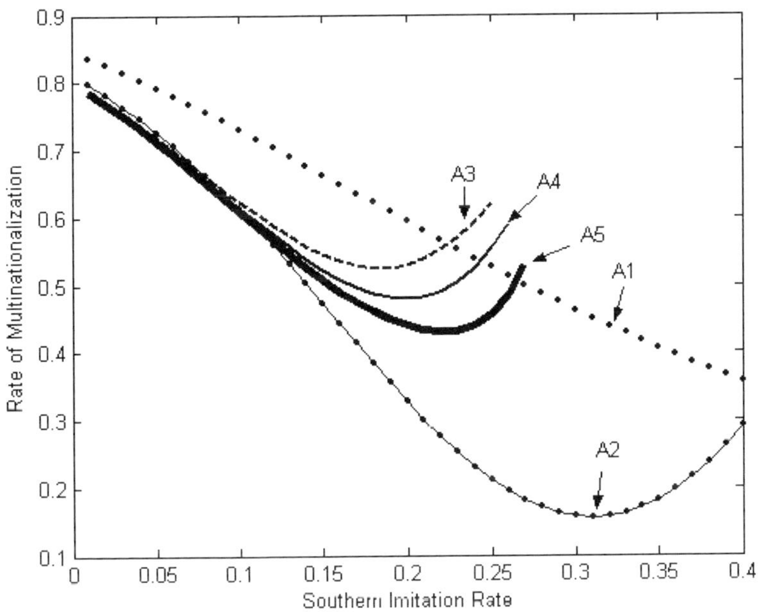

Figure 5. Multinationalization rate with export and FDI imitation.
Note: Dotted line (A1): Benchmark, no export imitation. Dash-dot line (A2): No spillover, no export imitation. Dashed line (A3): High reverse spillover, no export imitation. Light-solid line (A4): High reverse spillover with low export imitation. Dark-solid line (A5): High reverse spillover with medium export imitation.

this scenario also generates a U-shaped relationship between multinationalization rates and imitation propensities, suggesting that our model is robust to this change. The dark-solid curve (A5) repeats the analysis for the goods-imitation rate set at 33 percent of the FDI-imitation rate.

The fact that multinationalization rates are now lower than in the FDI-only imitation case reflects a balance between two forces. First, the second imitation channel raises competition for all Northern varieties, tending to push more firms to become multinationals. Second, the additional imitation reduces the Southern labor force available for production within multinationals. The second effect dominates in this scenario and reduces the multinationalization rate.

Finally, we note that in this case the innovation growth rate retains its U-shape with respect to imitation under our two channels.[12] However, the growth rates are smaller than in the case where imitation of exports is even less rapid (at 20 percent of the FDI imitation rate). This is broadly consistent with Lai's (1998) theoretical result that, with a sufficiently small export-imitation rate, a strengthened Southern patent regime raises Northern innovation.

6. Conclusion

We present a dynamic, general-equilibrium model in which the benchmark case permits endogenous FDI choice but there is no Southern innovation. In this case the more FDI activities that exist, the greater is the allocation of Southern labor to production and the greater is the allocation of Northern labor to innovation for any given imitation rate. This outcome represents an efficient allocation of labor in the two regions and generally maximizes the potential for global innovation growth. We find also in the model that as the Southern imitation rate declines there is an increase in multinationalization and higher global innovation. Thus, policy efforts in the South to reduce domestic imitation actually would expand international innovation.

When we next introduce the possibility of endogenous but high-cost Southern innovation, a complex set of tradeoffs emerges. A primary result is that the more Southern labor allocated to innovation, the slower is global growth because it implies relatively greater specialization by both regions in inefficient activities: the South in innovation and the North in production. In this case, however, there is a non-monotonic relationship between imitation and multinationalization. It is possible for increases in imitation risk (due perhaps to weaker patents) to expand MNC investment through a competition effect.

Next, a substantial spillover of Southern knowledge into improved Northern innovation implies at least partial reversal of the inefficient specialization. As a result, global innovation growth is intermediate between the no-spillover case and the benchmark case, and may

[12]The diagram for these simulations is available on request.

exceed the latter at high rates of imitation. A similar relationship arises for the rate of multinationalization. It is interesting that the combination of endogenous Southern innovation and knowledge spillovers to the North expands both FDI and global innovation at higher rates of imitation. In this sense the learning from South to North helps restore an efficient international resource allocation. Finally, our results remain intact when we permit a second channel of imitation through exports arriving in the South.

We emphasize that these results emerge in a complex and stylized model. However, they do highlight a possibility that has not been recognized widely in the literature. Specifically, technology transfer through multinational investment tends to go up in an environment of lower imitation risk, perhaps due to strengthened intellectual property protection. This multinationalization may kick off a process in the South in which local imitation and learning-by-doing establish the possibility of domestic innovation as the costs of R&D fall. In equilibrium, however, Southern innovation and investment in multinational subsidiaries must pay the same economic return and cover both the innovation costs and the FDI setup cost. This fact implies that Southern innovation remains expensive relative to its Northern counterpart. As a result, inefficient specialization can reduce the extent of FDI and international knowledge accumulation. To counter this inefficiency a Southern policy of strengthening IPP and reducing the costs of inward investment can expand multinationalization and growth, an effect enhanced by the reverse spillover. As noted in Figure 2, global innovation is maximized under weak Southern imitation risk.

Our analysis turns up another novel feature, however. Beyond a certain range, increases in the Southern imitation rate may raise the arrival rate of multinational enterprises due to a competition effect arising from lower-cost Southern innovation and the reverse technology spillover. This factor can reverse the impact of inefficient specialization and ultimately increase global innovation growth. In this context, the relationship between growth and patent protection is U-shaped. Thus, in cases where the reverse spillover from South to North is sufficiently strong, world innovation may benefit from a relaxation in local IPP.

References

Baldwin, R. E., H. Braconier and R. Forslid. 2005. Multinationals, Endogenous Growth and Technological Spillovers: Theory and Evidence. *Review of International Economics* 13: 945–963.

Baldwin, R. E. and R. Forslid. 2000. Trade Liberalization and Endogenous Growth: a Q-Theory Approach. *Journal of International Economics* 50: 497–517.

Barro, R. J. and J.-W. Lee. 2001. International Data on Educational Attainment: Updates and Implications. *Oxford Economic Papers* 53: 541–563.

Barro, R. J. and X. Sala-i-Martin. 1995. *Economic Growth* (New York: McGraw-Hill).

Benkard, L. 2000. The Dynamics of Aircraft Production. *American Economic Review* 90: 1034–1054.

Broda, C. and D. E. Weinstein. 2006. Globalization and the Gains from Variety. *Quarterly Journal of Economics* 121: 541–585.

Carr, D. L., J. R. Markusen and K. E. Maskus. 2001. Estimating the Knowledge-Capital Model of the Multinational Enterprise. *American Economic Review* 91: 693–708.

Chen, Y. and T. Puttitanun. 2005. Intellectual Property Rights and Innovation in Developing Countries. *Journal of Development Economics* 78: 474–493.

Chui, M., P. Levine and J. Pearlman. 2001. Winners and Losers in a North–South Model of Growth, Innovation and Product Cycles. *Journal of Development Economics* 65: 333–365.

Cooper, R. and A. Johri. 2002. Learning-by-Doing and Aggregate Fluctuations. *Journal of Monetary Economics* 49: 1539–1566.

Glass, A. J. and K. Saggi. 2002. Intellectual Property Rights and Foreign Direct Investment. *Journal of International Economics* 56: 387–410.

Glass, A. J. and X. Wu. 2007. Intellectual Property Rights and Quality Improvement. *Journal of Development Economics* 82: 393–415.

Grossman, G. M. and E. Helpman. 1991a. *Innovation and Growth in the Global Economy* (London: MIT Press).

Grossman, G. M. and E. Helpman. 1991b. Trade, Knowledge Spillovers, and Growth. *European Economic Review* 35: 517–526.

Helpman, E. 1993. Innovation, Imitation, and Intellectual Property Rights. *Econometrica* 61: 1247–1280.

Irwin, D. A. and P. J. Klenow. 1994. Learning-by-Doing Spillovers in the Semiconductor Industry. *Journal of Political Economy* 102: 1200–1227.

Jaffe, A. B., M. Trajtenberg and R. Henderson. 1993. Geographic Localization of Knowledge Spillovers as Evidenced by Patent Citations. *Quarterly Journal of Economics* 108: 577–598.

Keller, W. 2002. Geographic Localization of International Technology Diffusion. *American Economic Review* 92: 120–142.

Keller, W. 2010. International Trade, Foreign Direct Investment, and Technology Spillovers. In B. Hall and N. Rosenberg, editors, *Handbook of the Economics of Innovation* (Amsterdam: Elsevier-North Holland).

Keller, W. and S. R. Yeaple. 2009. Multinational Enterprises, International Trade, and Technology Diffusion: a Firm-Level Analysis of the Productivity Effects of Foreign Competition in the United States. *Review of Economics and Statistics* 91: 821–831.

Krugman, P. R. 1979. A Model of Innovation, Technology Transfer, and the World Distribution of Income. *Journal of Political Economy* 87: 253–265.

Lai, E. L.-C. 1998. International Intellectual Property Rights Protection and the Rate of Product Innovation. *Journal of Development Economics* 55: 133–153.

Linden, G. 2003. Optical Storage in China: a Study in Strategic Industrial Policy. ISIC Report 2003–01, the Information Storage Industry Center, Graduate School of International Relations and Pacific Studies, University of California.

Maskus, K. E. 2004. Encouraging International Technology Transfer. Issue Paper number 7, UNCTAD-ICTSD Project on Intellectual Property Rights and Sustainable Development, International Centre for Trade and Sustainable Development, Geneva.

Maskus, K. E. and M. Penubarti. 1995. How Trade-Related are Intellectual Property Rights? *Journal of International Economics* 39: 227–248.

Organization of Economic Co-operation and Development. 2005. *OECD Science, Technology and Industry Outlook* (Paris: OECD Publishing).

Park, W. G. 2008. Intellectual Property Rights and International Innovation. In K.E. Maskus, editor, *Intellectual Property, Growth and Trade: Frontiers of Economics and Globalization* (Amsterdam: Elsevier-North Holland).

Park, W. G. and S. Wagh. 2002. Index of Patent Rights. In *Economic Freedom of the World: 2002 Annual Report*, 233–241 (Fraser Institute: Calgary, Canada).

Segerstrom, P. S., T. C. A. Anant and E. Dinopolous. 1990. A Schumpeterian Model of the Product Life Cycle. *American Economic Review* 80: 1077–1091.

Sjöholm, F. 1996. International Transfer of Knowledge: the Role of International Trade and Geographic Proximity. *WeltwirtschaftlichesArchiv* 132: 97–115.

Vernon, R. 1966. International Investment and International Trade in the Product Cycle. *Quarterly Journal of Economics* 80: 190–207.

World Intellectual Property Organization. 2007. *WIPO Patent Report: Statistics on Worldwide Patent Activity* (Geneva: WIPO).

Xu, J.-L. 2006. Analyzing China's Patent Strategies. In *Guangming Daily* (Beijing, China, Guangming Daily Pressing, 16 February).

Appendix: Parameter Values for Simulation

The reverse knowledge spillover from South to North (λ) may be measured by the geographic pattern of learning. Jaffe *et al.* (1993) find that 60–75 percent of knowledge through citations of U.S. patent in the 1970s and 1980s is learned by domestic firms while 25–40 percent is absorbed by foreign firms. Thus, we take the value of our knowledge spillover parameter to be either 1/3 or 2/3.

Another important parameter is the Southern rate of imitation (i). We calculate it from two components: the relative strength of IPP and the relative learning ability of the FDI-recipient country. The stronger are patent rights and the lower is the learning ability, the lower would be the imitation rate. The strength of IPP is calculated from the Ginarte–Park (GP) patent index in the year 2000 (Park and Wagh, 2002). The average index in a selection of developing countries was 2.84 while that of developed regions was 4.05. Thus, the strength of IPP in developing regions was about 70 percent of the level in developed regions, suggesting an imitation margin of 30 percent. The learning ability of developing countries relative to developed nations is calculated from data on the education attainment of total population of age 25 and over in Barro and Lee (2001). The average years of schooling in developed nations was 9.4 and that in developing countries was 6.1. Using these data, workers in developing countries have about 65 percent of the learning ability of those in developed regions. Multiplying the relative imitation possibility from the weakness of IPP (0.3) and the relative average learning ability (0.65), developing countries are calibrated to have an imitation rate of 0.2 in the benchmark. Our lower bound for this parameter is 0.0, while we take 0.4 as the upper bound. These values are used also for the imitation rate of multinational varieties (i^m) in the extended model. The parameter τ, which captures the proportionately lower imitation rate for Northern exports than for multinationals ($i^e = \tau i^m$), takes a benchmark value of 0.2 and we also consider 0.0 and 0.33.

The relative investment barrier in the South (Γ) is calculated from the investment cost index developed by Carr *et al.* (2001). The index measures the cost of investing in a country by averaging several indexes of impediments to foreign operations as reported in the *World Competitiveness Report* published by the World Economic Forum. In our simulations, Γ is defined as the percentage difference in the average investment cost index in developing countries compared to that in developed countries. This difference suggested that the average additional investment cost was around 65 percent and we take our benchmark value to be 0.65. The minimum is set to zero and the maximum is set to 1.3, which doubles the benchmark value.

The value of the Northern LBD spillover parameter (μ) is taken from the literature. Irwin and Klenow (1994) estimate the LBD coefficient to be 0.2 in the semiconductor industry. Benkard (2000) concludes that if knowledge does not depreciate then the learning rate is roughly 18 percent in the commercial aircraft industry. Cooper and Johri (2002) present estimates of coefficients for LBD spillovers (measured by various output-based measures of organizational capital) ranging from 0.22 to 0.38 on the plant level. Therefore, we set our benchmark value of Northern LBD to 0.2 and allow it to change between 0.0 and 0.4.

The LBD spillover in the South (θ) reflects the learning ability of workers from watching local production processes. Recalling that workers in developing countries have about 65 percent of the learning ability of workers in developed countries, the benchmark value of the LBD spillover in the South is set to be 0.13, which is 65 percent of the benchmark value of the Northern LBD spillover. We permit the Southern parameter to range from 0.0 to 0.2, implying that the South could have no LBD or it could enjoy the same LBD as their Northern counterpart.

The remaining parameters do not have direct policy relevance and we simply fix them at a benchmark value. Based on Broda and Weinstein's (2006) finding that the mean elasticity of substitution at the Standard International Trade Classification (SITC) 5-digit level is 6.6, the elasticity of substitution between differentiated varieties (ε) is set to 6.5. The time preference parameter (ρ) is set to be 0.02, suggested by Barro and Sala-i-Martin (1995). The global expenditure share of manufacturing goods (Φ) is assumed to equal the value added of industrial goods divided by GDP. Taken from the 2007 *World Development Indicators* database of the World Bank, the value of this ratio for high-income and middle-income regions is calculated to be 0.3. Finally, define the "North" to be comprised of countries with ratios of gross enterprise R&D to GDP greater than 2.0 percent in 2007, taken from the OECD's *SourceOECD* online database.

This amounts to a population of approximately 650 million in 2008, leaving a population in the "South" of around 6.51 billion. This ratio of 10 to 1 is not quite sufficient to support innovation, multinational production, and local production in the South in steady-state equilibrium. Thus, we raise the benchmark value of Southern population to 12 (with the Northern population at 1) to permit comparing the benchmark with extended models.

All of our parameter values are summarized in Table 2.

Chapter 10

How National and International Financial Development Affect Industrial R&D*

Keith E. Maskus, Rebecca Neumann and Tobias Seidel[†]

1. Introduction

One of the most fundamental issues in economic development is how important the maturation and deepening of financial markets are for growth. This issue has been much debated since its reintroduction into the literature by King and Levine (1993). According to Levine (2005), a consensus has been reached that more developed domestic financial markets stimulate economic growth. Presumably this linkage stems from firms gaining better access to credit, which allows easier financing of investment projects, in turn leading to productivity growth. At a deeper level, however, it is important to study the specific channels through which access to finance enhances economic growth. In this paper we consider one particular type of investment, namely research and development (R&D), and its association with national and international financial market development. It is prominently argued in the literature that more investment in R&D stimulates technological progress, which provides the foundation for the bulk of economic growth (Griliches, 1998; Acemoglu, 2009).

*Reprinted from *European Economic Review*, Vol. 56, No. 1, pp. 72–83. Copyright 2012, with permission from Elsevier.
[†]Much of this research was conducted while Neumann and Seidel were visiting scholars at the Department of Economics at the University of Colorado. Seidel gratefully acknowledges financial support from the Fritz Thyssen Foundation. We are grateful to two referees for insightful comments.

Our specific focus is to study how endogenous research intensities, defined as industry-level R&D as a share of output, respond to country-wide financial development in industries that rely heavily on external finance or have limited tangible assets. This approach sheds light on whether financially dependent companies are more likely to have higher R&D intensities when national capital markets develop or when there is greater access to international financial markets. If so, financial development and financial access provide key reasons why enterprises in countries with deeper markets invest more in R&D than do their counterparts in other nations. Additionally, industries with low proportions of tangible capital tend to be those with higher proportions of intellectual capital. If the R&D intensities of those sectors also rise with capital-market sophistication then financial development supports higher innovation in knowledge-based endeavours.

Prior work on financing investments at the firm level helps motivate our study (Aghion *et al.*, 2004; Hall and Lerner, 2010). This research has demonstrated that firms first tap internal funds in order to maintain control rights over their innovations. As they need additional capital to fund R&D expenditures, however, they turn to external funds, first accessing bank credit and then equity markets.[1] This apparent hierarchy of finance specific to innovation motivates our explicit consideration of both credit and equity measures, such as liquid liabilities, private credit, stock market and private bond market capitalization.

We also include variables that describe international financial market development (portfolio investment, foreign direct investment (FDI), and foreign debt), which has been ignored in this line of research. This is an important consideration as recent work has documented a positive link between the integration of financial markets across countries and economic growth (Eichengreen, 2001; Kose *et al.*, 2009). Nevertheless, very little focus has been placed on whether openness to international capital markets may affect innovation and growth through financing R&D. The standard conception is that FDI is a direct source of technology transfer and may induce greater local

[1]See Myers (1984), Myers and Majluf (1984) and Leland and Pyle (1977) regarding this pecking-order logic for general investments. Harris and Raviv (1991) offer a useful review.

innovation through learning spillovers (He and Maskus, 2012; Keller and Yeaple, 2010). However, there has been considerable growth in the internationalization of R&D as multinational firms establish research affiliates abroad (Gammeltoft, 2006). Further, affiliates may be financed via a mix of ownership and debt (Kesternich and Schnitzer, 2010). It is therefore of interest to study how access to international financial sources may affect innovation.

Financing constraints may be particularly restrictive for R&D relative to other forms of investment.[2] According to Brown *et al.* (2009), innovative firms (those with high R&D expenditures) tend to have few tangible assets that can serve as collateral for credit. R&D expenditures largely go to salaries and wages for scientists and researchers. These human-capital investments cannot be collateralized. Further, firms may wish to protect their proprietary information over innovation, and thus may be unable or unwilling to offer sufficient signals about the effectiveness of their intended R&D programs to credit providers. These issues motivate the inclusion of a measure of asset tangibility into our analysis.

The relationships between innovation and economic growth, and between financial development and economic growth, have been explored in a wide swath of literature. However, there are far fewer studies of the effects of financial development on innovation as a specific channel by which the former may stimulate growth. Aghion *et al.* (2010) provide one such link with a theoretical model that explores the impact of financing constraints on the composition of investment. Financial frictions may limit economic growth by reducing long-term investment in R&D during economic downturns. Aghion *et al.* (2008) provide related firm-level evidence. They show that R&D investment is pro-cyclical, rising with firms' sales in the presence of firm-specific credit constraints. This link is particularly pronounced for enterprises with greater external financial dependence and fewer tangible assets.

Using aggregate data, Carlin and Mayer (2003) study the relationship between R&D intensity at the industry level and national institutional variables describing the structure of countries' financial

[2]See Hall (1992), Himmelberg and Petersen (1994), Mulkay *et al.* (2001), and Bond *et al.* (2005).

systems. In particular, these authors interact accounting standards, bank concentration and control of voting rights with equity finance dependence, bank finance dependence, and skill dependence to look at the effects on growth, fixed investment and R&D investment. They provide initial evidence on the relationship between domestic financial institutions and R&D expenditures, showing broadly that better accounting standards and more developed credit markets positively impact investment in R&D for those industries that rely more on external equity. They find little impact on R&D from the development of equity markets.

Our paper contributes to this literature in several ways. First, we deploy alternative measures of financial development to describe more fully the impacts on R&D intensities of such different dimensions as private credit, stock markets, private bond markets, portfolio investments and foreign direct investment. Second, we derive novel findings based on two classifications of the data: a differentiation between national and international sources of finance, and a categorization of financial systems as bank-based or market-based. Third, we calculate the implied impacts of financial development on R&D propensities.

To preview, our findings point to a strong association between domestic financial market development and R&D intensity. However, among sources from the international capital markets, only FDI seems to be a major factor in financing research and development. We find significant effects for varieties of both bank-based and market-based financial systems. However, we obtain the strongest effect for market-based (direct) forms of finance. Quantitatively, the R&D intensity of an industry that is heavily dependent on external sources of finance in a country with well developed domestic financial markets is more than 0.4 percentage points higher than that of an industry that relies less on external sources of funding in a country with poorly developed financial markets. This difference is about 20 percent of the average R&D intensity in our industry sample. The largest difference arises as the private bond market becomes larger, with our estimates suggesting as much as a 45-percent expansion of R&D intensity in industries located in a country with extensive bond-market capitalization versus those located in one with little capitalization.

A fourth contribution is that, to our knowledge, this is the first paper to include a measure of asset tangibility in the context of financial market development and R&D investments. We find that industries with fewer tangible assets generally benefit more from financial market development than industries that are endowed with more tangible assets.

In the next section we formally introduce the hypotheses we test and lay out the econometric methodology to do so. We then describe the data we use in Section 3 before we discuss our results in Section 4. In Section 5 we report the results of several robustness checks of our benchmark specification. We offer concluding remarks in Section 6.

2. Hypotheses and Econometric Approach

We study the different impacts of financial market development on R&D intensity in industries that (i) depend on external finance to different extents and (ii) are characterized by varying degrees of tangible assets in their overall balance sheets. Our hypotheses are that more developed financial markets should be associated with greater R&D in industries that (i) rely more on external finance and (ii) have less tangible assets to use as collateral.

The first industry characteristic — external financial dependence — captures the industry's ability to generate cash flow to finance investment projects, such as R&D. As Rajan and Zingales (1998) argue, this measure should be determined to a large extent by industry-specific factors that are external to the firm. For instance, it is plausible to argue that projects differ across industries with respect to "initial project scale, the gestation period, the cash harvest period, and the requirement for continuing investment" (Rajan and Zingales, 1998, p. 563). In our regressions, we expect a positive coefficient on the interaction term between financial development and sectoral external financial dependence.

The idea behind the second industry characteristic — asset tangibility — is that industries with a high degree of tangible assets, such as equipment and plants, should be able to access credit more easily, and at lower cost, because they have more collateral with which to guarantee such credit. Hence, we expect a negative coefficient on the interaction term between financial development and asset tangibility because industries with a greater presence of intangible assets, such as intellectual capital, should gain relatively greater access to capital as financial markets deepen. Like external dependence, tangibility is an industry characteristic largely external to the firm.

The approach we use follows the influential work by Rajan and Zingales (1998) in interacting industry variables (external financial dependence and asset tangibility) with several country variables capturing the degree of financial market development.[3] We thus regress R&D intensity on these interaction terms, the industry share in GDP, financial market development, and a set of country, industry and year dummies. The model we estimate is

$$\begin{aligned} R\&D\ intensity_{j,k,t} &= \beta_0 + \beta_1(external\ financial\ dependence_k \\ &\quad \times financial\ market\ development_{j,t}) \\ &\quad + \beta_2(tangibility_k \times financial\ market\ development_{j,t}) \\ &\quad + \beta_3(industry\ share_{j,k,t}) \\ &\quad + \beta_4(financial\ market\ development_{j,t}) \\ &\quad + \eta_j + \eta_k + \eta_t + \varepsilon_{j,k,t}, \end{aligned} \tag{1}$$

where j denotes country, k represents industry, t is the year index, and $\varepsilon_{j,k,t}$ describes the error term. As Rajan and Zingales (1998) point out, the interaction terms deliver predictions about within-country differences across industries after controlling for country-, industry- and year-specific effects. We do not estimate the direct effect of external financial dependence or tangibility as they are captured by our industry indicators.

Time-invariant country characteristics are controlled for by our country indicator η_j. The year dummies capture all effects that countries and industries face in common, such as global booms and busts and international trade liberalization. Furthermore, we use figures on industry output shares to control for different industry patterns

[3]Those authors interact indicators of financial development with external financial dependence to explore the relationship between financial development and economic growth.

across countries. This is an important inclusion since the industry shares control for national variations in comparative advantage that could affect the relationships between industry characteristics, financial development and R&D intensity. In contrast to Rajan and Zingales (1998), and more closely related to Manova (2008), we do not take average annual growth rates but instead use the annual levels for the years 1990–2003. Hence, identification of β_1 and β_2 results from variation across countries and time of the financial development indicators and variation across industries in external financial dependence and asset tangibility. According to our hypotheses, we expect β_1 to be positive and β_2 to be negative.

There are potential concerns about endogeneity and causality in equation (1), which we now address. As a basic observation, it is difficult to envision factors that would be general determinants of financial development and also drive variations in R&D intensity at the industry level. In addition, there are several technical reasons for arguing that causality problems are likely less important in our context. Note first that asset tangibility and external dependence measures are industry characteristics that do not vary over time or across nations. Moreover, they are calculated using U.S. data from 1990–1999. Thus, time-varying R&D expenditures at the industry level, which change annually over 1990–2003 for each country in the sample, are not likely to be causal for asset tangibility or external dependence computed for U.S industries. Still, we leave the United States out of the regression analysis to eliminate any feedback effect from U.S. R&D expenditures to these industry variables.

Second, financial development is endogenous to at least macroeconomic factors, which has been a paramount question in the literature analyzing the impacts of market deepening on growth. For example, the extent of economic expansion surely influences the demand for financial instruments and the capacity to invest in national financial development. In turn, any measures of financial development, including our own, are affected by other factors linking growth to financial market structure. In prior work these influences are often instrumented by such variables as an index for the rule of law, efficient governance, and bank regulation. In our case, however, we argue that this concern is considerably lessened since it seems unlikely that R&D intensity at the industry level is causal to the development of broad financial markets (though it might be for narrower measures, such as venture capital).

Third, the interaction terms between financial development and industry characteristics could suffer from a potential causality problem. Specifically, it is reasonable to suppose that the more developed are financial markets the less problematic are unobservable credit constraints, permitting firms to raise external finance without frictions. Suppose also that industries requiring higher amounts of outside capital also happen to have better investment prospects than those with lower dependence on external finance. Then the data will show both high national financial development and greater R&D propensities in sectors with high external dependence. A positive interaction coefficient then might not demonstrate a causal impact of credit constraints, which in this situation are absent in financially developed economies.[4]

As a practical matter this issue is a concern only if external dependence and R&D intensities become more correlated as credit constraints are relaxed. We check this possibility by ranking the six countries with the highest, and six with the lowest, indicators for each measure of financial development, then computing correlations between R&D ratios and external dependence across these country groups. These correlations are all positive and range from 0.24 to 0.44. However, there is no systematic pattern across types of financial development and in most cases there is no significant difference between the higher-ranked nations and the lower-ranked nations. Thus, there is no evidence in the data that sectors with higher external dependence also are more R&D-intensive in countries with well-developed financial markets.

With respect to asset tangibility, in our data this variable is negatively correlated with external dependence at the industry level. Thus, the concern expressed in the prior paragraphs does not exist, for if firms in industries with fewer tangible assets have better innovation opportunities, and therefore higher R&D intensities, this situation would not be conflated with the absence of credit constraints in

[4]We are grateful to a referee for pointing this out.

highly financially developed economies. Our hypothesized negative coefficients on the interaction terms between asset tangibility and financial development would imply that the availability of collateral does matter for the ability to raise external capital, suggesting there is a causal effect of credit constraints on R&D. Nonetheless, we compute the correlations between asset tangibility and R&D intensities in our high-development and low-development sub-samples. These correlation coefficients are all negative and vary in a narrow range between -0.31 and -0.48. More importantly, there is again no systematic pattern in these correlations across countries grouped by levels of financial development.

For all these reasons, we argue that it is unnecessary in our specification to deploy an instrumental-variables approach. However, we undertake several robustness checks to raise confidence in our results. For example, using lagged financial development variables in our baseline specification gave nearly identical results to those discussed below.

3. Data

We use R&D expenditure data for 22 manufacturing industries in 18 OECD countries for the years 1990 to 2003.[5] R&D intensity is calculated from OECD data as industry-level R&D expenditures as a share of industry output in each country.[6] Industry share is defined as each industry's production relative to GDP. GDP data are from the World Bank's *World Development Indicators 2007*.

We use the measure of external financial dependence from Klapper *et al.* (2006) defined as the industry-level median (across firms) of the ratio of capital expenditures minus cash flow from operations divided by capital expenditures.[7] The idea is that higher cash flow makes a firm less dependent on external finance. This proxy

[5]The countries we consider are Australia, Belgium, Canada, Czech Republic, Denmark, Finland, France, Germany, Ireland, Italy, Japan, Korea, Netherlands, Norway, Poland, Spain, Sweden, and the United Kingdom. For more detailed exposition of the data sources and descriptive statistics, see Maskus, Neumann, and Seidel (2011).

[6]An alternative measure, R&D expenditures as a share of value added, provides similar results.

[7]We thank Luc Laeven for providing us with the external financial dependence data.

is calculated for each industry with figures for U.S. companies over 1990–1999 using Standard and Poor's Compustat database. Following Braun (2003), we calculate tangibility as each industry's share of physical assets in total capital stock using U.S. data from Compustat.[8] The measures for both external dependence and tangibility are calculated from U.S. data with the idea that they capture technological differences across industries and can thus be used to rank industries. Rajan and Zingales (1998) provide the central argument for using U.S. data. Specifically, because the United States has well-developed financial markets, its levels of sectoral external dependence and tangibility are those that would exist under complete markets. Hence, these measures provide proxies for external dependence and tangibility that industries in other countries would achieve in the absence of other financial market frictions. It is important to note, however, that it is the ranking rather than the level that matters for identifying the coefficients in our regressions. Technological differences within the same industry across countries are likely to be small so that U.S. data can be used as a proxy for the ranking of industries in other countries.

Indicators for domestic financial development are taken from the World Bank's *Financial Structure Database 2007* and comprise liquid liabilities, private credit by deposit money banks, stock-market capitalization, and private bond-market capitalization. The international variables are taken from Lane and Milesi-Ferretti (2006) and include total external assets and liabilities, portfolio equity, FDI equity, and foreign debt. All financial development indicators are expressed relative to GDP.

4. Results

We discuss the role of domestic and international financial markets sequentially in the following two subsections, focusing attention on our two main hypotheses: (i) R&D intensity should be higher in industries that depend more on external finance in countries with

---

[8]We thank Nathalie Moyen for assistance with the Compustat data to generate the tangibility measure.

more developed financial markets, and (ii) R&D intensity should be lower in industries that have more tangible assets in countries with more developed financial markets. Hence, estimates of β_1 should be positive while estimates of β_2 should be negative.

4.1. *Domestic Financial Market Development*

Table 1 reports baseline results for specification (1) using liquid liabilities, private credit, stock-market capitalization, and private bond-market capitalization (all relative to GDP) as indicators of domestic financial market development. Each of these variables is interacted with external dependence and tangibility. We observe first that β_1, the coefficient on the interaction term using external dependence, has a positive sign and is statistically significant in all cases. Second, the coefficient on the tangibility interaction term, β_2, has a negative sign and is significant in all cases, except for stock-market capitalization. Generally, the statistical significance is lower for the tangibility interactions than for the external financial dependence interactions. Thus, we find broad support for our two hypotheses.

Third, the estimates for external dependence interacted with liquid liabilities, private credit, or stock-market capitalization are very similar in magnitude across specifications. This is true as well for asset tangibility interacted with liquid liabilities and private credit. The sizes of the private bond market coefficients differ from these levels and exert the biggest effect on R&D intensity. The estimate is larger by a factor of three to four for external dependence and by a factor of two for tangibility. Finally, it is notable that we do not find a statistically significant direct effect of financial market development on R&D intensity. For completeness we retain the level variables in the regressions but note that the coefficients on the interaction terms are little changed by leaving out these direct effects.

To gauge the economic significance of the estimates, we undertake the following thought experiment. We examine by how much the R&D intensity in the 75th percentile of external dependence (furniture; manufacturing n.e.c.) exceeds the R&D intensity in the 25th percentile of external dependence (pulp, paper and paper products) if the respective financial development measure is raised from the

Table 1. Regression of R&D intensity on domestic financial development interacted with external financial dependence and tangibility. The table illustrates OLS panel regression results for 22 manufacturing industries in 18 OECD countries for the years 1990–2003. Each column represents a separate regression using a different measure of domestic financial market development (as indicated in the first row) interacted with both external financial dependence and asset tangibility. The dependent variable is R&D intensity calculated as R&D expenditures as a share of industry output. Industry share in GDP is defined as industry production relative to GDP. Robust standard errors are in parentheses: * significant at 10%; ** significant at 5%; *** significant at 1%.

	Liquid liabilities	Private credit	Stock market capitalization	Private bond market capitalization
Industry share in GDP	−0.1347 (0.0189)***	−0.1386 (0.0193)***	−0.138 (0.0210)***	−0.1331 (0.0183)***
Liquid liabilities	0.0039 (0.0086)			
External dependence* Liquid liabilities	0.0175 (0.0049)***			
Tangibility* Liquid liabilities	−0.0193 (0.0111)*			
Private credit		0.0025 (0.0059)		
External dependence* Private credit		0.0183 (0.0052)***		
Tangibility* Private credit		−0.0243 (0.0127)*		
Stock market capitalization			−0.0044 (0.0046)	
External dependence* Stock market capitalization			0.0233 (0.0057)***	
Tangibility* Stock market capitalization			−0.0047 (0.0119)	
Private bond market capitalization				0.0088 (0.013)

(*Continued*)

Table 1. (*Continued*)

	Liquid liabilities	Private credit	Stock market capitalization	Private bond market capitalization
External dependence* Private bond market cap				0.0674 (0.0069)***
Tangibility* Private bond market cap				−0.0375 (0.0112)***
Constant	0.0172 (0.0051)***	0.0195 (0.0041)***	0.0155 (0.0034)***	0.0147 (0.0034)***
Country-, Industry- and Year-Dummies	YES	YES	YES	YES
Observations	3643	3643	3667	3626
R-squared	0.5	0.5	0.48	0.49

25th to the 75th percentile of the distribution. Put differently, we ask by how much the R&D intensity of a high-dependence sector would exceed that of a low-dependence sector if both were moved from a country with low financial development to one with high financial development. We do the same comparison for physical-capital tangibility by comparing the industry at the 75th percentile of tangibility (rubber and plastics products) with the one at the 25th percentile of tangibility (chemicals and chemical products).

Table 2 provides a summary of the economic magnitudes for the financial measures that are significant in the baseline regressions, computing the implied percentage-point change in R&D intensity and relating that value to the average R&D intensity. It is evident that the estimated effects can be quite large, with a difference in R&D intensity of up to 45 percent of the mean R&D intensity for the interaction between the size of the private bond market and external financial dependence. Specifically, the industry at the 75th percentile of external dependence has an R&D intensity that is about one percentage point higher in France (the country at the 75th percentile of bond-market capitalization) than does the industry at the

Table 2. Magnitudes of estimates. The first line reports the difference in R&D intensity in percentage points between an industry at the 75th percentile of external financial dependence in a country at the 75th percentile of the respective financial development and an industry at the 25th percentile of external financial dependence in a country at the 25th percentile of financial development. The second line relates the percentage point difference to the mean R&D intensity. Lines 3 and 4 show magnitudes for the same thought experiment undertaken for asset tangibility.

Differential in R&D intensity	Liquid liabilities	Private credit	Stock market capitalization	Private bond market capitalization	FDI equity
External financial dependence	0.40	0.46	0.56	0.99	0.44
As percentage of R&D intensity	18.1	20.9	25.2	45.0	19.7
Tangibility	−0.37	−0.51	—	−0.49	−0.38
As percentage of R&D intensity	−16.5	−22.9	—	−22.3	−17.4

25th percentile of external dependence in Canada (the country at the 25th percentile of bond-market capitalization). This difference makes up 45 percent of the average R&D intensity, indicating that deeper bond-market capitalization would increase the R&D intensity of the high-dependence sector relative to the low-dependence sector by 45 percent of the mean industry value. For liquid liabilities, private credit, and stock-market capitalization, the magnitudes are between 18 and 25 percent of the mean R&D intensity.

Tangibility also is a substantial determinant of the cross-industry variation in R&D as financial markets get deeper. An increase in the relative size of private credit and private bond markets from the 25th-percentile to the 75th-percentile nation are both estimated to reduce R&D intensity by around 22 to 23 percent in comparing the industry at the 75th percentile of tangibility to that at the 25th percentile. The effect of more developed financial markets as measured by liquid liabilities is 16.5 percent. Put differently, the data suggest that as these markets grow relative to GDP, relatively more R&D occurs in sectors with lower stocks of physical capital. One interpretation is that improvements in credit markets (measured broadly as liquid liabilities or more specifically finance supplied through banks or the bond market) tend to benefit those industries with fewer tangible assets, thus loosening credit constraints for R&D investment.

It is interesting to note that there is not a significant effect via tangibility when financial development is measured by the size of the stock market. That is, R&D intensities are not different between levels of tangibility as stock markets grow.

The coefficient on the industry share in GDP is negative and significant throughout in Table 1, with similar magnitudes across the different regressions. Thus, R&D intensity is lower in larger industries, which is similar to the finding in Rajan and Zingales (1998) that larger industries grow more slowly than smaller industries.

4.2. *International Financial Market Development*

While the link between domestic financial market development and R&D intensity turns out to be strong, there is evidence for just one influential channel of international capital. We interact four measures of international integration — total external assets and liabilities, portfolio equity, FDI equity, and foreign debt — with both the external financial dependence and tangibility variables. As Table 3 shows, FDI equity seems to be the crucial international financial development dimension that affects R&D intensity via the interaction terms. Total assets and liabilities also show a (weakly) significant effect when interacted with external dependence, which is likely picking up the impact of FDI as a major component of total assets and liabilities.

The coefficient on the interaction between FDI equity and external dependence is highly significant while that on the tangibility interaction is less precisely estimated, though still significant. Both coefficients are of comparable sizes to those on the domestic measures for liquid liabilities, private credit, and stock-market capitalization. Turning to the magnitudes of the effects of FDI and external dependence in Table 2, we find that they are similar to the liquid liabilities and private credit variables. Specifically, the industry at the 75th percentile of external dependence has an R&D intensity that is 0.44

Table 3. Regression of R&D intensity on international financial development interacted with external financial dependence and tangibility. The table illustrates OLS panel regression results for 22 manufacturing industries in 18 OECD countries for the years 1990–2003. Each column represents a separate regression using a different measure of international financial market development (as indicated in the first row) interacted with both external financial dependence and asset tangibility. The dependent variable is R&D intensity calculated as R&D expenditures as a share of industry output. Industry share in GDP is defined as industry production relative to GDP. Robust standard errors are in parentheses: * significant at 10%; ** significant at 5%; *** significant at 1%.

	Total external assets and liabilities	Portfolio equity	FDI equity	Foreign debt
Industry share in GDP	−0.1401 (0.0232)***	−0.1378 (0.0241)***	−0.1496 (0.0223)***	−0.1302 (0.0214)***
Total external assets and liabilities	−0.0004 (0.0006)			
External dependence ∗ Total external assets and liab.	0.0012 (0.0007)*			
Tangibility ∗ Total external assets and liab.	−0.0008 (0.0019)			
Portfolio equity		−0.0002 (0.0025)		
External dependence ∗ Portfolio equity		0.0037 (0.003)		
Tangibility ∗ Portfolio equity		−0.003 (0.0087)		
FDI equity			−0.0023 (0.0046)	
External dependence ∗ FDI equity			0.0194 (0.0052)***	
Tangibility ∗ FDI equity			−0.0191 (0.0115)*	
Foreign debt				−0.0009 (0.0008)
External dependence ∗ Foreign debt				0.0006 (0.0009)
Tangibility ∗ Foreign debt				0.0011 (0.0022)

(*Continued*)

Table 3. (*Continued*)

	Total external assets and liabilities	Portfolio equity	FDI equity	Foreign debt
Constant	0.0174	0.0168	0.02	0.0158
	(0.0035)***	(0.0035)***	(0.0034)***	(0.0033)***
Country-, Industry- and Year-Dummies	YES	YES	YES	YES
Observations	3795	3795	3795	3795
R-squared	0.47	0.47	0.48	0.47

percentage points higher in France (the country at the 75th percentile of FDI) than does the industry at the 25th percentile of external dependence in Germany (the country at the 25th percentile of FDI). This difference makes up 19.7 percent of the average R&D intensity. For tangibility, this thought experiment delivers a difference in R&D intensity of -0.38 percentage points or -17.4 percent of average R&D intensity. The interaction between tangibility and FDI shows a similar magnitude (both in percentage points and relative to mean R&D intensity) as the liquid liability measure. Thus, greater stocks of FDI are associated with greater R&D investments for industries with lower proportions of tangible assets.

4.3. *Discussion*

In this section we relate our empirical findings regarding R&D investments to established theories in the finance literature. We structure our discussion according to the main contributions of this paper: (i) the role of tangibility; (ii) national versus international measures of financial development; and (iii) market-based versus intermediated finance.

4.3.1. *Tangibility — Debt versus Equity*

Generally, when a firm has few tangible assets, it is less able to take advantage of credit markets since it has little real capital and equipment on which to collateralize the debt. This is particularly true for R&D investment, where much of the expenditures are on wages and salaries for scientists and other researchers. We find that the results via the tangibility channel are particularly strong for the various forms of credit markets, with the largest effects for private credit and private bond-market capitalization. This is consistent with the idea that R&D expenditures face financing constraints based on the lack of tangible assets. Thus, development of such financial markets plays a particularly large role in loosening such constraints.

As noted above, the size of the stock market, interacted with tangibility, does not have a significant effect on R&D expenditures. This finding indicates that development of the stock market plays little role in relaxing credit constraints associated with low stocks of capital and equipment. This makes sense since industries with fewer tangible assets are more inclined to use equity financing, requiring no collateralization, for their R&D programs even at lower levels of financial development. Firms in those sectors may therefore find little additional stimulus to R&D from the development of domestic stock markets.

In contrast, there is a negative impact of FDI equity interacted with tangibility. That is, FDI can allow for greater financing of R&D even for firms with few physical-capital assets. This finding is consistent with evidence that industries with high reliance on intellectual capital tend to be more multinational in scope (Markusen, 2002), indicating that FDI may loosen any capital constraints facing their R&D programs.

4.3.2. *National versus International Measures of Financial Development*

The basic results show that financial development at the international level may be just as important as financial development at the domestic level. The significantly positive impacts of FDI (via the external finance channel) indicate that openness to international capital can increase R&D expenditures by amounts that are similar to the impacts of development of domestic credit markets. Further, the tangibility results indicate that FDI may by a useful form of financing for firms with less capital and equipment. Note that FDI offers greater potential to monitor local-affiliate R&D activities, which seems consistent with the notion that sectors with low tangibility (and higher intellectual capital) would resort to this form of international capital.

Indeed, it is not surprising that the magnitude of FDI in an economy is the primary international variable that provides significant results for relative R&D expenditures. There are a number of reasons to expect FDI to be more effective in supporting R&D activity than other forms of international finance, as discussed by Kose *et al.* (2009). First, innovative activities often entail greater uncertainty than physical-capital investments (Hall and Lerner, 2010). It thus takes a greater commitment to invest in firms with high R&D intensity. FDI might serve this requirement quite well. Neither portfolio equity nor foreign debt offers the monitoring activities that are embodied in FDI, and thus provide a poor alternative form of financing for those firms with fewer tangible assets.

Second, the close relationships that are often established between multinational firms and their subsidiaries may reduce the need for tangible assets as collateral to support R&D investments. Further, subsidiaries may be able to access the internal credit resources of the multinational parent.[9] Third, FDI is a considerably more stable source of funding than are short-term financial flows, providing the ability to finance innovative activities with longer time horizons.

More fundamentally, our finding that FDI is a primary source of R&D finance complements prior studies of the growth effects of international capital liberalization. In that literature, positive growth impacts are often found for FDI liberalization but not necessarily for openness to other forms of portfolio equity or debt flows (Kose *et al.*, 2009). Our results add another layer to this analysis by showing that FDI can spur R&D investments, presumably because FDI supports more relationship-based finance and less arm's-length lending.

---

[9]See Desai *et al.* (2004) and Manova *et al.* (2009).

4.3.3. *External Financial Dependence — Market-Based versus Intermediated Finance*

As an alternative focus, we may consider a division of results into direct or market-based finance (e.g., stock-market capitalization and bond-market capitalization) versus intermediated finance (e.g., liquid liabilities and private credit). In the literature on the financing of innovative activities, one primary argument is that once firms must access external funds, market-based financing may provide a preferred form of financing relative to bank-based financing. The notion is that banks are effective at monitoring specific firm activities where the outcomes are relatively well defined but are less able to monitor innovative activities with highly uncertain returns. Further, bank lending decisions are typically made by a single bank manager, who forms an opinion about the potential returns and may be less likely to finance highly uncertain projects.

Market-based financing, on the other hand, relies on a wide variety of opinions, with this diversity allowing for investment in uncertain activities. Allen and Gale (2000) point out that market-based systems may be particularly significant when opinions vary about the potential outcome of risky R&D investments. In this case, a diverse set of financiers obtain information and invest based on their individual priors, which may differ from other investors' expectations. Thus, development of market-based external finance is likely to be important for innovative enterprises. Conversely, bank-based systems may have a comparative advantage in monitoring borrower behavior and managing risk when there is greater agreement about the potential outcomes of investment.

This line of thinking is consistent with our results, which show that the positive impacts on R&D intensities are strongest for the stock-market capitalization and private bond-market capitalization variables (interacted with external dependence). Interestingly, we find that liquid liabilities and private credit have impacts on R&D intensity that are similar to those of stock-market capitalization. Thus, we may be picking up both the importance of market-based financing and the relative preference for debt over equity in terms of maintaining control rights. Of the international measures, FDI may incorporate aspects of both direct finance (through its equity component) and intermediated finance (since FDI entails some control over management).

Finally, note that Denis and Mihov (2003) find a negative relationship between R&D expenditures and the likelihood of issuing bonds. Those firms that do issue bonds (relative to private credit) tend to be larger, more profitable, have a higher proportion of fixed assets, and spend less on R&D. In contrast, Altunbaş *et al.* (2009) argue that bond markets may be particularly useful for financing activities that embody forward-looking expectations. In a study of large European firms, they find that firms with more growth opportunities (measured by higher and more visible capital investment spending, which they describe as R&D investment) prefer the bond market over syndicated loans. Thus, if large firms with higher credit ratings tend to undertake relatively more R&D investment, then this may explain the larger impacts of the private bond-market interaction terms in our regressions. Our data do not allow us to infer firm details such as size or credit ratings but this line of thought may be suggestive of the differential effects.

In summary, our findings on the determinants of relative R&D investments across industries and countries complement these various ideas. Development of domestic financial markets and openness to FDI clearly raise R&D intensities in sectors with higher degrees of external dependence and lower proportions of physical assets. Since the latter are largely industries with much intellectual capital, these forms of market growth favor R&D expenditures in high-technology sectors. This finding may offer one clue about why growth and productivity are enhanced as found in the prior literature.

5. Sensitivity Analysis

An important issue is whether there is an omitted variable bias from only considering one financial development variable at a time. Due to high correlations of financial development measures, we are unable to include all of these covariates at once. Nevertheless, we consider specific combinations that measure different aspects of financial market development and hence are less correlated. Thereby, we clarify whether the development of different types of financial variables may substitute for one another. In particular, we examine combinations of the credit variables and the stock-market variables. Further, since the bond market appears to play a large role, we focus on combinations that include the private bond-market variable.

Table 4 shows the results of these combinations of financial development variables. Generally, the overall results are little changed. Domestic private credit appears to complement domestic stock-market development as the interactions with the stock-market variable remain significant when included along with interaction terms with liquid liabilities or bank credit (columns 1 and 2). The coefficients on the interactions with external dependence are somewhat smaller (and less precise for private credit and liquid liabilities). The stock-market interaction with external dependence remains highly significant. The results on the tangibility interactions also remain largely the same, with R&D intensity responding to the two credit variables and not the stock-market size. The stock-market capitalization interaction with external dependence is also robust to including FDI and private bonds. In each case the effects of the size of stock markets come through the external dependence channel and not through the tangibility channel.

The results on the private bond-market variable for both the interactions with external financial dependence and tangibility are quite robust, with little change in either the magnitude or significance of the interaction terms when it is included along with any of the other measures of domestic or international financial development. The coefficients on the interaction terms with liquid liabilities or private credit (columns 3 and 4) are both slightly smaller and somewhat less significant, while the coefficient on the stock-market variable (column 5) interacted with external financial dependence is somewhat larger and more significant. Thus, joint development of the private bond market and domestic credit or equity markets plays a role in facilitating higher R&D intensity.

We next consider whether domestic and international financial development measures interact with one another. Columns 6–9 of Table 4 show that the impacts of the international-equity variable (FDI) via the external dependence channel remain significant and

Table 4. Combinations of financial development variables. The table illustrates OLS panel regression results for 22 manufacturing industries in 18 OECD countries for the years 1990–2003. Each column represents a different combination of various measures of financial development. The financial development variables are each interacted with external financial dependence and tangibility. The dependent variable is R&D intensity calculated as R&D expenditures as a share of industry output. Industry share in GDP is defined as industry production relative to GDP. Robust standard errors are in parentheses: * significant at 10%; ** significant at 5%; *** significant at 1%.

	Liquid liabilities and stock market capitalization	Private credit and stock market capitalization	Liquid liabilities and private bond market capitalization	Private credit and private bond market capitalization	Stock market cap and private bond market cap	Liquid liabilities and FDI	Private credit and FDI	Stock market capitalization and FDI	Private bond market capitalization and FDI
Industry share in GDP	−0.1422 (0.0206)***	−0.1462 (0.0211)***	−0.1354 (0.0181)***	−0.1372 (0.0185)***	−0.1436 (0.0201)***	−0.1539 (0.0226)***	−0.1539 (0.0226)***	−0.1526 (0.0233)***	−0.1515 (0.0210)***
Liquid liabilities	0.0046 (0.0091)		0.0014 (0.0092)			0.0034 (0.0085)			
Private credit		0.0041 (0.0062)		−0.001 (0.0068)			0.0029 (0.006)		
Stock market capitalization	−0.0058 (0.0043)	−0.0066 (0.0043)			−0.0059 (0.0052)			−0.0055 (0.0042)	
Private bond market capitalization			0.01 (0.0151)	0.012 (0.0166)	0.0066 (0.0148)				0.0077 (0.0136)
FDI equity						−0.0021 (0.0045)	−0.0029 (0.0048)	−0.0001 (0.0046)	−0.0034 (0.005)

(*Continued*)

Table 4. (*Continued*)

	Liquid liabilities and stock market capitalization	Private credit and stock market capitalization	Liquid liabilities and private bond market capitalization	Private credit and private bond market capitalization	Stock market cap and private bond market cap	Liquid liabilities and FDI	Private credit and FDI	Stock market capitalization and FDI	Private bond market capitalization and FDI
External dependence* Liquid liabilities	0.0126 (0.0051)**		0.0147 (0.0050)***			0.0175 (0.0049)***			
Tangibility* Liquid liabilities	−0.0193 (0.0111)*		−0.0122 (0.0115)			−0.0211 (0.0113)*			
External dependence* Private credit		0.0121 (0.0055)**		0.0142 (0.0053)***			0.016 (0.0052)***		
Tangibility* Private credit		−0.0251 (0.0121)**		−0.0138 (0.0138)			−0.0231 (0.0122)*		
External dependence* Stock market capitalization	0.0172 (0.0058)***	0.0161 (0.0057)***			0.0282 (0.0063)***			0.0173 (0.0052)***	

Table 4. (*Continued*)

	Liquid liabilities and stock market capitalization	Private credit and stock market capitalization	Liquid liabilities and private bond market capitalization	Private credit and private bond market capitalization	Stock market cap and private bond market cap	Liquid liabilities and FDI	Private credit and FDI	Stock market capitalization and FDI	Private bond market capitalization and FDI
Tangibility* Stock market capitalization	0.0036 (0.011)	0.0068 (0.0103)			0.0007 (0.0125)			0.0072 (0.0098)	
External dependence* Private bond market capitalization			0.0661 (0.0072)***	0.0655 (0.0070)***	0.0757 (0.0076)***				0.0671 (0.0067)***
Tangibility* Private bond market capitalization			−0.0385 (0.0112)***	−0.0377 (0.0114)***	−0.0329 (0.0128)**				−0.0362 (0.0111)***
External dependence* FDI equity						0.0185 (0.0057)***	0.0164 (0.0055)***	0.0108 (0.0048)**	0.0186 (0.0054)***
Tangibility* FDI equity						−0.0182 (0.012)	−0.0144 (0.0113)	−0.022 (0.0097)**	−0.0126 (0.0115)

Table 4. (*Continued*)

	Liquid liabilities and stock market capitalization	Private credit and stock market capitalization	Liquid liabilities and private bond market capitalization	Private credit and private bond market capitalization	Stock market cap and private bond market cap	Liquid liabilities and FDI	Private credit and FDI	Stock market capitalization and FDI	Private bond market capitalization and FDI
Constant	0.0166	0.0186	0.016	0.0178	0.0143	0.0217	0.0226	0.0181	0.0184
	(0.0056)***	(0.0044)***	(0.0045)***	(0.0036)***	(0.0039)***	(0.0056)***	(0.0044)***	(0.0035)***	(0.0035)***
Country-, Industry- and Year-Dummies	YES	YES	YES	YES	YES	YES	YES	YES	YES
Observations	3522	3522	3467	3467	3491	3643	3643	3667	3619
R-squared	0.5	0.5	0.52	0.52	0.5	0.5	0.5	0.48	0.5

are only slightly smaller in magnitude when included with any of the domestic credit or stock-market variables. Further, the FDI interactions have little effect on the size of the external dependence interactions with the credit variables but attenuate the size of the external dependence interactions with stock-market capitalization. One interesting finding is that the tangibility interaction with FDI generally loses significance except where stock-market capitalization is included, in which case it takes on a larger magnitude and is more significant. Thus, with respect to financing R&D investment, there appears to be a notable complementarity between development of the stock market and growth in FDI for sectors with low levels of physical assets.

Another major concern about our results could be that overall economic development and institutional quality are the silent drivers behind our financial development interaction terms. To rule this out, we control for real GDP per capita (in logs) and a "quality of governance" index interacted with external dependence and asset tangibility.[10] When these two controls are included, the estimates of the coefficients on interactions between financial development and external dependence and tangibility are little changed. Thus, our conclusions regarding the impacts of financial market development on R&D investments remain robust.

Among the interaction terms that include GDP per capita, only tangibility exerts a statistically significant effect. Specifically, we find that higher income has a positive effect on R&D investments in industries with more tangible assets. The coefficient on institutional quality by itself is positive and significant, suggesting a stimulative direct effect on R&D intensity. In this case, both interaction terms between institutional quality and industry characteristics are significant and have the expected sign as well. These results indicate that institutional development plays both direct and indirect roles in encouraging investments in research and development. The main point, however, is that including domestic institutions does not alter

[10]We do not report regression tables for this and subsequent robustness checks for the sake of brevity. However, they may be obtained from the authors upon request. Also see Maskus, Neumann, and Seidel (2011) for more details.

the importance of the different types of financial development we have already discussed.

Although our baseline regressions above controlled for relative industry size, the results may be still driven by comparative advantage that is not captured by this covariate. For example, systematic differences among economies in economic environments or relative factor endowments generate distinct comparative advantages in industries across countries. We thus control for national characteristics, including country size (as measured by real GDP) and three factor endowments (labor force, real capital stock and the share of the labor force with tertiary education).[11] These time-varying country variables do not exert a statistically significant effect on R&D intensity at the industry level and, more importantly, leave unaffected the estimates of the interaction terms.

To account for comparative advantage at the industry level, we compute sector-level capital-labor ratios across all countries, and interact these ratios with both the real capital stock and the labor force. It is notable that our basic results are robust to controlling for both the direct and interacted effects of these additional variables. Compared to our baseline results displayed in Tables 1 and 3, estimates of the interaction terms between financial development and external dependence remain positive and highly significant. The coefficients on tangibility retain their signs but are somewhat less significant. Overall, however, the main pattern of the effects of financial development on R&D investments remains undisturbed.

Finally, we stress that our results are robust to alternative measures of financial development available from the World Bank's *Financial Structure Database 2007* and to excluding less advanced economies (the Czech Republic and Poland) from our sample.

6. Conclusions

In this paper we examine the impacts of financial market development on R&D intensity in 22 manufacturing industries in 18 OECD countries for the period 1990–2003. As financing constraints for R&D

[11]All data are taken from the World Bank's World Development Indicators.

may be particularly tight due to the intangible nature of R&D assets, access to external funds can be expected to play an important role in stimulating innovation and eventually output growth.

We find statistically and economically strong effects for most domestic indicators of financial market development. In particular, the largest impacts arise from the development of the bond market for firms relying more on external finance. The other measures of domestic financial market development are smaller and similar to each other. For international capital flows, FDI has by far the strongest impacts on R&D, operating both through sectoral financial dependence and asset tangibility. It is remarkable that neither portfolio investments nor foreign debt seem to be related to R&D intensity at the industry level. We interpret these results as suggesting that it requires strong monitoring and managerial commitment (which is present with FDI but not with more short-term and volatile portfolio investment) to support more risky investments in research and development.

The largest impacts occur in response to private bond-market growth. This outcome may be related to the size and credit ratings of firms that undertake R&D investment, although we cannot examine this question specifically due to data limitations. Alternatively, it may be that the development of private bond markets has larger effects for firms relying more on external finance because those markets allow for a greater diversity of opinion among financiers, while maintaining some control rights (in the event of bankruptcy) that stock markets do not permit.

In terms of asset tangibility, we find similar magnitudes of both credit markets and FDI, with insignificant results for stock markets. This may indicate an important role of monitoring through relationship lending for those firms with limited tangible assets, which generally correspond to a high degree of intellectual capital established through R&D spending.

In summary, we find that development of different forms of financing may stimulate R&D investment to different degrees. Policymakers in both advanced and emerging economies are keen to promote domestic private R&D spending, which is a direct input into innovation and growth. Thus, they offer tax advantages, direct subsidies and infrastructural support for R&D programs. Our results suggest that a key factor in encouraging R&D investments is access to finance of various forms. Thus, authorities may wish to remove impediments to the growth of domestic financial markets, which seem to promote higher R&D intensities in sectors with high external dependence or low asset tangibility. It also seems important to relax restrictions on FDI flows, which have similar effects and seem particularly significant for building R&D in sectors with high intellectual capital. Of course, these findings do not necessarily imply that these policies will maximize overall welfare. Rather, they underscore the need to ensure access to financial resources if R&D investments are an important policy concern.

References

Acemoglu, D., 2009. *Introduction to Modern Economic Growth.* Princeton University Press, Princeton.

Aghion, P., Angeletos, G.-M., Banerjee, A., and Manova, K., 2010. Volatility and growth: Credit constraints and the composition of investment. *Journal of Monetary Economics* 57, 246–265.

Aghion, P., Askenazy, P., Berman, N., Cette, G., and Eymard, L., 2008. Credit constraints and the cyclicality of R&D investment: Evidence from France. Paris School of Economics Working Paper No. 2008-26.

Aghion, P., Bond, S., Klemm, A., and Marinescu, I., 2004. Technology and financial structure: Are innovative firms different? *Journal of the European Economic Association* 2, 277–288.

Allen, F. and Gale, D., 2000. *Comparing Financial Systems.* The MIT Press, Cambridge, MA.

Altunbaş, Y., Kara, A., and Marqués-Ibáñez, D., 2009. Large debt financing: syndicated loans versus corporate bonds. European Central Bank Working Paper Series, No. 1028.

Beck, T., Demirgüç-Kunt, A., and Levine, R., 2000. A new database on financial development and structure. *World Bank Economic Review* 14, 597–605.

Bond, S., Harhoff, D., and van Reenen, J., 2005. Investment, R&D and financial constraints in Britain and Germany. *Annales d'Economie et de Statistique* 79/80, 435–463.

Braun, M., 2003. Financial contractibility and asset hardness. Unpublished working paper, Harvard University.

Brown, J.R., Fazzari, S.M., and Petersen, B.C., 2009. Financing innovation and growth: Cash flow, external equity, and the 1990s R&D boom. *The Journal of Finance* 64, 151–185.

Carlin, W. and Mayer, C., 2003. Finance, investment, and growth. *Journal of Financial Economics* 69, 191–226.

Denis, D.J. and Mihov, V.T., 2003. The choice among bank debt, non-bank private debt, and public debt: Evidence from new corporate borrowings. *Journal of Financial Economics* 70, 3–28.

Desai, M.A., Foley, C.F., and Hines, J.R., 2004. A Multinational perspective on capital structure choice and internal capital markets. *Journal of Finance* 59, 2451–2488.

Eichengreen, B., 2001. Capital account liberalization: What do cross-country studies tell us? *World Bank Economic Review* 15, 341–365.

Gammeltoft, P., 2006. Internationalisation of R&D: trends, drivers, and managerial challenges. *International Journal of Technology and Globalisation* 2, 177–199.

Griliches, Z., 1998. *R&D and Productivity: The Econometric Evidence.* The University of Chicago Press, Chicago.

Hall, B.H., 1992. Investment and research and development at the firm level: Does the source of financing matter? Cambridge, MA. NBER Working Paper No. 4096.

Hall, B.H. and Lerner, J., 2010. The financing of R&D and innovation. In: Hall, B.H. and Rosenberg, N. (Eds.), *The Elsevier Handbook of the Economics of Innovation, Vol. 1.* Elsevier, Amsterdam, 609–639.

Harris, M. and Raviv, A., 1991. The theory of capital structure. *Journal of Finance* 46, 297–355.

He, Y. and Maskus, K.E., 2012. Southern innovation and reverse knowledge spillovers: a dynamic FDI model. *International Economic Review* 53, 281–304.

Himmelberg, C.P. and Petersen, B.C., 1994. R&D and internal finance: A panel study of small firms in high-tech industries. *Review of Economics and Statistics* 76, 38–51.

Keller, W. and Yeaple, S.R., 2010. Multinational enterprises, international trade, and productivity growth: firm-level evidence from the United States. *Review of Economics and Statistics* 91, 821–832.

Kesternich, I. and Schnitzer, M., 2010. Who is afraid of political risk? Multinational firms and their choice of capital structure. *Journal of International Economics* 82, 208–218.

King, R. and Levine, R., 1993. Finance and growth: Schumpeter might be right. *Quarterly Journal of Economics* 108, 717–737.

Klapper, L., Laeven, L., and Rajan, R.G., 2006. Entry regulation as a barrier to entrepreneurship. *Journal of Financial Economics* 82, 591–629.

Kose, M.A., Prasad, E.S., Rogoff, K., and Wei, S.-J., 2009. Financial globalization: A reappraisal. *IMF Staff Papers* 56, 8–62.

Kose, M.A., Prasad, E.S., and Terrones, M.E., 2008. Does openness to international financial flows contribute to productivity growth? Berlin. IZA Discussion Paper No. 3634.

Lane, P.R. and Milesi-Ferretti, G.M., 2006. The external wealth of nations mark II: Revised and extended estimates of foreign assets and liabilities, 1970-2004. Washington DC. IMF Working Paper 06/69.

Leland, H. and Pyle, D., 1977. Information asymmetries, financial structure, and financial intermediation. *Journal of Finance* 32, 371–387.

Levine, R., 2005. Finance and growth: Theory and evidence. In: Aghion, P. and Durlauf, S. (Eds.), *Handbook of Economic Growth, vol. 1.* Elsevier, Amsterdam, pp. 865–994.

Manova, K., 2008. Credit constraints, equity market liberalizations and international trade. *Journal of International Economics* 76, 33–47.

Manova, K., Wei, S.-J., and Zhang, Z., 2009. Firm exports and multinational activity under credit constraints. Unpublished manuscript, Stanford University.

Markusen, J.R., 2002. *Multinational Firms and the Theory of International Trade.* MIT Press, Cambridge, MA.

Maskus, K.E., Neumann, R., and Seidel, T., 2011. How national and international financial development affect industrial R&D. Munich. CESifo Working Paper No. 3480.

Mulkay, B., Hall, B.H., and Mairesse, J., 2001. Investment and R&D in France and in the United States. In Deutsche Bundesbank (Ed.), *Investing Today for the World of Tomorrow.* Springer Verlag, Berlin.

Myers, S.C., 1984. The capital structure puzzle. *Journal of Finance* 39, 575–592.

Myers, S.C. and Majluf, N.S., 1984. Corporate financing and investment decisions when firms have information that investors do not have. *Journal of Financial Economics* 13, 187–221.

Organization for Economic Cooperation and Development, 2005. STAN Structural Analysis Database STAN Industry Vol. 2005 release 05.

Organization for Economic Cooperation and Development, 2006. STAN Structural Analysis Database ANBERD — R&D Expenditure in Industry (ISIC Rev.3) Vol. 2006 release 01.

Rajan, R.G. and Zingales, L., 1998. Financial dependence and growth. *American Economic Review* 88, 559–586.

PART III

TRADE IN FACTORS AND TECHNOLOGY

Chapter 11

A Test of the Heckscher–Ohlin–Vanek Theorem: The Leontief Commonplace*

Keith E. Maskus†

1. Introduction

The Heckscher–Ohlin–Vanek (HOV) multi-factor content theorem of international trade remains open to empirical verification, despite the abundance of literature purporting to test it in some fashion. The primary reason for the uncertainty regarding the theorem's empirical validity is that it has rarely been subjected to a logically complete test. Such a test requires independent measures of a country's trade, its factor intensities in production, and factor endowments both of the country and the rest of the world. The first two measures, for example, could be used to predict the structure of factor endowments under the HOV assumptions. Large deviations of the observed endowments from these predicted endowments would cast doubt on the theorem as an empirical proposition.

Unfortunately, virtually all studies of the relationship between factor inputs and trade have computed only factor intensities and

*Reprinted from *Journal of International Economics*, Vol. 19, No. 1, pp. 201–212.

†This paper was partially written while the author was a Visiting Scholar at the Federal Reserve Bank of Kansas City. I am grateful to Renee Lindsey and Terry Fitzgerald for computational assistance, and I thank Sven W. Arndt, Jagdish Bhagwati, Robert M. Stern, Alan V. Deardorff and two anonymous referees for helpful comments. Any remaining errors are solely my responsibility.

trade patterns across industries without also developing data on endowments. Instead, the results have often been used to infer, possibly erroneously, the structure of factor endowments as revealed by the factor content of trade. Indeed, Leontief's (1953) celebrated results were considered paradoxical because the theorem was generally accepted without question; Leontief had, therefore, apparently demonstrated labor abundance and capital scarcity in the United States. Two competing explanations — that Leontief's calculations were inappropriate or that the theorem itself was invalid — were largely ignored.

Despite this fundamental oversight in the empirical literature on the HOV theorem, very few studies have sought explicitly to relate the factor content of a country's trade to the country's endowment structure. Examples are Clifton and Marxsen (1984), Bowen (1983), and Balassa (1979) but, for various reasons, the results in each of these articles are only partial and indirect. Considerable room remains for systematic testing of the HOV theorem.[1]

This paper provides such a test by relating the labor and capital contents of U.S. net exports to measures of U.S. and world endowments of these factors in 1958 and 1972. The tests are based on the observations by Leamer (1980) and Brecher and Choudhri (1982) that traditional calculations of the factor contents of exports and imports separately are inappropriate for inferring revealed abundance rankings even if the HOV theorem were valid. The next section notes the revealed abundance rankings of several factors under the appropriately specified HOV criterion. Section 3 performs the tests of the HOV theorem. In nearly all cases, the HOV theorem fails to hold in some versions. Thus, the Leontief paradox may in fact be a commonplace in the data. Section 4 concludes by noting briefly that these results are likely due to the restrictive assumptions of the HOV theorem.

---

[1]After this paper was written, my attention was called to the manuscript by Bowen, Leamer, and Sveikauskas (1984) in which rather similar calculations to my own are made for a variety of countries in 1967. Their results are substantially complementary to those reported here.

2. Revealed Factor Abundance and the Leontief Paradox

Since Leontief's study, it has been common in the literature at least implicitly to infer relative factor endowment rankings for a country by comparing the ratios of factor requirements in the country's representative bundle of exports relative to its bundle of competitive-import replacements. Thus, for example, Baldwin (1971) noted the continued existence of the Leontief paradox in 1958 in the United States, while Stern and Maskus (1981) suggested the paradox may have evaporated by 1972.[2] Both studies went on to compute the factor requirements of trade for factors other than labor and capital, indicating that for all factors a comparison of their embodiments in exports relative to imports could be useful in inferring the factor endowment structure.

Leamer (1980) pointed out the inadequacy of this procedure in the multi-factor context. Based upon Vanek's (1968) generalization of the Heckscher–Ohlin model to n goods and factors, Leamer showed that a larger capital–labor ratio in exports than in imports is an unambiguous indicator of capital abundance and labor scarcity only if capital services are exported on net and labor services are imported on net. Since in a world of more than two factors a country may be a net exporter of both capital and labor, the appropriate comparison is between capital per worker embodied in net exports and capital per worker in production (or consumption). Equivalently, one may compare the share of capital in trade with the share of labor going to trade. Leamer used these notions to show that the Leontief paradox did not exist in 1947, although the paradox could not be refuted for 1958.

Following Leamer, a country may be considered abundant in one factor relative to a second factor if the country's share of the world endowment of the first factor exceeds its share of the world supply of the second factor. The HOV model, as specified in Vanek (1968), implies that for any factor F,

$$F_x - F_m = F - (C/C_w)F_w \tag{1}$$

---

[2]The latter study also supported its claim by examining some of the more appropriate ratios discussed below.

where F_x and F_m represent, respectively, the total (direct plus indirect) amount of the factor embodied in U.S. exports and imports, F and F_w are the factor endowments in the United States and the world, respectively, and C and C_w are aggregate consumption in the United States and the world, respectively. C/C_w is, accordingly, the U.S. share in world consumption.[3] Using Eq. (1), Leamer develops three criteria for ranking factors according to their relative abundance. First, any factor net exported is more abundant than any factor net imported. Second, if two factors, say capital (K) and labor (L), are both exported on net, then

$$(K_x - K_m)/K > (L_x - L_m)/L \tag{2}$$

is sufficient to demonstrate capital abundance relative to labor. That is, the share of domestic capital exported must exceed the share of labor exported under capital abundance. Finally, if both factors are imported on net, then Eq. (2) again suffices to reveal relative capital abundance, so that the share of capital imported is less than the share of labor imported. These conditions together imply that abundance rankings among factors may be established by ranking the factors according to their total contents in net exports relative to production.[4]

For purposes of computing these rankings, data were assembled on U.S.production, factor inputs, and trade across industries in 1958 and 1972. These data were concorded to the 79-sector input–output tables for those years, thereby including all available industries. Thus, all factor contents are measured appropriately as total requirements of trade or production on an economywide basis [see Hamilton and Svensson (1983)]. Production is measured by deliveries

---

[3]Leamer's version of the model assumes identical constant-returns-to-scale technologies in all countries, identical homothetic preferences across countries, the existence of n freely traded commodities and n internationally immobile factors, perfect competition, and factor price equalization. Retention of the final assumption allows the number of factors and commodities to differ, however, as in Vanek's paper.

[4]Equivalent rankings may be established by comparing the factor contents of net exports with consumption, as in Sveikauskas (1983), or by comparing the capital–labor ratio in net exports with the capital–labor ratio in consumption [Stern and Maskus (1981)] or production.

to final demand in the input–output tables for the years covered. It should be noted that reliable data on net exports for the service industries are very difficult to obtain. The service sectors, therefore, are treated here as nontradeables so that all services output goes to consumption, as in Sveikauskas (1983).[5] Nominal values for all variables are expressed in 1967 dollars. Additional details on data construction are listed in the Appendix.

Table 1 presents factor abundance rankings for the United States in 1958 and 1972. The factors considered are employed engineers

Table 1. Total factor contents of U.S. production (F) and net exports ($F_x - F_m$), with factor abundance rankings, 1958 and 1972.

Factor	F	$F_x - F_m$	$(F_x - F_m)/F$	Rank
1958				
Engineers and scientists (mill. man years)	1.168	0.047	0.0402	1
Production labor (mill. man years)	42.339	0.784	0.0185	4
Other labor (mill. man years)	12.990	0.351	0.0270	2
Total labor	56.497	1.182	0.0209	—
Gross physical capital ($ mill.)	$750,178	$9,942	0.0133	5
Human capital ($ mill.)	$1,323,410	$28,388	0.0215	3
1972				
Engineers and scientists (mill. man years)	1.249	0.015	0.0120	1
Production labor (mill. man years)	46.705	−0.374	−0.0080	5
Other labor (mill. man years)	12.243	0.024	0.0020	2
Total labor	60.197	−0.335	−0.0056	—
Gross physical capital ($ mill.)	$1,340,929	−$9,593	−0.0072	4
Human capital ($ mill.)	$1,783,938	−$11,526	−0.0065	3

[5] Of course, trade occurs in some services, so some error is introduced by this procedure. The indirect contributions of the various factors to trade through the production of services remain in the calculations through the use of the full input–output tables. Such contributions are relevant to the factor content of trade that is covered.

and scientists as a measure of highly skilled labor across industries,[6] production labor as a proxy for unskilled labor, other labor, gross physical capital stocks, and human capital stocks as measured by discounted industry wage differentials. Factor contents of overall labor are reported as well but no ranking is given because this factor obviously overlaps with the individual labor groups. It is possible, however, to rank total labor vis-á-vis physical and human capital.

It is clear from Table 1 that labor skills are revealed by trade to be the most abundant factors in the United States according to the criterion in Eq. (2). In both years, engineers and scientists were revealed to be most abundant, followed by other labor (consisting of nonproduction labor other than engineers and scientists). Human capital ranked third in both years, indicating the importance of education in the determination of trade performance. Note that despite being net imported in 1972, human capital retained its abundance relative to physical capital in contrast to the findings in Stern and Maskus (1981) of a reversal in the relative rankings of human and physical capital. Unskilled labor and physical capital were at the bottom of the rankings in both years.[7]

The status of the Leontief paradox is revealed by the abundance rankings of labor and capital. In 1958, capital was revealed by trade to be scarce relative to labor, reaffirming the paradox by this criterion. Although capital had apparently become more abundant than production (unskilled) labor by 1972, it was still scarce relative to total labor. Hence, the scarce capital paradox had not evaporated

[6]Data on employment of engineers and scientists by industry were developed from 1960 and 1970 census figures. Thus, the direct coefficients for this variable are not strictly applicable to the 1959 and 1972 interindustry tables. It is unlikely that this problem would impart any systematic biases across industries in comparison to actual industry employment of engineers and scientists.

[7]In order to compare these results with the rankings from the basic Leontief criterion, the factor requirements of representative bundles of exports and imports were computed as well. Interestingly, the Spearman rank correlations between the Leontief rankings and the Leamer rankings were 0.9 in both years. The sample sizes (5) were too small to attach much confidence to these results, however.

by 1972, despite earlier reports to the contrary (Stern and Maskus, 1981).[8]

The factor abundance rankings in Table 1 are accurate only under the condition that the HOV theorem is empirically valid. As mentioned above, the validity of the HOV theorem remains an open question. The next section, therefore, provides some explicit tests of the theorem.

3. Tests of the HOV Theorem

Equation (1) provides the basis for the tests of the theorem. As Brecher and Choudhri (1982) pointed out, Eq. (1) implies:

$$F_x - F_m > 0 \quad \text{iff } C/F < C_w/F_w. \tag{3}$$

That is, a country may be a net exporter of factor F under the HOV assumptions if and only if its aggregate expenditure per unit of that factor is less than world expenditure per unit of the factor. Intuitively, a country's net export of a factor should be related to a large endowment, causing a relatively small C/F ratio in Eq. (3), given identical homothetic preferences. Thus, for example, the United States could be a net exporter of labor in 1958, as it was, if and only if U.S. expenditures per worker were less than in the world. The inequality in Eq. (3) will be referred to here as the 'weak' HOV prediction.

A second test is suggested by the ranking procedure in Eq. (2). In particular, if the HOV model is valid the revealed endowment rankings must duplicate the actual endowment rankings computed from independent data. This will be called the HOV 'rank' prediction; it is somewhat stronger than the weak prediction because it requires an ordering across factors.[9]

---

[8]There are two essential differences between this study and the Stern–Maskus study. First, the Stern–Maskus paper measured output as value of shipments, which is inappropriate in an interindustry context because it counts intermediate inputs double. Thus, the production figures in that paper are much larger than the current ones. Second, the Stern–Maskus paper allowed for trade in services.

[9]I am grateful to a referee for suggesting a different version of the rank prediction test.

Finally, the HOV theorem makes a very strong prediction as well. Equation (1) may be rewritten as:

$$C_w/F_w = C/F[1/(1 - (F_x - F_m)/F)]. \tag{4}$$

Thus, if U.S. trade in a factor is balanced, expenditure per unit of endowment of that factor must be the same in the United States and the world. Positive (negative) U.S. net exports of the factor imply a larger (smaller) world expenditure per unit as given in Eq. (4). This condition is called here the 'strong' HOV prediction.

To test these hypotheses, the factor contents of U.S. actual net exports and production were computed in 1958 and 1972 for three factors for which U.S. and world endowment measures were available. These were skilled labor as measured by professional, technical, and kindred workers, other labor as a proxy for unskilled labor, and gross physical capital.

A basic data source for this exercise is Bowen (1980), which provides data on labor force, skilled labor, and net capital stocks in 1963 and 1972 for the United States and 33 of its trading partners.[10] The bulk of U.S. nonoil trade is with these countries, so this 34-country definition of the world is reasonably comprehensive. Figures for labor and skilled labor for 1958 were generated by linear extensions from 1963 of Bowen's data. This procedure proved unacceptable for the capital stocks, so 1963 capital stock data were retained for use with 1963 expenditure data. Net stocks were converted to gross stocks by a procedure detailed in the Appendix.

The results of the weak HOV tests are presented in columns (1) and (3) through (6) of Table 2. Given the positive net exports of all three factors in 1958, U.S. expenditure per unit of each factor should be less than its world counterpart. National expenditure for each country was computed as GNP less net exports, expressed in dollars at prevailing exchange rates. It is evident that in 1958 U.S. expenditures relative to both its skilled and unskilled labor force were much higher than the corresponding world ratios. Under the HOV assumptions, therefore, the United States could not have been a net

[10]There are actually 34 countries besides the United States in Bowen's data, but Hong Kong is excluded here.

Table 2. Tests of the HOV theorem in three factors using a 34-country world, 1958 and 1972.

Factor	(1) $(F_x - F_m)/F$	(2) Rank	(3) C/F[a]	(4) Weak HOV Prediction	(5) C_w/F_w[b]	(6) HOV Holds	(7) U.S. Share	(8) Rank	(9) Strong HOV Prediction	(10) Percent[c] Deviation
1958										
Prof., technical & kindred workers	0.0165	2	$60,157	$<$	$37,697	No	0.3189	2	$61,166	38.4
Other labor	0.0181	1	$7,287	$<$	$1,752	No	0.1223	3	$7,421	76.4
Gross physical capital	0.0133	3	0.6303	$<$	0.6237[d]	Yes	0.4222[d]	1	0.6084	2.5
1972										
Prof., technical & kindred workers	−0.0011	1	$95,616	$>$	$64,322	Yes	0.2693	2	$95,511	32.6
Other labor	−0.0075	3	$16,229	$>$	$4,857	Yes	0.1198	3	$16,108	69.8
Gross physical capital	−0.0072	2	0.7270	$>$	0.6283	Yes	0.3460	1	0.7218	13.0

Notes: [a]U.S. aggregate expenditure per unit of factor endowment. [b]World (U.S. plus 33 countries) aggregate expenditure per unit of factor endowment. [c]Absolute deviation as percentage of predicted ratio. [d]Computed from 1963 data.

exporter of these factor services. Since it clearly was a net exporter, the HOV theorem failed to satisfy even its weak prediction in these cases. The theorem did hold in this weak sense in physical capital in 1958, however.[11] It also held in all three factors in 1972, when the net imports of the factor services reversed the inequality in the prediction. It appears, therefore, that the theorem fares rather well in satisfying the simple inequality criterion in comparing U.S. and foreign factor endowments.[12]

Columns (1), (2), (7), and (8) note the results of the rank prediction test. According to the actual endowment shares in column (7), the United States was more abundant in gross capital than in skilled labor in both years, while unskilled labor was the most scarce factor. The revealed rankings in column (2) suggest a rather different factor abundance structure, however, again depicting the inability of trade data to rank endowments accurately. On this score, therefore, the HOV theorem is evidently denied empirical support.

The results of the tests of the strong HOV predictions are listed in columns (5), (9), and (10). Obviously, any failure to satisfy the weak condition implies failure in the strong case also. Consider, therefore, the bottom four rows of Table 2, which relate to the successful cases in the weak test. In each case, the observed world expenditure-to-endowment ratio diverged from the ratio predicted by Eq. (4), leading apparently to a rejection of the theorem. Unfortunately, the nature of these computations does not allow the development of significance tests to apply to the differences between columns (9) and (5).[13] Because it is unreasonable to expect these ratios to coincide exactly, however, the percentage deviations are reported in column (10). While this is a crude measure, it does suggest the strong

[11]Recall that the world expenditure-to-capital-endowment ratio was actually computed for 1963, so it should be viewed with caution.

[12]Although note that if there were no systematic relationship, the probability of two or fewer 'nos' in six random observations, assuming a 50 percent probability of a 'no', would be 0.344. That is, 34.4 percent of the time, chance would support the HOV theorem to the extent demonstrated in Table 2. This far exceeds conventional significance levels. A similar result was noted in Clifton and Marxsen (1983).

[13]Bowen, Leamer, and Sveikauskas (1984) make an interesting attempt to explain the nature of similar differences in their results, ascribing their differences primarily to measurement errors in the data.

condition is violated in every case but gross physical capital in 1958. Overall, therefore, the HOV theorem does not receive support from the data used here.[14] Indeed, since each of the negative conclusions in Table 2 represents some variant of the Leontief paradox, it appears that paradoxical outcomes may be the rule rather than the exception.

As a final test of the theorem, U.S. expenditure-to-endowment ratios were compared to similar ratios for a more restrictive world consisting of the United States and five major OECD countries. Data on gross physical capital stocks in 1958 and 1972 were directly available from published sources for these countries, so the capital stock comparisons should be somewhat more accurate than the earlier ones based on converted net stocks.

It is evident from Table 3 that the HOV theorem fared no better in the smaller definition of the world. The theorem failed to meet its weak prediction in four of the six cases. Precisely the same ranking problems surfaced here as well. The HOV theorem also failed in all of the strong tests, although the error in the 1972 capital stock estimates was small enough to shed some doubt on both the strong and weak test conclusions.

4. Concluding Remarks

There has been considerable uncertainty in the literature about the empirical validity of the HOV theorem, despite the large volume of research devoted to that question. Most of the uncertainty stems from the widespread use of inappropriate testing techniques that may yield contradictory results.

Application of the appropriate techniques has shown here that the HOV theorem is inconsistent with available data on factor endowments, factor intensities, and trade, at least for the United States. Efforts to reveal factor abundance rankings by computing the factor contents of trade and production, for example, are therefore unlikely

[14] As noted by a referee, this strong test may be unreasonable given the crudity of international data comparisons. It may be more useful to focus on the two weaker tests in further testing of the HOV model. Of course, even these tests led frequently to rejection of HOV in this paper.

Table 3. Tests of the HOV theorem in three factors using a six-country world, 1958 and 1972.

Factor	(1) $(F_x - F_m)/F$	(2) Rank	(3) C/F[a]	(4) Weak HOV Prediction	(5) C_w/F_w[b]	(6) HOV Holds	(7) U.S. Share	(8) Rank	(9) Strong HOV Prediction	(10) Percent[c] Deviation
1958										
Prof., technical & kindred workers	0.0165	2	$60,157	<	$45,730	No	0.4983	2	$61,166	25.2
Other labor	0.0181	1	$7,287	<	$3,977	No	0.3795	3	$7,421	46.4
Gross physical capital	0.0133	3	0.6003	<	0.4941	No	0.5395	1	0.6084	18.8
1972										
Prof., technical & kindred workers	−0.0011	1	$95,616	>	$84,169	Yes	0.4743	2	$95,511	11.9
Other labor	−0.0075	3	$16,229	>	$11,158	Yes	0.3704	3	$16,108	30.7
Gross physical capital	−0.0072	2	0.7270	>	0.7314	No	0.5420	1	0.7218	1.3

Notes: [a]U.S. aggregate expenditure per unit of factor endowment. [b]World (U.S. plus 5 countries) aggregate expenditure per unit of factor endowment. [c]Absolute deviation as percentage of predicted ratio.

to produce reliable findings. Attention should perhaps be directed to developing direct measures of factor endowments.

These conclusions do not necessarily imply that differences in relative factor endowments across countries do not play an important role in the determination of world trade patterns. The empirical literature has clearly demonstrated the fairly consistent and suggestive links between trade performance and factor intensities, for example. Rather, the most reasonable interpretation is that the HOV assumptions are simply too restrictive for that version of the factor endowments theory to hold in an empirical context. Accordingly, it is important for trade economists to continue the search for versions of the theory that hold under more general circumstances, as in Deardorff (1982) and Helpman (1984), with a view toward allowing for empirical verification.

References

Balassa, Bela, 1979. The changing pattern of comparative advantage in manufactured goods. *Review of Economics and Statistics* 61, 259–266.

Baldwin, Robert E., 1971. Determinants of the commodity structure of U.S. trade. *American Economic Review* 61, 126–146.

Bowen, Harry P., 1980. Resources, technology and dynamic comparative advantage: A cross-country analysis of the product cycle theory of international trade (unpublished Ph.D. dissertation, UCLA).

Bowen, Harry P., 1983. Changes in the international distribution of resources and their impact on U.S. comparative advantage. *Review of Economics and Statistics* 65, 402–414.

Bowen, Harry P., Edward E. Leamer, and Leo Sveikauskas, 1984. Multicountry, multifactor tests of the factor abundance theory (unpublished manuscript, NYU).

Brecher, Richard A. and Ehsan U. Choudhri, 1982. The Leontief Paradox: Continued. *Journal of Political Economy* 90, 820–823.

Clifton, David S. and William P. Marxsen, 1984. An empirical investigation of the Heckscher-Ohlin theorem. *Canadian Journal of Economics* 17, 32–38.

Deardorff, Alan V., 1982. The general validity of the Heckscher-Ohlin theorem. *American Economic Review* 72, 683–694.

Hamilton, Carl and Lars E. O. Svensson, 1983. Should direct or total factor intensities be used in tests of the factor proportions hypothesis? *Weltwirtschaftliches Archiv* 119, 453–463.

Helpman, Elhanan, 1984. The factor content of foreign trade. *Economic Journal* 94, 84–94.

Leamer, Edward E., 1980. The Leontief Paradox, reconsidered. *Journal of Political Economy* 88, 495–503.

Leontief, Wasily W., 1953. Domestic production and foreign trade: The American capital position re-examined. *Proceedings of the American Philosophical Society* 97, 331–349.

Stern, Robert M. and Keith E. Maskus, 1981. Determinants of the structure of U.S. foreign trade, 1958–1976. *Journal of International Economics* 11, 207–224.

Sveikauskas, Leo, 1983. Science and technology in United States foreign trade. *Economic Journal* 93, 542–554.

Vanek, Jaroslav, 1968. The factor proportions theory: Then-factor case. *Kyklos* 21, 749–756.

Appendix: Data Sources

The Leontief inverse matrices for 1958 and 1972 were adapted from the 79-sector input–output matrices published in *Survey of Current Business* (September, 1965, and April, 1979, respectively). Production was measured as deliveries to final demand within these tables.

For purposes of computing factor–output ratios, output was measured as value of shipments by industry. Within manufacturing, these data were compiled by three-digit SIC industry from various issues of the Bureau of the Census, *Annual Survey of Manufactures.* Labor, production labor, and nonproduction labor figures were taken from these sources as well. Capital-output coefficients for 1958 were computed using constant dollar (1967) gross capital-stock data by three-digit SIC industry in U.S. General Services Administration, Office of Emergency Preparedness, *Capital Stock Figures* (*Final*), *Input–Output Sectors*, *SIC Level with Vintage* (May, 1974). Those for 1972 came from *Annual Survey of Manufactures: Industry Profiles* (1976). The human capital measure was constructed as the excess of the industry wage over the wage of unskilled labor (the median income of persons aged 25 and over with eight or less years of schooling), capitalized at 10 percent, multiplied by industry employment. Wages in manufacturing were taken from *Annual Survey of Manufactures*, while the unskilled wage was taken from various issues of the U.S. Bureau of the Census, *Current Population Report: Consumer Income*, Series P-60. The number of engineers and scientists for both manufacturing and nonmanufacturing for the 1960 and 1970 census years were taken from U.S. Bureau of Census, *Occupations of*

Employed Persons by Industry. Data on professional, technical, and kindred workers by industry were taken from this source as well.

Data on the values of exports and imports by industry were obtained from the Bureau of the Census, *U.S. Commodity Exports and Imports as Related to Output.* These data, along with shipments, production, and wages, were converted to constant (1967) dollars by means of price deflators constructed for each three-digit SIC industry. These deflators were constructed from U.S. Bureau of Labor Statistics, *Wholesale Prices and Price Indexes.*

The various data for the nonmanufacturing sectors were constructed from a variety of sources, primarily U.S. Bureau of the Census, *Statistical Abstract of the United States*, various issues. Details are available on request.

The assembled factor–output coefficients and trade data were then classified on a basis comparable to the industry coverage of the Leontief inverse matrices. These classifications were based on details reported in *Survey of Current Business.*

Resource endowment data were taken from three sources. First, for the 34-country experiments, Bowen (1980) provided figures on labor force, professional, technical, and kindred workers, and net capital stocks in 1963 and 1972. For the former two factors, 1958 data were generated through linear extensions from 1963. Net capital stocks in 1963 and 1972 were converted to gross stocks by applying the U.S. ratios of gross-to-net stocks in both years. This assumes equal depreciation rates across countries, which corresponds to Bowen's procedure. Second, for the six-country experiments, gross physical capital stocks were taken from Organization for Economic Cooperation and Development, *Flows and Stocks of Fixed Capital, 1955–1980.* Third, data on labor and professional, technical, and kindred workers for the six-country model were developed from International Labor Organization, *Yearbook of Labor Statistics*, various issues.

Finally, data on GNP, aggregate net exports, and exchange rates for the United States and selected countries were taken from various issues of International Monetary Fund, International Financial Statistics.

Chapter 12

Development-Related Biases in Factor Productivities and the HOV Model of Trade*

Keith E. Maskus and Shuichiro Nishioka†

1. Introduction

Early tests of the Heckscher–Ohlin–Vanek (HOV) model of international factor trade demonstrated that it failed to predict trade better than a coin toss (Maskus, 1985; Bowen, Leamer and Sveikauskas, 1987). As noted by Maskus (1985), the assumptions of the strict HOV model are too unrealistic to expect them to generate actual data.[1] Later tests relaxed many of these assumptions to generate extended HOV models that were more consistent with data (Trefler, 1995; Davis *et al.*, 1997; Davis and Weinstein, 2001; Hakura, 2001). Much of this analysis has focused on the unrealistic assumptions of internationally identical technologies and factor price equalization (FPE).

Trefler (1993) made a first important step to integrate international differences in factor-prices into the HOV model. He introduced a simple Hicks-neutral (HN) productivity modification at the individual factor level to measure endowments in productivity-equivalent units. For example, if the labor supplies of the United States and Brazil were the same, but U.S. workers were twice as productive, the

*Reprinted from *Canadian Journal of Economics*, Vol. 42, No. 2, pp. 519–553. Copyright 2009, with permission from John Wiley and Sons.

†We thank Ron Balvers, Keith Head, Wolfgang Keller, James Markusen and two referees for helpful comments. All errors remain our responsibility.

[1]The strict version assumes: (1) identical constant returns to scale (CRS) technology and factor price equalization; (2) perfectly competitive markets in goods and factors; (3) identical and homothetic preferences; (4) factor endowment differences; and (5) free trade in goods but not factors.

former nation would have twice as much labor at the productivity-equivalent level.[2] At the same time, the wage of U.S. workers would be twice that of Brazilian workers and ratios of factor prices could be used to infer relative productivities. This modification is consistent with the HOV model after adjusting for international differences in factor productivity.

Davis and Weinstein (2001) argued that Trefler's productivity modification is incomplete because it fails to introduce general differences in technology. With step-by-step relaxations of the standard HOV assumptions, they found substantial improvements in prediction power when national technologies are modified according to factor abundance measures.

Though both studies focused on modifying FPE, the conceptual distinction between their empirical approaches is important. Is it differences in productivity of factors or underlying technology that is responsible for factor price disparity? If it is because of factor-productivity differences, the HOV model is fundamentally acceptable, for its failures would come from the inability to measure factors in productivity-equivalent units. However, if the failures occur because of general technology differences, both the standard HOV model and FPE break down.

Several papers analyzing factor abundance have relied on Trefler's method to justify the introduction of productivity adjustments (Trefler, 1995; Antweiler and Trefler, 2002; Debaere, 2003; Debaere and Demiroglu, 2003; Fitzgerald and Hallak, 2004). However, the validity of his results has been questioned. Gabaix (1997), for example, showed that Trefler's adjustment to labor productivity (capital productivity) merely reflects differences in GDP/labor (or GDP/capital) due to his method of deriving productivities. There is surely a strong correlation between GDP per unit of factor and factor prices that may not be solely the result of differences in productivity. Thus, Trefler's claim of strong support for the standard HOV model is questionable unless factor productivities are estimated appropriately.

In this paper we introduce a different approach to estimate factor-specific productivities based solely on a constant returns to scale

[2] This was Leontief's (1953) conjecture to explain his celebrated paradox.

(CRS) production function.[3] This methodology permits extending Trefler's approach to HOV testing without facing the argument made by Gabaix. Using a newly constructed dataset covering 29 countries, we find evidence for the notion of basic factor-augmenting technology differences. Moreover, incorporating estimated factor productivities raises the fit of the standard HOV signs test from 46.6 percent to 72.4 percent and increases the variance ratio from 0.024 to 0.436.

More fundamentally, we also find that the estimated productivities are strongly correlated with aggregate factor abundance in a relative sense. For example, workers in Japan, which is capital abundant, are productive (relative to Japanese capital) because they have access to machines and computers that make them efficient. This systematic correlation between labor-productivity and capital-abundance, which is consistent with general principles of the factor-proportions model of trade without FPE, was previously discussed by Dollar, Wolff, and Baumol (1988).

Moreover, the idea of factor-specific productivity is strongly related to the literature on skill-biased technological change (Krusell *et al.*, 2000; Caselli and Coleman II, 2006). Because the rapid growth of physical capital interacts differently with different types of labor, capital productivity and labor productivity evolve differently with the stages of economic development. In particular, capital-skill complementarity could play a key role because the efficient operation of highly productive capital in developed countries requires skilled labor.[4] Its importance here is that the empirical success of Trefler's basic model can be attributable to systematic productivity differences across factors that cannot be obtained from the Hicks-neutral form. Thus, similar to Davis and Weinstein (2001) who adjusted national technologies according to factor abundance,

[3]Our approach builds on those of Davis and Weinstein (2001) and Maskus and Webster (1999). Maskus and Webster (1999) were concerned with ranking endowments across the United States and the United Kingdom, assuming the HOV model to be valid, rather than testing the trade model itself.

[4]Krusell *et al.* (2000) defined capital-skill complementarity as the situation where the elasticity of substitution between capital and unskilled labor is higher than that between capital and skilled labor. In this paper, using the data from Trefler (1993), we show that skilled-labor productivity and capital productivity co-move with economic development but these productivities move differently from unskilled-labor productivity.

our results also indicate the important link between technology, productivity, and factor abundance.[5] However, rather than general technology differences we consider only factor-augmenting, industry-neutral productivity variations. This more restricted specification achieves considerable success for HOV.

Our finding points out a potential danger in applying strict HOV-type models in a relative (i.e., bilateral) sense because factor productivity interacts systematically with factor abundance. For example, Debaere (2003) demonstrated that the HOV equation holds better for South–North country pairs than for North–North country pairs. But this finding raises the question of whether South–North factor-productivity gaps or South–North factor-abundance differences drive support for the factor contents of trade. In fact, we show that the likely reason Debaere found strong evidence only for South–North country pairs of particular factor combinations is systematic South–North differences in factor productivity. Because the abundant factor (unskilled labor) has limited access to capital in the South, its labor productivities are systematically lower than those of capital. However, Northern data do not entail this feature. This productivity gap makes it hard to conclude that the success of the relative factor-abundance model is purely derived from South–North differences in adjusted factor endowments. Rather, both differences in factor productivities and factor endowments are responsible, with the balance of each element being unclear.

Our analysis lies within the HOV framework of incomplete specialization adjusted for productivity differences. It does not test for the possibility of multiple specialization cones that can arise due to large endowment differences (Debaere and Demiroglu, 2003; Schott, 2003). While we do not dispute that possibility, our primary concern is to analyze whether a straightforward specification of technological differences, appropriately estimated, can achieve a significant improvement in HOV performance. Thus, in contrast to the complicated specifications in Davis and Weinstein (2001), we find that the HOV equation performs well with the simple correction of factor-augmenting productivity.

---

[5]Davis and Weinstein relied on the Dornbusch, Fischer, and Samuelson (1980) model and Helpman's (1999) multiple-cone model of specialization to motivate this correlation.

In addition, Schott (2003) argues that the standard use in HOV testing of U.S. technologies for all countries suppresses evidence of specialization. Here we incorporate consistent technology matrices from many countries and discover evidence supporting the extended HOV model, suggesting that the bias he mentions may not be significant at the industry level. Moreover, this dataset avoids some of the problems in data construction facing those authors. For example, Debaere and Demiroglu (2003) incorporate factor prices as proxies for productivity differences, but this procedure is likely to overstate those variations. We directly estimate these variables from international technology data.

We organize the paper as follows. In Section 2 we revisit Trefler's (1993) model and the criticism in Gabaix (1997). In Section 3 we set out our empirical results from the estimation of factor-augmenting productivities and relate them to Trefler's approach. In addition, we study the characteristics of estimated productivities, particularly the correlation between productivity and factor abundance. In Section 4 we examine the potential biases from ignoring factor productivities in the context of Debaere's (2003) relative factor-abundance model. Concluding remarks are offered in the final section.

2. The HOV Model and Factor-Augmenting Productivity

We begin by deriving the basic HOV prediction in a world with F factors, C countries, and N products (sectors). Assume that all countries have identical constant returns to scale production technology; markets for goods and factors are perfectly competitive; there are no barriers to trade and zero transportation cost; factors move freely within a country but do not move across countries; and the distribution of factors is consistent with integrated equilibrium so that factor prices are equalized across countries.

For each country c, the net-export vector can be obtained as the difference between net production and the final consumption:

$$\mathbf{T}_c = (\mathbf{I} - \mathbf{B}_c)\mathbf{Q}_c - \mathbf{C}_c \tag{1}$$

where $\mathbf{T}_c$ is an $N \times 1$ vector of net exports, $\mathbf{Q}_c$ is an $N \times 1$ vector of gross output, and $\mathbf{C}_c$ is an $N \times 1$ vector of final consumption. $\mathbf{B}_c$ is an $N \times N$ input–output (indirect) matrix for the unit intermediate requirements so that $(\mathbf{I} - \mathbf{B}_c)\mathbf{Q}_c$ equals the net output vector $\mathbf{Y}_c$.

Let $\mathbf{A}_c$ be the $F \times N$ direct technology matrix and its elements (a'_{cif}) represent the amount of a factor needed to produce one unit of gross output in sector i. Pre-multiplying Equation (1) by direct and indirect technology matrix $\mathbf{A}_c(\mathbf{I} - \mathbf{B}_c)^{-1}$, and applying the factor-exhaustion assumption $\mathbf{A}_c\mathbf{Q}_c = \mathbf{V}_c$ where $\mathbf{V}_c$ is an $F \times 1$ vector of factor endowments, we have that a country's factor contents of trade are the difference between factors absorbed in production $(\mathbf{A}_c\mathbf{Q}_c = \mathbf{V}_c)$ and factors absorbed in final consumption $(\mathbf{A}_c(\mathbf{I} - \mathbf{B}_c)^{-1}\mathbf{C}_c)$:

$$\mathbf{A}_c(\mathbf{I} - \mathbf{B}_c)^{-1}\mathbf{T}_c = \mathbf{V}_c - \mathbf{A}_c(\mathbf{I} - \mathbf{B}_c)^{-1}\mathbf{C}_c. \tag{2}$$

Assuming identical and homothetic preferences, along with identical prices of goods and services, the final consumption vector is proportional to the world net output vector $(\mathbf{Y}_W)$:

$$\mathbf{C}_c = s_c\mathbf{Y}_W \tag{3}$$

where s_c is a scalar representing the share of country c in world expenditure. Because the production technology is identical worldwide and there is FPE, the technologies of the United States may be used to derive the following standard HOV equation:

$$\mathbf{F}_c = \mathbf{V}_c - s_c\mathbf{V}_W \tag{4}$$

where $\mathbf{F}_c = \mathbf{A}_{US}(\mathbf{I} - \mathbf{B}_{US})^{-1}\mathbf{T}_c$ is measured factor contents of trade with U.S. technologies and $\mathbf{V}_c - s_c\mathbf{V}_W$ is predicted factor contents of trade.[6] Thus, the HOV model tells us that measured factor contents of trade for any country can be predicted by that country's factor endowments, world factor endowments, and final consumption shares.

To integrate factor productivity into the HOV model, Trefler (1993) introduced coefficients (π_{cf}) with the interpretation that if V_{cf} is the factor endowment of country c, then $V^*_{cf} = \pi_{cf}V_{cf}$ is the corresponding factor endowment measured in productivity-equivalent units. Let w_{cf} be the price per unit of V_{cf} and let w^*_{cf}

[6]To focus on predictions from the restrictive HOV model, we use U.S. technology to obtain measured factor contents of trade throughout this paper.

be the price per unit of V^*_{cf}. Since one unit of V_{cf} provides π_{cf} productivity-equivalent units of service, $1/\pi_{cf}$ units of V_{cf} provide one productivity-equivalent unit service priced at $w^*_{cf} = w_{cf}/\pi_{cf}$. Assuming identical technologies at the productivity-equivalent level and normalizing factor productivities of the United States to unity, we have this system of equations:

$$F_{cf} = \pi_{cf}V_{cf} - s_c \sum_{g=1}^{G} \pi_{gf}V_{gf} \tag{5}$$

$$w_{cf}/\pi_{cf} = w_{USf}/\pi_{USf} \Leftrightarrow w_{cf}/w_{USf} = \pi_{cf}/\pi_{USf} \tag{6}$$

where Equation (5) captures the elements of $\mathbf{F}_c = \mathbf{A}_{US}(\mathbf{I} - \mathbf{B}_{US})^{-1}$ $\mathbf{T}_c, \pi_{USf} = 1$, and g is the index of countries in the dataset.

2.1. *Trefler's Derivation of Factor-Productivity*

Trefler (1993) built the extended HOV model, adjusted by factor productivities, with the system of Equations (5) and (6), using a dataset for 33 countries. Equation (5) is the HOV model with productivity-equivalent factors and Equation (6) indicates that FPE holds when international factor productivities are adjusted. To estimate factor productivities (π_{cf}), Trefler derived Equation (7) from Equation (5):

$$\mathbf{F}_f = \mathbf{X}_f\mathbf{\Pi}_f$$

where

$$\mathbf{F}_f = \begin{bmatrix} F_{1f} \\ F_{2f} \\ \vdots \\ F_{Cf} \end{bmatrix},$$

$$\mathbf{X}_f = \begin{bmatrix} (1-s_1)V_{1f} & -s_1V_{2f} & \cdots & -s_1V_{Cf} \\ -s_2V_{1f} & (1-s_2)V_{2f} & \cdots & -s_2V_{Cf} \\ \vdots & \vdots & \ddots & \vdots \\ -s_CV_{1f} & -s_CV_{2f} & \cdots & (1-s_C)V_{Cf} \end{bmatrix}, \quad \mathbf{\Pi}_f = \begin{bmatrix} \pi_{1f} \\ \pi_{2f} \\ \vdots \\ \pi_{Cf} \end{bmatrix}. \tag{7}$$

Normalizing the productivities in terms of the United States, the π_{cf} parameters may be estimated by ordinary least squares (OLS). However, once these estimated factor productivities are introduced into Equation (5) it is inappropriate to apply standard testing procedures of the HOV model because fitted values for predicted factor contents of trade are identical to measured factor contents of trade. That is, all the HOV test statistics automatically would indicate a perfect fit.

To deal with this issue, Trefler set out two methods to demonstrate the validity of his estimated factor productivities. One was to check the signs of the productivity parameters, with all expected to be positive. The other was to study the correlation between relative price (w_{cf}/w_{USf}) and relative productivity (π_{cf}/π_{USf}) in Equation (6) for each factor, with the correlation expected to be unity. Trefler noted that the productivities estimated from Equation (7) were positive and that Equation (6) performed well, with the correlation for labor being 0.90 and that for physical capital being 0.68.

While the approach generated a number of comments, Gabaix (1997) in particular criticized this methodology for deriving the estimation method (Equation (7)) testing factor productivities. His reasoning came from the "missing trade" phenomenon analyzed in Trefler (1995). Missing trade is the finding that measured factor contents of trade are generally very small relative to predicted factor contents of trade. Thus, if the vector of measured factor contents of trade were virtually zero in Equation (5), we would have:

$$0 = \pi_{cf} V_{cf} - s_c V^*_{wf} \Leftrightarrow \pi_{cf} = s_c \frac{V^*_{wf}}{V_{cf}} = \frac{Y_c}{Y_w} \frac{V^*_{wf}}{V_{cf}} = \frac{Y_c}{V_{cf}} R \qquad (8)$$

where world aggregates ($V^*_{wf} = \Sigma_g \pi_{gf} V_{gf}$ and $R = V^*_{wf}/Y_w$) are essentially independent of the data from country c. In the case of labor, for example, the estimated productivity of labor would equal GDP per worker. Therefore, it is not surprising that Trefler's estimated productivities were positive and correlated strongly with factor prices. In this context, Trefler's approach offered no independent validation for the empirical success of his productivity modification of HOV.

Although Gabaix's criticism invalidates Trefler's methodology and statistical evidence, it does not necessarily mean the rejection of Trefler's model *per se*. Rather, if it were possible to estimate factor-productivity parameters independently of the equation system, incorporating them would not make HOV a truism and standard testing procedures would be valid. To this end, we develop unit total factor requirements (a_{cif} is each element of total technologies $\mathbf{A}_c(\mathbf{I}-\mathbf{B}_c)^{-1}$) for each country and estimate factor productivities for each country across sectors. These estimated parameters are then incorporated to test Equations (5) and (6). This procedure escapes the problems Gabaix (1997) pointed out.

2.2. *The Modified Approach*

Within Trefler's framework, countries share identical production technologies at the productivity-equivalent level, making adjusted unit factor requirements identical across countries for each factor: $a_{USif} = a^*_{cif}$ for country c and factor f where a^*_{cif} is $\pi_{cf}a_{cif}$. If firms minimize unit cost functions with CRS technology, the quantity of factor f required in sector i for country c divided by corresponding output is the unit factor requirement to produce one unit of gross output. Following Davis and Weinstein (2001), we estimate the productivity parameters by regressing the logarithm of unit factor requirements on country fixed effects and common unit factor requirements.[7]

We begin with Hicks-neutral productivity differences. With constant returns to scale, a single productivity (π_c) augments each factor at the same rate ($V^*_{cL} = \pi_c V_{cL}$ and $V^*_{cK} = \pi_c V_{cK}$). We estimate the following equation, which is the same as technology specification (P3) of Davis and Weinstein:

$$\ln a_{cif} = \theta_c + \ln \bar{a}_{if} + \varepsilon_{cif} \tag{9}$$

where $\ln a_{cif}$ is the log of measured unit factor requirements, $\ln \bar{a}_{if}$ is the log of cross-country average unit factor requirements, and θ_c is

[7]We also estimate factor-augmenting productivities, using the method of Maskus and Webster (1999), by regressing the unit factor requirements of the United States against those of each country across sectors. This approach confirmed that our results with Davis and Weinstein's technique are robust.

a vector of coefficients on country dummies. A conventional measure of national total factor productivity (TFP) is then $\pi_c = 1/\exp(\theta_c)$. We normalize these coefficients so that the U.S. measure is equal to unity ($\pi_{US} = 1$). Note that heteroskedasticity is likely to be present and so all technology regressions are weighted by $\bar{a}_{if}\sqrt{\ln(VA_{ci})/\bar{a}_f}$, where VA_{ci} indicates value added, as in Davis and Weinstein (2001).[8]

We next consider the more general case of factor-augmenting productivity adjustments by allowing $\theta_{cf} \neq \theta_{cf'}$:

$$\ln a_{cif} = \theta_{cf} + \ln \bar{a}_{if} + \varepsilon_{cif}. \tag{10}$$

Here, $(1/\exp(\theta_{cf}))$ is consistent with Trefler's (1993) and Maskus and Nishioka's (2006) estimations of factor-augmenting productivity with the U.S. normalization. To compare the performance of Equation (10) with various technology specifications with and without FPE in Davis and Weinstein (2001), we also estimate their technology specifications (P4), (P5), and (P5′) and present the results in Table 1.[9]

Using the estimated factor-productivities from Equation (10), we test both Equations (5) and (6). This implicitly assumes that no other sources of international differences in unit factor requirements (technologies) exist after international factor productivities are adjusted. Because the factor-productivities are estimated solely from unit factor requirements, it is possible to examine Equations (5) and (6) separately and apply standard testing procedures of the HOV model to Equation (5). In addition, we can separately assess Equation (6) in terms of the correlations between price and productivity for each factor.

In testing HOV we consider both the aggregate specification in (5) and the pair-wise HOV model (Staiger, Deardorff, and Stern, 1987; Hakura, 2001). The primary advantage of the pair-wise HOV model is that the testing equation does not include world aggregates. Because our dataset consists only of 29 countries, there is some question about data sums representing world aggregates. To derive the pair-wise model, apply Equation (4) to two arbitrarily chosen

---

[8]Simple corrections with robust standard errors do not change the results of these estimations.

[9]In Appendix B we describe these specifications.

Table 1. Results of technology estimations.

	Specifications from Davis and Weinstein (2001)					
	Hicks-Neutral (P3)	DFS with FPE (P4)	Helpman No-FPE (P5)	Unrestricted Helpman No-FPE (P5′)	Factor-Augmenting Productivity	
	Equation (9)	Equation (B.1)	Equation (B.2)	Equation (B.3)	Equation (10)	
	Implied TFP	Implied TFP	Implied TFP	Implied TFP	Labor Productivity	Capital Productivity
Australia	0.782	0.782	0.783	0.776	0.719	0.852
Austria	0.763	0.763	0.763	0.762	0.738	0.788
Belgium	0.829	0.830	0.830	0.837	0.859	0.802
Brazil (D)	0.561	0.554	0.556	0.492	0.233	1.312
Canada	0.866	0.866	0.867	0.864	0.812	0.925
Czech Republic (D)	0.609	0.607	0.609	0.578	0.415	0.889
Denmark	0.858	0.857	0.858	0.851	0.760	0.966
Finland	0.693	0.692	0.693	0.690	0.648	0.739
France	0.870	0.870	0.869	0.875	0.854	0.885
Germany	0.707	0.707	0.707	0.710	0.711	0.703
Greece (D)	0.696	0.694	0.696	0.670	0.514	0.940
Hungary (D)	0.606	0.603	0.606	0.566	0.371	0.982
Indonesia (D)	0.526	0.517	0.525	0.451	0.211	1.291
Ireland	0.828	0.826	0.828	0.805	0.707	0.963
Israel	0.984	0.982	0.982	0.979	0.877	1.100
Italy	0.836	0.836	0.836	0.835	0.834	0.837

(Continued)

Table 1. *(Continued)*

	Specifications from Davis and Weinstein (2001)					
	Hicks-Neutral (P3)	DFS with FPE (P4)	Helpman No-FPE (P5)	Unrestricted Helpman No-FPE (P5′)	Factor-Augmenting Productivity	
	Equation (9)	Equation (B.1)	Equation (B.2)	Equation (B.3)	Equation (10)	
	Implied TFP	Implied TFP	Implied TFP	Implied TFP	Labor Productivity	Capital Productivity
Japan	0.559	0.556	0.556	0.562	0.631	0.488
Korea (D)	0.573	0.573	0.573	0.570	0.529	0.620
Netherlands	0.797	0.797	0.798	0.792	0.712	0.892
New Zealand	0.854	0.854	0.856	0.847	0.754	0.968
Norway	0.821	0.824	0.824	0.831	0.736	0.919
Poland (D)	0.608	0.605	0.609	0.554	0.310	1.186
Portugal (D)	0.649	0.648	0.650	0.625	0.508	0.828
Slovak Republic (D)	0.524	0.522	0.524	0.493	0.334	0.817
Spain	0.745	0.745	0.745	0.737	0.694	0.799
Sweden	0.797	0.796	0.797	0.785	0.727	0.873
Turkey (D)	0.615	0.608	0.610	0.542	0.242	1.514
United Kingdom	0.857	0.856	0.858	0.833	0.706	1.039
United States	1.000	1.000	1.000	1.000	1.000	1.000
γKT		0.450 (0.019)	0.429 (0.019)			
γKN			0.560			
γLT		−0.450 (−0.019)	−0.472 (−0.019)			
γLN			−0.517 (−0.021)			

(Continued)

Table 1. (*Continued*)

	Specifications from Davis and Weinstein (2001)					
	Hicks-Neutral (P3)	DFS with FPE (P4)	Helpman No-FPE (P5)	Unrestricted Helpman No-FPE (P5′)	Factor-Augmenting Productivity	
	Equation (9)	Equation (B.1)	Equation (B.2)	Equation (B.3)	Equation (10)	
	Implied TFP	Implied TFP	Implied TFP	Implied TFP	Labor Productivity	Capital Productivity
Observations (T)	1676	1676	1676	1676	1676	
Parameters (k)	86	87	89	137	114	
-Log L (LL)	−890	−645	−242	−106	−218	
SIC	1.443	1.155	0.683	0.760	0.766	
AIC	1.164	0.874	0.395	0.297	0.397	

Notes: (1) Standard errors are reported in parentheses. (2) Hicks-neutral and factor-augmenting productivities are estimated from $1/\exp(\theta)$ with the normalization of U.S. = 1. (3) There are 29 industries and 2 factors (sector 1 agriculture is dropped from the sample). (4) The number of observations is not 1,682 ($= 29 * 29 * 2$). We have to drop 6 observations of unit factor requirement (a_{cif}) because they are zeros. (5) Developing countries (the South) are marked with (D). We sort countries by labor compensation per worker.

countries. For example, take the ratio of the United States ($c = 1$) and Japan ($c=2$) and cancel the net world output ($\mathbf{Y}_W$) in Equation (3) ($\mathbf{C}_1 = s_1/s_2\mathbf{C}_2 = \alpha\mathbf{C}_2$). Then, with appropriate subtraction, the pair-wise HOV model follows:

$$\mathbf{F}_1 - \alpha\mathbf{F}_2 = \mathbf{V}_1 - \alpha\mathbf{V}_2 \tag{11}$$

where $\mathbf{F}_1 - \alpha\mathbf{F}_2$ is the measured relative factor contents of trade with U.S. technology ($\mathbf{F}_1 = \mathbf{A}_{US}(\mathbf{I} - \mathbf{B}_{US})^{-1}\mathbf{T}_1$ and $\mathbf{F}_2 = \mathbf{A}_{US}(\mathbf{I} - \mathbf{B}_{US})^{-1}\mathbf{T}_2$) and $\mathbf{V}_1 - \alpha\mathbf{V}_2$ is the predicted relative factor content of trade.[10]

Using Equations (10) and (11), the pair-wise HOV model with factor-productivity adjustment follows:

$$\mathbf{F}_1 - \alpha\mathbf{F}_2 = \mathbf{\Pi}_1\mathbf{V}_1 - \alpha\mathbf{\Pi}_2\mathbf{V}_2. \tag{12}$$

Here $\mathbf{\Pi}_1$ is a diagonal $F \times F$ matrix with elements that are the corresponding productivity coefficients of country c (π_{cf}) estimated from Equation (10). Note that the measured relative factor contents of trade in the pair-wise model do not change due to the introduction of factor-augmenting productivity adjustments. Testing procedures are the same as for the basic HOV model.

3. Data and Empirical Results

Estimating factor-productivity parameters from Equation (10) requires data on actual technologies for multiple countries. Thus, we assembled a comprehensive dataset for a group of 29 countries in the year 2000 as described in Appendix A. There are two factors (physical capital and aggregate labor) and 30 industrial sectors. The dataset is similar to that in Davis and Weinstein (2001), who developed a 35-sector dataset of 10 advanced members of OECD. However, our data spans both developed and developing countries, with 19 being in the former category according to our sorting procedure.[11]

[10]This specification is more restrictive than that in Hakura (2001) because she used each country's actual technology to measure factor contents of trade.

[11]Country names may be found in Table 1. We sort the countries according to labor compensation per worker (average wage). Countries with average wage less than \$20,000 U.S. dollars (2000, PPP) are developing countries.

3.1. *Factor-Augmenting Productivity Estimates*

The final two columns of Table 1 report the estimated factor-productivity parameters and associated statistics for Equation (10), where factor efficiencies are defined relative to the United States. All coefficients measuring factor productivities are positive and statistically significant at the one-percent level. The coefficients on aggregate labor for all 29 countries are lower than unity, suggesting that the United States has the highest levels of labor productivity. Regarding physical capital, some developing countries (Brazil, Indonesia, Poland and Turkey) and some developed countries (Israel and the United Kingdom) are more productive than those in the United States.

For comparison purposes, the first four columns present national TFP estimates using Equation (9) of this paper and various technology specifications in Davis and Weinstein (2001). According to the Schwarz Information Criteria (SIC),[12] the technology estimation with factor-augmenting productivity adjustment is better than those with Hicks-neutral productivity adjustment (Equation (9)) and with Dornbusch–Fischer–Samuelson technology adjustment (Equation (B.1)) but is slightly worse than those with Helpman's specifications without FPE (Equations (B.2) and (B.3)). Thus, for estimating technology differences our factor-productivity adjustment performs at least as well as Davis and Weinstein's successful specifications.

It is also of interest to compare the national factor-productivity parameters developed using Trefler's (1993) method (Equation (7)) and those using our method in Equation (10). In Table 2 we list the parameters computed from Table 1 (the final two columns) and those in Trefler's paper (the next pair of columns). The correlations between the corresponding factors are very high, at 0.830 for physical capital and 0.987 for aggregate labor. Thus, Trefler's estimated factor-productivities are very similar to those obtained from estimation based only on unit factor requirements.

In addition, we compare these factor productivities with our TFP estimates from Equation (9) in the initial column of Table 1.

[12]$\text{SIC} = -2(LL)/(T) + k\ln(T)/T$ where LL is the log likelihood, T is the observation, and k is the number of independent variables.

Table 2. Estimated factor-augmenting productivity.

		Maskus and Nishioka (2008)		Trefler (1993)		
		(1) Labor	(2) Capital	(3) Labor	(4) Capital	TFP (5)
Australia		0.719	0.852	0.729	0.907	0.782
Austria		0.738	0.788	0.720	0.777	0.763
Belgium		0.859	0.802	0.865	0.800	0.829
Brazil (D)		0.233	1.312	0.234	1.186	0.561
Canada		0.812	0.925	0.784	0.882	0.866
Czech Republic (D)		0.415	0.889	0.463	0.935	0.609
Denmark		0.760	0.966	0.772	1.021	0.858
Finland		0.648	0.739	0.699	0.838	0.693
France		0.854	0.885	0.842	0.768	0.870
Germany		0.711	0.703	0.728	0.671	0.707
Greece (D)		0.514	0.940	0.595	1.065	0.696
Hungary (D)		0.371	0.982	0.444	1.149	0.606
Indonesia (D)		0.211	1.291	0.149	1.245	0.526
Ireland		0.707	0.963	0.741	1.302	0.828
Israel		0.877	1.100	0.916	1.102	0.984
Italy		0.834	0.837	0.847	0.812	0.836
Japan		0.631	0.488	0.642	0.547	0.559
Korea (D)		0.529	0.620	0.524	0.529	0.573
Netherlands		0.712	0.892	0.692	0.822	0.797
New Zealand		0.754	0.968	0.811	1.117	0.854
Norway		0.736	0.919	0.772	1.115	0.821
Poland (D)		0.310	1.186	0.328	1.163	0.608
Portugal (D)		0.508	0.828	0.500	0.895	0.649
Slovak Republic (D)		0.334	0.817	0.407	1.069	0.524
Spain		0.694	0.799	0.690	0.780	0.745
Sweden		0.727	0.873	0.723	0.948	0.797
Turkey (D)		0.242	1.514	0.283	1.262	0.615
United Kingdom		0.706	1.039	0.674	1.026	0.857
United States		1.000	1.000	1.000	1.000	1.000
	(1)	1.000	−0.457	0.987	−0.409	0.894
	(2)	—	1.000	−0.468	0.830	−0.037
Correlations	(3)	—	—	1.000	−0.369	0.885
	(4)	—	—	—	1.000	−0.009
	(5)	—	—	—	—	1.000

The correlation between TFP and our estimated labor-augmenting productivity is very high (0.894) but there is no correlation between TFP and capital productivity. In addition, the values of TFP generally lie between those for capital and labor. This would suggest that the empirical success of the factor-productivity adjustments in Trefler (1993) is attributable to systematic productivity differences across factors that the Hicks-neutral form (e.g., TFP) cannot account for. It also confirms previous findings in the literature that Hicks-neutral adjustments do not overturn failures of the HOV equation.

3.2. *Performance of the HOV Models*

Table 3 shows the results of testing the HOV model. The basic model, without factor-productivity adjustments, performs poorly as expected (Panel A). The sign fit is 46.6 percent for our two factors, the slope coefficient is 0.093, and the variance ratio is 0.024. Though the sign fit is marginally better than a coin-flip for a subset of developing countries, the slope and variance ratio tests indicate significant missing trade. The pair-wise tests do not achieve much better results. Neither is there a noticeable gain when a Hicks-neutral adjustment is added.[13] Therefore, the results strongly reject both the standard and Hicks-neutral HOV models regardless of the level of economic development.

However, once the estimated factor productivities are introduced, these numbers improve considerably. For the standard HOV specification, as shown in Panel B, the sign fit improves to 72.4 percent, the slope coefficient rises to 0.506, and the variance ratio increases to 0.436. The improvement for developed countries (the North) is particularly noteworthy. The sign test improves from 36.8 percent to 73.7 percent, the slope from 0.100 to 0.529, and the variance ratio from 0.023 to 0.431.

Regarding the pair-wise HOV model, Panel B also shows a considerable improvement when factor-productivity adjustments are incorporated. The sign test improves from 55.8 percent to 78.9 percent and the variance ratio improves from 0.100 to 0.786. These positive improvements from the pair-wise models suggest that the effective performance of Trefler's factor-productivity adjustments was not simply spurious.

For further perspective, the correlation between factor productivities and factor prices as specified in Equation (6) are very high, with

[13] To conserve space we do not report these results but they are available upon request.

Table 3. Results of the HOV models.

Panel A: Production Factors without Productivity Adjustments

		All (29 countries)			North (19 developed countries)			South (10 developing countries)			North–South Pairs		
		Total	Labor	Capital	Total	Labor	Capital	Total	Labor	Capital	Total	Labor	Capital
Standard HOV Test (Trefler, 1995)	Observations	29	29	29	19	19	19	10	10	10	—	—	—
	Sign Test	0.466	0.345	0.586	0.368	0.263	0.474	0.650	0.500	0.800	—	—	—
	Slope Test	0.093	0.029	0.093	0.100	0.042	0.100	−0.050	0.012	−0.051	—	—	—
	Standard error	0.017	0.004	0.024	0.019	0.004	0.027	0.054	0.004	0.080	—	—	—
	R-squared	0.326	0.664	0.309	0.414	0.858	0.397	0.021	0.476	−0.002	—	—	—
	Variance Test	0.024	0.001	0.024	0.023	0.002	0.022	0.058	0.000	0.061	—	—	—
Pair-Wise HOV Test (Hakura, 2001)	Observations	406	406	406	171	171	171	45	45	45	190	190	190
	Sign Test	0.558	0.485	0.631	0.591	0.614	0.567	0.544	0.711	0.378	0.532	0.316	0.747
	Slope Test	0.051	0.005	0.051	−0.020	0.048	−0.020	−0.076	0.020	−0.077	−0.005	0.017	−0.005
	Standard error	0.011	0.002	0.015	0.021	0.017	0.030	0.034	0.003	0.049	0.015	0.001	0.021
	R-squared	0.026	0.026	0.026	−0.011	0.031	−0.025	−0.021	0.485	−0.104	−0.037	0.566	−0.078
	Variance Test	0.100	0.001	0.102	0.157	0.049	0.160	0.103	0.001	0.097	0.079	0.001	0.079
Debaere's Test (Debaere, 2003)	Observations	406	406	406	171	171	171	45	45	45	190	190	190
	Sign Test	0.628	—	—	0.474	—	—	0.289	—	—	0.847	—	—
	Slope Test	0.040	—	—	0.108	—	—	−0.112	—	—	0.059	—	—
	Standard error	0.008	—	—	0.042	—	—	0.017	—	—	0.009	—	—
	R-squared	0.050	—	—	0.033	—	—	0.478	—	—	−0.291	—	—
	Variance Test	0.031	—	—	0.327	—	—	0.026	—	—	0.051	—	—

(Continued)

Table 3. (*Continued*)

Panel B: Production Factors with Factor-Augmenting Productivity Adjustments

		All (29 countries)			North (19 developed countries)			South (10 developing countries)			North–South Pairs		
		Total	Labor	Capital	Total	Labor	Capital	Total	Labor	Capital	Total	Labor	Capital
Standard	Observations	29	29	29	19	19	19	10	10	10	—	—	—
HOV Test	Sign Test	0.724	0.759	0.690	0.737	0.789	0.684	0.700	0.700	0.700	—	—	—
(Trefler,	Slope Test	0.506	0.426	0.506	0.529	0.752	0.528	0.309	0.199	0.309	—	—	—
1995)	Standard error	0.058	0.061	0.083	0.066	0.061	0.095	0.147	0.021	0.213	—	—	—
	R-squared	0.562	0.636	0.551	0.624	0.894	0.613	0.171	0.905	0.151	—	—	—
	Variance Test	0.436	0.284	0.425	0.431	0.645	0.421	0.555	0.043	0.617	—	—	—
Pair-Wise	Observations	406	406	406	171	171	171	45	45	45	190	190	190
HOV Test	Sign Test	0.789	0.788	0.791	0.775	0.772	0.778	0.778	0.733	0.822	0.805	0.816	0.795
(Hakura,	Slope Test	0.587	0.226	0.587	0.698	0.703	0.698	0.737	0.191	0.737	0.376	0.190	0.376
2001)	Standard error	0.023	0.014	0.033	0.032	0.043	0.045	0.046	0.019	0.066	0.032	0.015	0.045
	R-squared	0.438	0.404	0.438	0.578	0.603	0.572	0.720	0.686	0.698	0.239	0.461	0.209
	Variance Test	0.786	0.129	0.786	0.827	0.809	0.820	0.719	0.055	0.701	0.513	0.078	0.495
Debaere's	Observations	406	406	406	171	171	171	45	45	45	190	190	190
Test	Sign Test	0.490	—	—	0.602	—	—	0.400	—	—	0.411	—	—
(Debaere,	Slope Test	0.228	—	—	0.685	—	—	−0.228	—	—	0.095	—	—
2003)	Standard error	0.041	—	—	0.055	—	—	0.064	—	—	0.062	—	—
	R-squared	0.071	—	—	0.475	—	—	0.201	—	—	−0.554	—	—
	Variance Test	0.726	—	—	0.978	—	—	0.237	—	—	0.468	—	—

the correlation for aggregate labor being 0.97 and that for physical capital being 0.74.[14]

3.3. *Characteristics of Factor Productivity*

While introducing factor-productivity parameters (π_{cf}) is a convenient method to modify the HOV model, their interpretation is not entirely clear. Suppose that workers in the United States have the highest labor productivity. Then it could be that workers in the United States simply work harder than workers in other countries,[15] American workers are no more industrious as workers elsewhere but they have access to technologies that make them more efficient, or the simple number of workers cannot account for the difference in each worker's efficiency occurring from educational attainment (i.e., the human-capital approach).

We are particularly interested in the second possibility, under which we expect that labor productivity correlates positively with capital abundance.[16] This feature characterizes the data strongly between the developing countries and the developed countries, but weakly among the developed countries in an absolute sense. While the simple correlations between factor-augmenting productivity and the capital–labor ratio are −0.72 for capital and 0.89 for labor across all 29 countries, those among the developed countries are −0.54 and 0.24, respectively. Shown in Table 4 are the slope coefficients from simple regressions, with the corresponding partial correlations being −0.49 and 0.63, respectively, across all countries and −0.55 and 0.16, respectively, for the developed economies. Using unadjusted input requirements, capital productivities decline with capital abundance and labor productivities rise with capital abundance. One reason

---

[14]We obtain the factor prices from total compensation for each factor divided by total amount of the corresponding factors. Compensation for physical capital is derived from gross operating surplus, which is net operation surplus plus consumption of fixed capital.
[15]It would be ideal to adjust the total number of employees by annual working hours. For most OECD countries such data are available, but not for most developing countries. Once we adjust by annual working hours the labor productivities for most Western European countries increase by 5–10 percent.
[16]This link was previously discussed by Dollar, Wolff, and Baumol (1988) who found a systematic correlation between labor productivity and capital abundance.

Table 4. Correlations between factor-augmenting productivity and factor abundance.

Dependent variable		Labor productivity			Capital productivity		
Explanatory variables		Constant	K/L	R-squared	Constant	K/L	R-squared
All (29 observations)	Coefficient	0.144	0.628	0.793	1.303	−0.489	0.515
	(t-statistics)	(4.320)	(11.195)		(14.626)	(−4.667)	
North (19 observations)	Coefficient	0.609	0.162	0.060	1.387	−0.546	0.292
	(t-statistics)	(4.965)	(1.122)		(7.205)	(−2.494)	
South (10 observations)	Coefficient	0.164	0.476	0.860	1.481	−1.042	0.792
	(t-statistics)	(5.528)	(5.335)		(15.838)	(−6.208)	

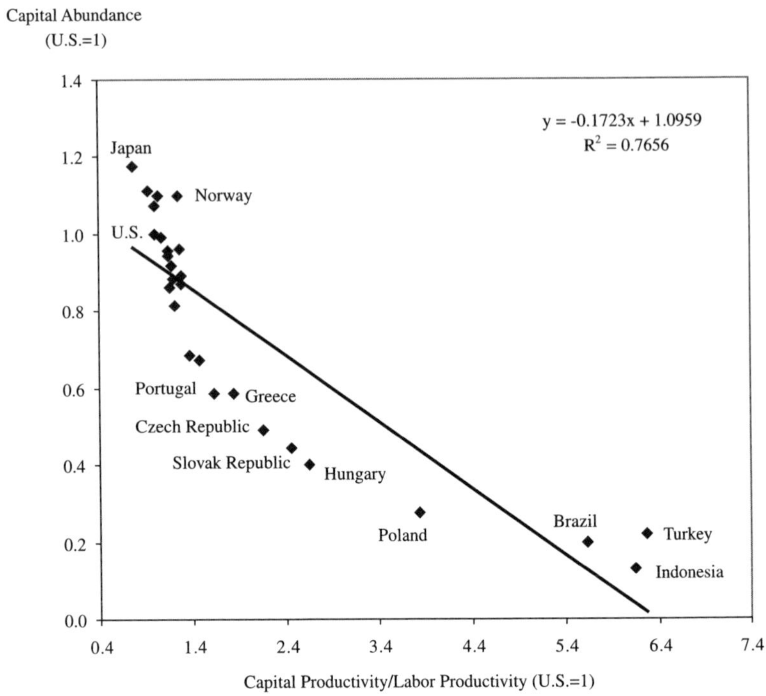

Figure 1. Relative factor productivity and relative factor abundance.

for these weak correlations among developed countries might be the limitation of our data to just two factors, with other elements such as knowledge capital and human capital being partially responsible for varying productivities.

However, when we incorporate the adjusted productivity ratios (π_{cL}/π_{cK}), they correlate strongly with corresponding factor endowments as shown in Figure 1. For example, capital-abundant Japanese workers are productive relative to Japanese capital because they have access to abundant capital (machines and computers). It seems that Trefler's original explanation holds well in this "relative" sense. This observation suggests that, similar to the approach of Davis and Weinstein (2001), who adjusted technologies according to factor abundance, our adjusted factor productivities also capture the link between technology, productivity, and factor-abundance that the Hicks-neutral form cannot accommodate.

4. Relative Factor Abundance Model and Factor Productivities

The strong correlation between factor abundance and factor productivity is particularly relevant to the relative factor-abundance model of Debaere (2003). Debaere developed a prediction of the factor content of trade for the HOV model that relates bilateral differences in endowments to bilateral differences in factor trade. Our objective here is to reexamine his conclusion that, based on the sign test, the trade of South–North country pairs is consistent with HOV but that of North–North country pairs is not. We show that Debaere's result is caused not only by South–North differences in factor endowments, which is the issue he emphasized, but also by South–North differences in factor productivity. Specifically, because unskilled labor, the abundant factor in the South, has limited access to skilled labor and physical capital, the productivity of unskilled workers there is systematically lower than that in the North. This difference is an additional important reason that only South–North country-pairs perform well in his examination of HOV.

4.1. *Endowment-Related Productivity Biases*

To develop Debaere's relative factor abundance model, take Equation (2) with U.S. technologies and impose identical and homothetic preferences:

$$\mathbf{F}_c = \mathbf{V}_c - \mathbf{A}_{US}(\mathbf{I} - \mathbf{B}_{US})^{-1} s_c \mathbf{Y}_W. \tag{13}$$

Divide both sides of Equation (13) by the scalar expenditure share s_c to obtain:

$$\mathbf{F}_c^* = \mathbf{V}_c^* - \mathbf{A}_{US}(\mathbf{I} - \mathbf{B}_{US})^{-1} \mathbf{Y}_W \tag{14}$$

where $\mathbf{F}_c = \mathbf{A}_{US}(\mathbf{I} - \mathbf{B}_{US})^{-1}\mathbf{T}_c$, $\mathbf{F}_c^* = \mathbf{F}_c/s_c$ and $\mathbf{V}_c^* = \mathbf{V}_c/s_c$. Now consider Equation (14) for two countries, c and c', and take the difference between their expressions:

$$\mathbf{F}_c^* - \mathbf{F}_{c'}^* = \mathbf{V}_c^* - \mathbf{V}_{c'}^*. \tag{15}$$

Equation (15) may be expressed for a particular factor (f) and divided by the sum of factor endowments, $V^*_{cf} + V^*_{c'f}$:

$$\frac{F^*_{cf} - F^*_{c'f}}{V^*_{cf} + V^*_{c'f}} = \frac{V^*_{cf} - V^*_{c'f}}{V^*_{cf} + V^*_{c'f}}. \tag{16}$$

Then, express Equation (16) for another factor (f') and again take differences:

$$\begin{aligned}
&\frac{F^*_{cf} - F^*_{c'f}}{V^*_{cf} + V^*_{c'f}} - \frac{F^*_{cf'} - F^*_{c'f'}}{V^*_{cf'} + V^*_{c'f'}} \\
&\quad= \frac{V^*_{cf} - V^*_{c'f}}{V^*_{cf} + V^*_{c'f}} - \frac{V^*_{cf'} - V^*_{c'f'}}{V^*_{cf'} + V^*_{c'f'}} \\
&\quad\Leftrightarrow \frac{F^*_{cf} - F^*_{c'f}}{V^*_{cf} + V^*_{c'f}} - \frac{F^*_{cf'} - F^*_{c'f'}}{V^*_{cf'} + V^*_{c'f'}} \\
&\quad= \frac{2V^*_{c'f}}{V^*_{cf'} + V^*_{c'f'}} \left(\frac{V^*_{c'f'}}{V^*_{c'f}} - \frac{V^*_{cf'} + V^*_{c'f'}}{V^*_{cf} + V^*_{c'f}} \right).
\end{aligned} \tag{17}$$

Here, the relative difference in measured factor content of trade is on the left-hand side and the relative difference in predicted factor content of trade is on the right-hand side.

For any two factors f and f', a country c is said to be relatively abundant in factor f compared to country c' when $V^*_{cf}/V^*_{cf'}$ is larger than $V^*_{c'f}/V^*_{c'f'}$. This statement is easily shown because the relative factor abundance relationship $V^*_{cf}/V^*_{cf'} > V^*_{c'f}/V^*_{c'f'}$ holds if and only if $V^*_{c'f'}/V^*_{c'f} > (V^*_{cf'} + V^*_{c'f'})/(V^*_{cf} + V^*_{c'f})$, which is the right-hand side of the second equation in (17). Therefore, the testing strategy is to check the sign concordance of measured and predicted relative differences in factor trade.[17] Debaere showed that Equation (17) holds for the case of Hicks-neutral productivity. We show in Appendix C that it holds also for the case of adjusted factor productivities.

Our tests using the 29-country dataset are reported in Table 3. As discussed in Debaere (2003), there are substantial differences in the

[17]Though Debaere (2003) reported only sign tests, we report slope tests and variance ratios as well.

performance of Equation (17) for varying pairs of countries, depending on the stages of economic development. In particular, the sign match of country pairs when picking one North and one South is 84.7 percent, which is significantly better than the cases with North–North (47.4 percent) and South–South (28.9 percent). This finding is consistent with Debaere's basic result using either unadjusted or Hicks-neutral-adjusted factor endowments and U.S. technology. However, as seen in Panel B, adjusting endowments by factor productivities makes a critical difference in the performance of the relative HOV model for pairs of developed countries. In those cases the sign match improves from 47.4 percent to 60.2 percent and the variance ratio improves from 0.327 to 0.978. However, once factors are adjusted by our estimated factor productivities, these testing statistics for North–South pairs worsen significantly from 84.7 percent to 41.1 percent for the sign test, even though the variance ratio improves from 0.051 to 0.468. The reduction in success of the sign test suggests implicitly that Debaere's finding was attributable, in an important degree, to productivity biases.

To explore further why the North–South pairs perform poorly under the relative HOV model with factor-augmenting productivity adjustments, we show that the relative factor abundance ratios ($V^*_{cf}/V^*_{cf'} > V^*_{c'f}/V^*_{c'f'}$) in Debaere's model can be decomposed into two parts: (1) relative factor-productivity ratios; and (2) productivity-equivalent relative factor abundance. Denote factor endowments at the productivity-equivalent level as $V'_{cf} = \pi_{cf} V^*_{cf}$ and rewrite relative factor abundance as:

$$\frac{V^*_{cf}}{V^*_{cf'}} > \frac{V^*_{c'f}}{V^*_{c'f'}} \Leftrightarrow \frac{V'_{cf}/\pi_{cf}}{V'_{cf'}/\pi_{cf'}} > \frac{V'_{c'f}/\pi_{c'f}}{V'_{c'f'}/\pi_{c'f'}} \Leftrightarrow \frac{V'_{cf}}{V'_{cf'}}\frac{\pi_{cf'}}{\pi_{cf}} > \frac{V'_{c'f}}{V'_{c'f'}}\frac{\pi_{c'f'}}{\pi_{c'f}}. \tag{18}$$

Equation (18) explains that the relative factor-abundance ratio without productivity adjustments ($V^*_{cf}/V^*_{cf'}$ or $V^*_{c'f}/V^*_{c'f'}$) is the product of the productivity-equivalent relative factor abundance ratio ($V'_{cf}/V'_{cf'}$ or $V'_{c'f}/V'_{c'f'}$) and the factor-productivity ratio ($\pi_{cf'}/\pi_{cf}$ or $\pi_{c'f'}/\pi_{c'f}$). If, as Debaere assumed, the Hicks-neutral form ($\pi_{cf'}/\pi_{cf} = \pi_{c'f'}/\pi_{c'f}$) is realistic, then relative factor abundance and productivity-equivalent factor abundance are identical and his basic conclusion holds. However, if productivity adjustments are more general, then both elements matter. For example, if f is labor (L) and f' is physical capital (K) for the South ($c = S$) and the North ($c' = N$), we might expect labor in the South to be less productive than in the North because it operates with a smaller relative capital endowment. As a result, we have an inequality in relative productivity ratios: $\pi_{SK}/\pi_{SL} > \pi_{NK}/\pi_{NL}$ or $\pi_{SK}/\pi_{NK} > \pi_{SL}/\pi_{NL}$. Here, inequality of the factor-productivity ratio might be the same as that of efficiency-adjusted factor abundance.

It is important, therefore, to study South–North differences in factor productivities in addition to relative factor endowments. For this purpose, using our 29-country dataset, we divide countries into the South and the North and develop the South–North productivity ratios for factors. These productivity figures for aggregate labor and physical capital are displayed in Figures 2.1, 2.2, and 2.3.

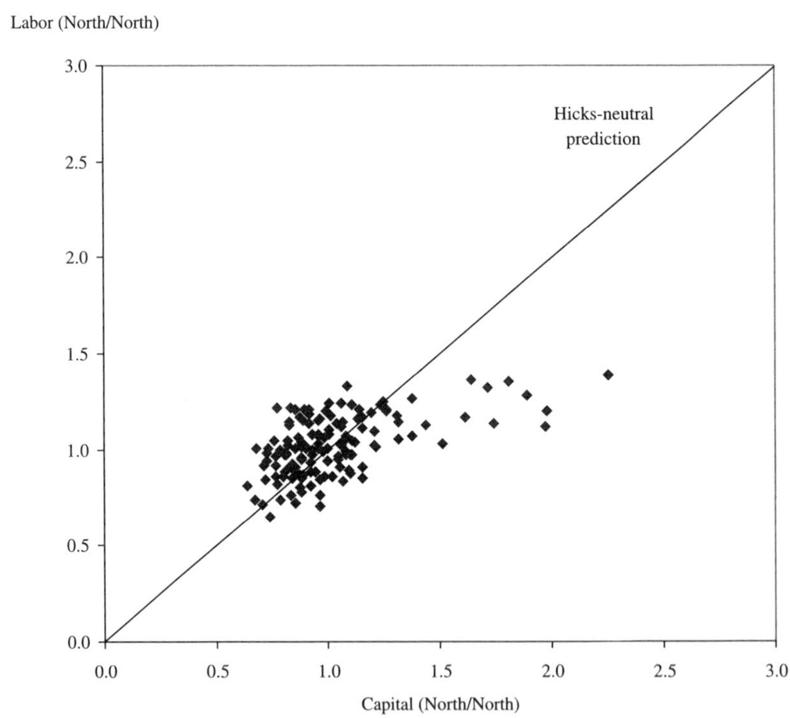

Figure 2.1. Productivity ratios of labor–capital (North–North pairs).

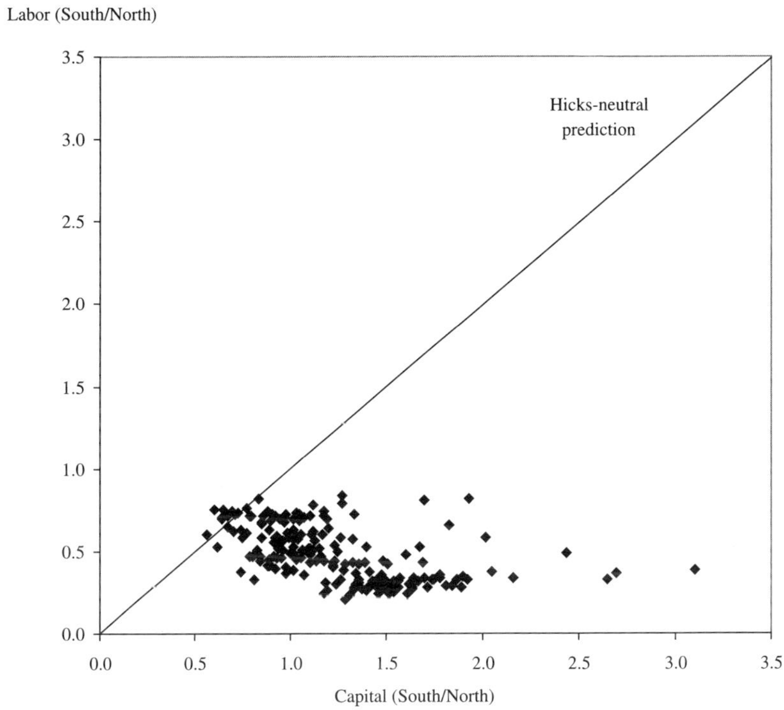

Figure 2.2. Productivity ratios of labor–capital (South–North pairs).

If Hicks-neutral productivity differences were realistic, we would expect these ratios to be identical across any factor pair ($\pi_{cf'}/\pi_{c'f'} = \pi_{cf}/\pi_{c'f}$). However, while this tendency is evident in North–North pairs (Figure 2.1), it is not the case for North–South pairs (Figure 2.2). Rather, we find the interesting tendency that the productivity ratio of the South to the North for aggregate labor (π_{SL}/π_{NL}) is nearly uniformly smaller than that for physical capital (π_{SK}/π_{NK}). Regarding South–South comparisons (Figure 2.3), there is a negative relationship between relative factor productivities. Therefore, systematic tendencies in factor productivities support the inequality in Equation (18) only for the South–North country pairs.

To conclude our analysis we investigate this issue using the same data as Debaere. Because his dataset is the same as that in Trefler (1993), it is not possible to estimate appropriate factor productivities by using bilateral simple regressions as in Equation (10). Our compromise is to use Trefler's method in Equation (7) to obtain

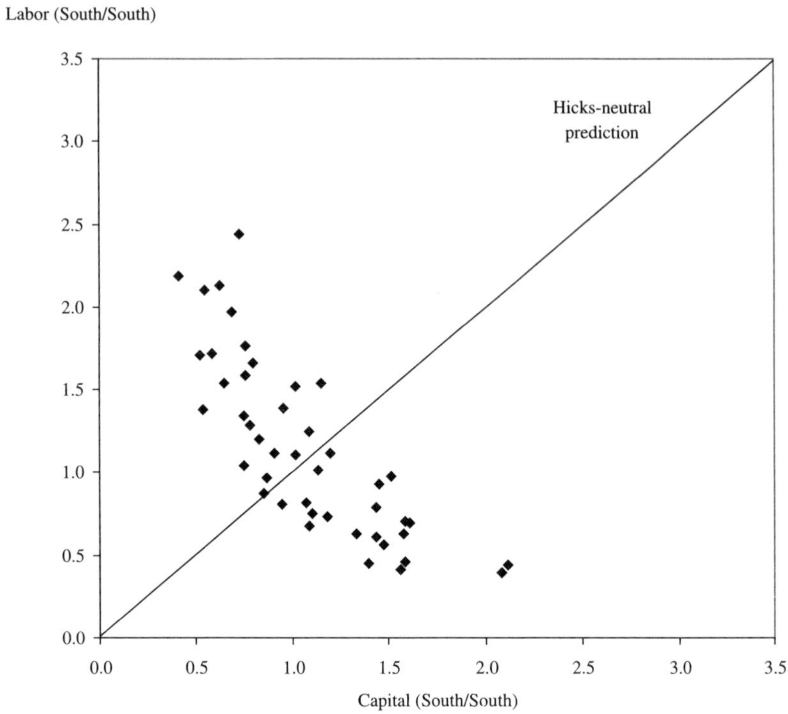

Figure 2.3. Productivity ratios of labor–capital (South–South pairs).

factor productivities. We divide countries into the South and the North according to Debaere (2003), and develop the South–North productivity ratios for factors. The parameters (π_{cf}) are obtained for physical capital, skilled labor, and unskilled labor. Similar to our previous results, we find that productivity ratios of the South to the North for unskilled labor are smaller than those for capital (Figure 3.4) and skilled labor (Figure 3.6). There is not a similar tendency for the North–North pairs of capital/skilled labor and unskilled labor/skilled labor, nor for any country pairs of skilled labor/capital. Thus, the inequality in Equation (18) is found in data only for the South–North country pairs of unskilled labor/skilled labor and unskilled labor/capital.[18]

[18]This same factor-productivity bias exists for the case of aggregate labor/capital in South–North pairs, but we do not report it here.

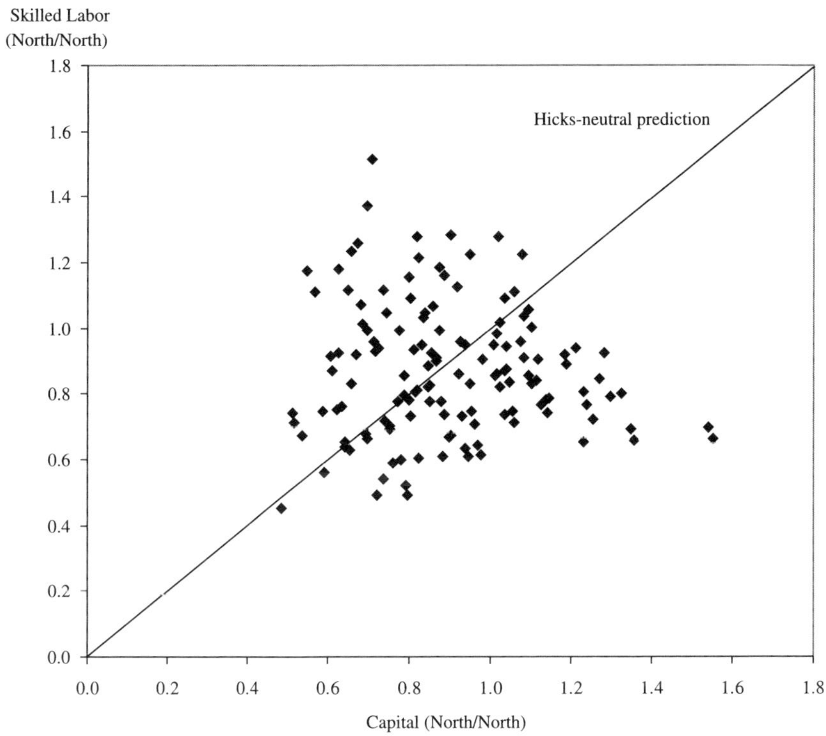

Figure 3.1. Productivity ratios of capital–skilled labor (North–North pairs).

In summary, the superior performance of the relative HOV model for North–South comparisons than for North–North and South–South pairs is mainly derived from these development-related biases in factor productivities. This evidence implies that Debaere's conclusion could be delivered by the interplay between endowment differences and factor-productivity differences.

4.2. *Relation to Other Explanations of Productivity Biases*

Here we have attributed relative factor-productivity biases solely to differences in relative factor abundance. However, there are other possible sources of these biases. In particular, capital-skill complementarity (Krusell *et al.*, 2000) and imperfect substitutability between

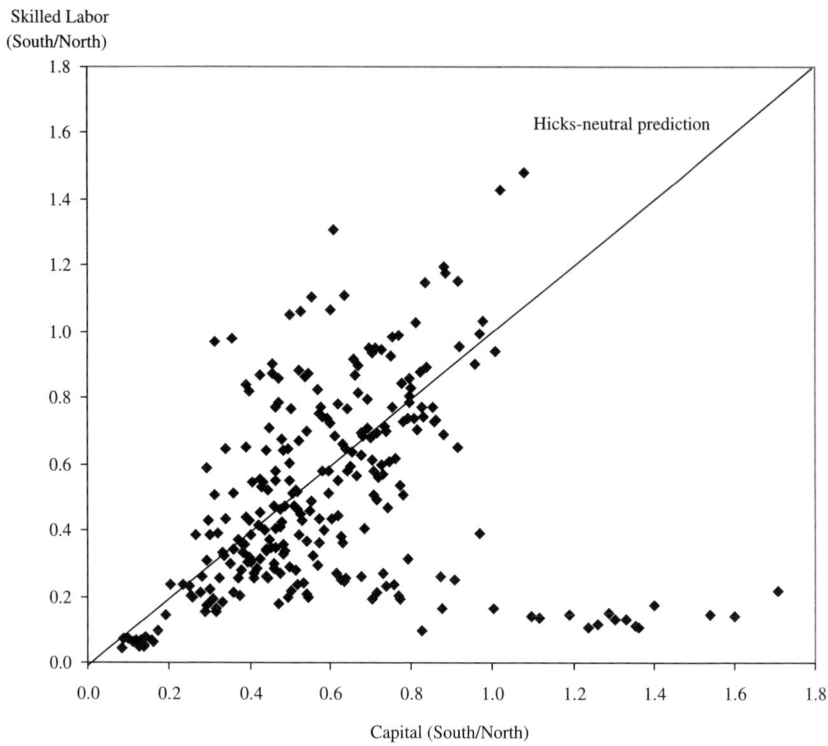

Figure 3.2. Productivity ratios of capital–skilled labor (North–South pairs).

skilled labor and unskilled labor (Caselli and Coleman II, 2006) have been cited as important possibilities. Regarding capital-skill complementarity, our analysis of Trefler's dataset unearths an implication similar to that of Krusell *et al.* (2000). Specifically, capital productivity and skilled-labor productivity co-move with economic development. As seen in Figure 3.2, where most observations are in the lower left-hand corner, developing countries have lower relative productivity in both skilled labor and capital. In Figure 3.1, however, the observations are concentrated in the center, suggesting convergence of these productivities among the developed countries. If capital and skilled labor are complements, the higher productivity performance of skilled labor in the North might be associated with higher quality and sophistication of capital in the North. Our approach does not amount to a test of such complementarity, however.

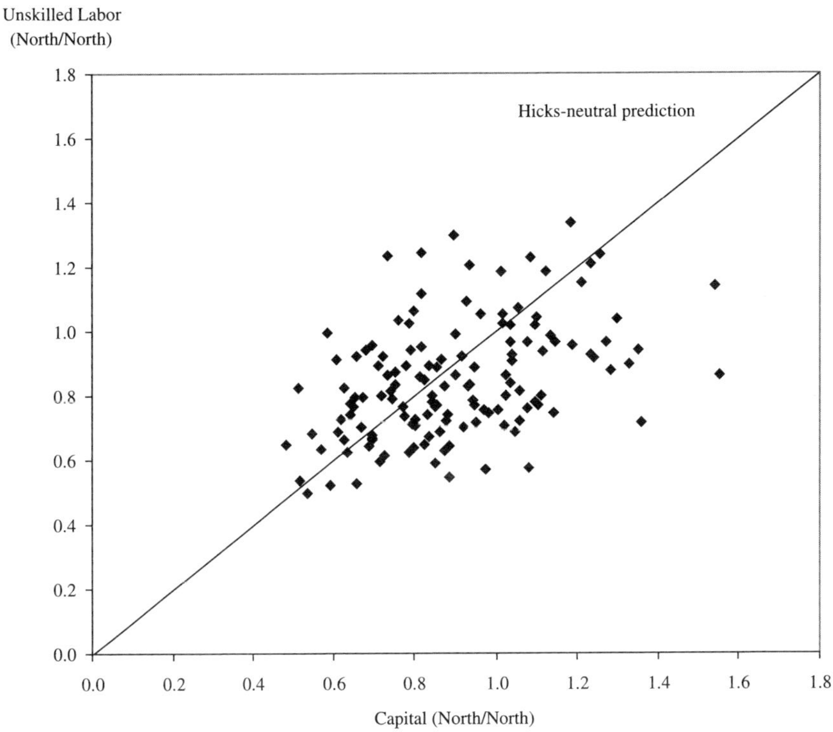

Figure 3.3. Productivity ratios of capital–unskilled labor (North–North pairs).

On the other hand, as seen in Figures 3.4 and 3.6, the South's usage of unskilled labor is unproductive relative to the North's usage, regardless of the relative productivities of capital and skilled labor. This tendency of skill bias is quite different from the finding in Caselli and Coleman II (2006).[19] They found evidence that developed countries use skilled labor more efficiently and developing countries use

[19]Strictly speaking, the calculation of factor productivities for Caselli and Coleman II (2006) is different from ours even though both assume constant return to scale production functions. Caselli and Coleman II estimated country-level production functions with a constant-elasticity-of-substitution (CES) aggregate of labor types: $y = K^{\alpha}[(AuLu)^{\sigma} + (AsLs)^{\sigma}]^{(1-\alpha)/\sigma}$ where Au and As are factor-augmenting productivities for unskilled-labor and skilled-labor, respectively. They calibrated these productivities by introducing the condition that skill premium equals relative marginal products of skills: $ws/wu = (As/Au)^{\sigma}\ (Ls/Lu)^{(\sigma-1)}$, and by using data measuring Ls, Lu, ws/wu, and parameters $\alpha = 1/3, 1/(1-\sigma) = 1.4$.

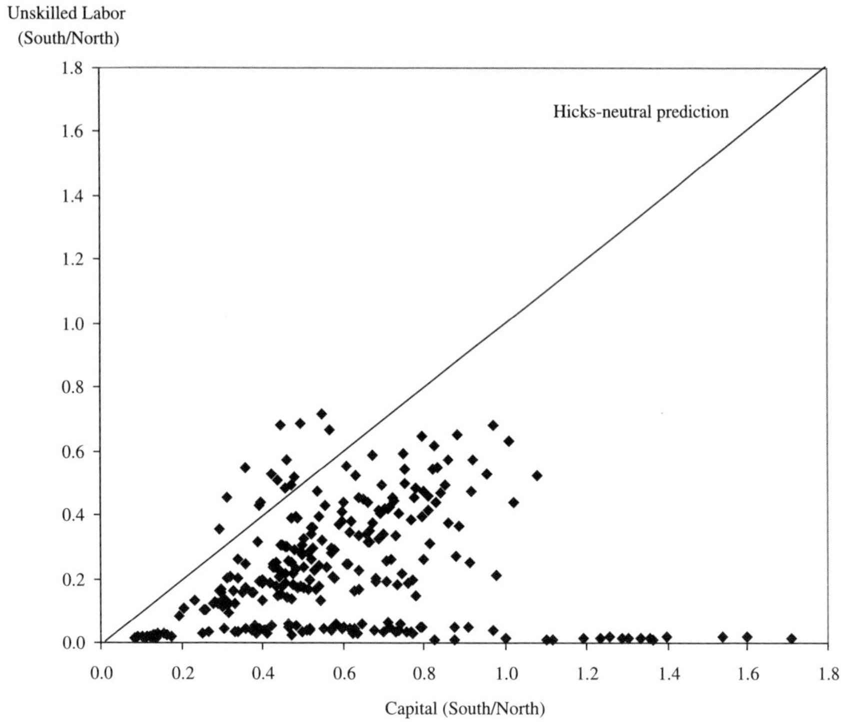

Figure 3.4. Productivity ratios of capital–unskilled labor (South–North pairs).

unskilled labor more efficiently. They explained this cross-country pattern of skill bias by imagining two different technologies to produce aggregate output. One is an assembly line where unskilled labor works with the supervision of just a few skilled workers. The other is a computer-controlled facility run solely by skilled workers while unskilled labor engages in janitorial work. They argue that the South chooses the first prototypical technology and the North chooses the second. Therefore, unskilled labor in the South is absolutely more productive.

We suggest two explanations why the dataset we use does not support this view, finding instead that unskilled labor is unproductive in the South. First, unskilled labor in Trefler's dataset consists of clerical, service, sales, agriculture, and production workers based on the International Standard Classification of Occupations. Therefore,

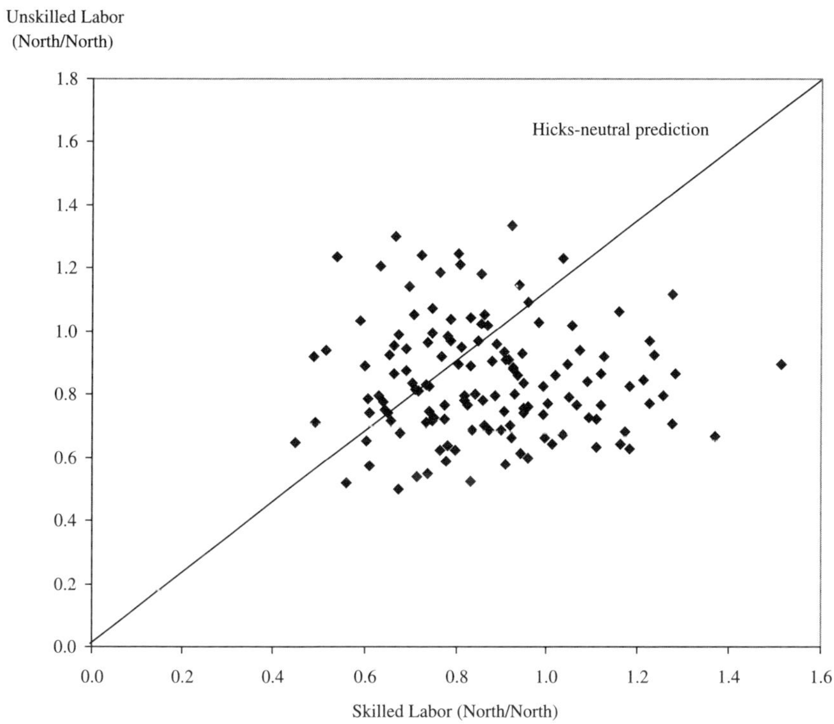

Figure 3.5. Productivity ratios of skilled–unskilled labor (North–North pairs).

Caselli and Coleman II's idea that unskilled workers operate solely as low-productivity support workers in the North is not applicable. As their preferred data show, their unskilled-labor measure amounts to less than three percent of skilled labor for the United States, indicating that unskilled workers are equivalent to the labor force that has virtually no educational achievement. Second, because of data limitations, we could not adjust Trefler's dataset to reflect international differences in educational attainment. Even in jobs for production facilities, workers with higher education might be employed in developed countries more than in developing countries. In contrast to the data we used, Caselli and Coleman II divided workers into skilled and unskilled labor from the categories of educational achievement set out by Barro and Jong-Wha Lee (2001). They weighted different types of education by relative wages so that the

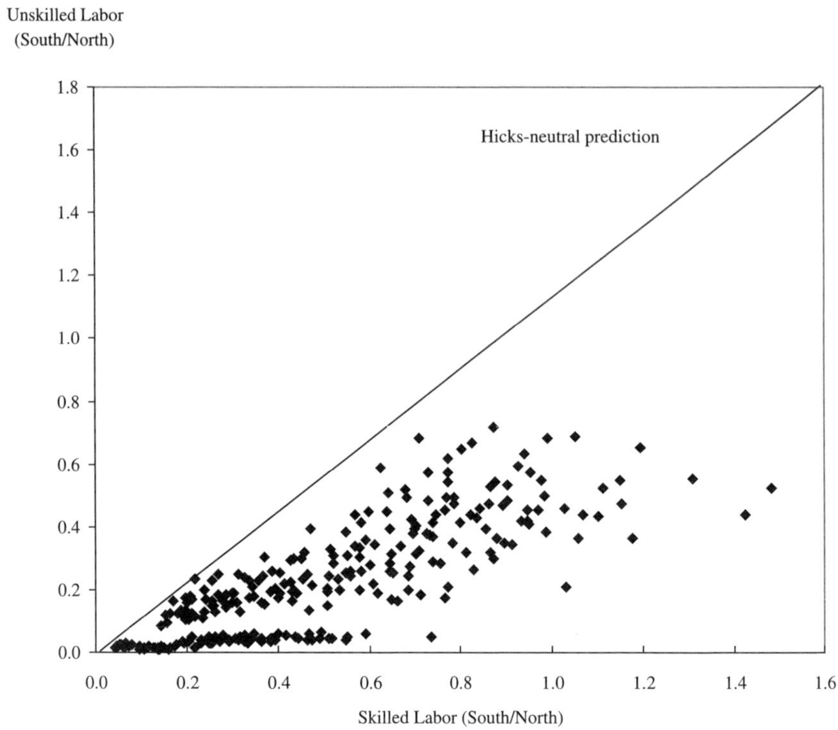

Figure 3.6. Productivity ratios of skilled–unskilled labor (South–North pairs).

number of unskilled labor (skilled labor) is equivalent to that of no-education labor (workers completing primary education). Our inability to adjust for education (human capital) might affect our estimates of factor productivities based on occupational differences. Thus, a useful subject for future research would be to combine the data on detailed educational achievement and wages with the HOV equation.

5. Concluding Remarks

In this paper we reexamine Trefler's (1993) basic factor-productivity model. Departing from his procedure, we estimate factor-productivity parameters from each country's actual technologies. This approach permits use of the standard evaluations of the HOV model (sign test, slope test, and variance ratio test). Using a dataset of 29 countries, we find evidence supporting the fundamental idea of factor-augmenting productivities, with both the sign concordance and the variance ratio increasing markedly. Our results indicate that factor-augmenting productivity differences are an appropriate modification of HOV models.

Prior studies that made technological specifications increasingly more flexible by using country-specific data also supported the extended HOV model (Hakura, 2001; Davis and Weinstein, 2001). Indeed, Davis and Weinstein (2001) established a strong fit of the HOV equations when technologies are modified across both industries and countries according to factor abundance. The contribution here is to show that a simpler modification — factor augmentation that is neutral across industries — can establish considerable gains in the predictive performance of the HOV model. More fundamentally, the analysis unearthed a particular feature of factor productivities in both our 29-country data from the year 2000 and Trefler's earlier dataset. Specifically, factor productivities are inversely correlated with own-factor endowments and positively correlated with other factor endowments, which is consistent with the neoclassical trade model without FPE due, say, to specialization within different cones. This is especially the case as regards labor in developing countries. As a result, Debaere's (2003) finding that South–North factor trade may be explained well by the relative HOV model with Hicks-neutral productivity differences needs to be supplemented by the interplay between relative endowments and factor productivities.

References

Antweiler, W. and D. Trefler, 2002, Increasing Returns and All That: A View from Trade, *American Economic Review* 92, 93–119.

Barro, R. J. and J.-W. Lee, 2001, International Data on Educational Attainment: Updates and Implications, *Oxford Economic Papers* 53, 541–563.

Bowen, H. P., E. E. Leamer and L. Sveikauskas, 1987, Multicountry, Multifactor Tests of the Factor Abundance Theory, *American Economic Review* 77, 791–809.

Caselli, F. and W. J. Coleman II, 2006, The World Technology Frontier, *American Economic Review* 96, 499–522.

Davis, D. R. and D. E. Weinstein, 2001, An Account of Global Factor Trade, *American Economic Review* 91, 1423–1453.

Davis, D. R., D. E. Weinstein, S. C. Bradford and K. Shimpo, 1997, Using International and Japanese Regional Data to Determine When the Factor Abundance Theory of Trade Works, *American Economic Review* 87, 421–446.

Debaere, P., 2003, Relative Factor Abundance and Trade, *Journal of Political Economy* 111, 589–610.

Debaere, P. and U. Demiroglu, 2003, On the Similarity of Country Endowments, *Journal of International Economics* 59, 101–136.

Dollar, D., E. N. Wolff and W. J. Baumol, 1988, The Factor-Price Equalization Model and Industry Labor Productivity: An Empirical Test Across Countries, in R. C. Feenstra, ed., *Empirical Methods for International Trade* (Cambridge, MA: MIT Press).

Dornbusch, R., S. Fischer and P. A. Samuelson, 1980, Heckscher–Ohlin Trade Theory with a Continuum of Goods, *Quarterly Journal of Economics* 95, 203–224.

Feenstra, R. C. and G. H. Hanson, 2000, Aggregation Bias in the Factor Content of Trade: Evidence from U.S. Manufacturing, *American Economic Review* 90, 155–160.

Fitzgerald, D. and J. C. Hallak, 2004, Specialization, Factor Accumulation and Development, *Journal of International Economics* 64, 277–302.

Gabaix, Xavier, 1997, The Factor Content of Trade: A Rejection of the Heckscher–Ohlin–Leontief Hypothesis, Harvard University, Department of Economics, manuscript.

Hakura, D. S., 2001, Why Does HOV Fail? The Role of Technological Differences Within the EC, *Journal of International Economics* 54, 361–382.

Helpman, E., 1999, The Structure of Foreign Trade, *Journal of Economic Perspectives* 13, 121–144.

Heston, A., R. Summers and B. Aten, 2006, Penn World Table Version 6.2, Center for International Comparisons of Production, Income and Prices at the University of Pennsylvania.

Keller, W., 2000, Do Trade Patterns and Technology Flows Affect Productivity Growth? *World Bank Economic Review* 14, 17–47.

Krusell, P., L. E. Ohanian, J.-V. Ríos-Rull and G. L. Violante, 2000, Capital-Skill Complementarity and Inequality: A Macroeconomic Analysis, *Econometrica* 68, 1029–1053.

Lai, H. and S. Chun Zhu, 2007, Technology, Endowments, and the Factor Content of Bilateral Trade, *Journal of International Economics* 71, 389–409.

Leamer, E. E., 1984, *Sources of International Comparative Advantage: Theory and Evidence* (Cambridge, MA: MIT Press).

Leontief, W., 1953, Domestic Production and Foreign Trade: The American Capital Position Re-examined, *Proceedings of the American Philosophical Society* 97, 332–349.

Maskus, K. E., 1985, A Test of the Heckscher–Ohlin–Vanek Theorem: The Leontief Commonplace, *Journal of International Economics* 19, 201–212.

Maskus, K. E. and A. Webster, 1999, Estimating the HOV Model with Technology Differences Using Disaggregated Labor Skill for the United States and the United Kingdom, *Review of International Economics* 7, 8–19.

Maskus, K. E. and S. Nishioka, 2006, Development-Related Biases in Factor Productivities and the HOV Model of Trade, University of Colorado, Department of Economics, manuscript.

Schott, P. K., 2003, One Size Fits All? Heckscher–Ohlin Specialization in Global Production, *American Economic Review* 93, 686–708.

Staiger, R. W., A. V. Deardorff and R. M. Stern, 1987, An Evaluation of Factor Endowments and Protection as Determinants of Japanese and American Foreign Trade, *Canadian Journal of Economics* 20, 449–463.

Trefler, D., 1993, International Factor Price Differences: Leontief was Right! *Journal of Political Economy* 101, 961–987.

Trefler, D., 1995, The Case of the Missing Trade and Other Mysteries, *American Economic Review* 85, 1029–1046.

Appendix A: Data Construction

A.1. *Input–Output Data*

Input–output (I-O) tables (total use) for Australia (1998–1999), Austria (2000), Belgium (2000), Brazil (2000), Canada (2000), Czech Republic (2000), Denmark (2000), Finland (2000), France (2000), Germany (2000), Greece (1999), Hungary (2000), Indonesia (2000), Ireland (1998), Israel (1995), Italy (2000), Japan (2000), Korea (2000), the Netherlands (2000), New Zealand (1995–1996), Norway (2001), Poland (2000), Portugal (1999), Slovak Republic (2000), Spain (2000), Sweden (2000), Turkey (1998), the United Kingdom (2000), and the United States (2000) are from the OECD input–output database for 2007. Four more countries (Argentina, China, India, and Taiwan) are available in the database but are excluded here, either because the data on gross operating surplus are unavailable (China and India) or there are no reliable factor input data sources across industries (Argentina and Taiwan).

The I-O tables from the OECD database employ ISIC Rev.3 classification containing 48 industrial groups but some countries do not report all 48 industries. Rather than estimating these unreported data, we aggregate into 30 industries (Table A.1) so that industrial classification is consistent across countries. Aggregation is inevitable but raises the risk of systematic bias in the HOV predictions, a problem in all such studies (Feenstra and Hanson, 2000).

The number of industrial groups is smaller than the 35 sectors used by Davis and Weinstein (2001) but is greater than the 23 sectors

Table A.1. Classification of industrial activities.

Sectors	Branches of activities	ISIC Rev.3
1	Agriculture, hunting, forestry, and fishing	1–2, 5
2	Mining and quarrying	10–14
3	Food products and beverages	15–16
4	Textiles	17–19
5	Wood and products of wood and cork	20
6	Paper and paper products	21–22
7	Coke, refined petroleum products and nuclear fuel	23
8	Chemicals and chemical products	24
9	Rubber and plastics products	25
10	Other non-metallic mineral products	26
11	Basic metals	27
12	Fabricated metal products	28
13	Machinery and equipment	29
14	Office, accounting and computing machinery	30
15	Electrical machinery and apparatus	31
16	Radio, television and communication equipment and apparatus	32
17	Medical, precision and optical instruments	33
18	Motor vehicles, trailers and semi-trailers	34
19	Other transport equipment	35
20	Other manufacturing (including recycling)	36–37
21	Electricity, gas and water supply	40–41
22	Construction	45
23	Wholesale and retail trade	50–52
24	Hotels and restaurants	55
25	Transport, storage and communications	60–64
26	Financial intermediation	65–67
27	Real estate, renting and business activities	70–74
28	Public administration and defense	75
29	Education	80
30	Health and social work	85, 90–93, 95, 99

used by Hakura (2001) and the 24 sectors used by Lai and Zhu (2007). The input–output matrices and final consumption, gross output, and net exports are derived from the I-O tables for the year 2000. Final consumption is the sum of final consumption of households, final consumption and investment of government, gross fixed capital formation, and changes in inventory. Therefore, the total use table of country c always satisfies the equation: $\mathbf{T}_c = (\mathbf{I} - \mathbf{B}_c)\mathbf{Q}_c - \mathbf{C}_c$ where

$\mathbf{B}_c$ is a 30 × 30 indirect technology matrix for the unit intermediate requirements and $(\mathbf{I} - \mathbf{B}_c)\mathbf{Q}_c$ vector equals net output $(\mathbf{Y}_c)$. $\mathbf{B}_c$ is obtained by taking input–output data from the I-O tables and dividing inputs in each sector by the corresponding sector's gross output.[20]

To convert the dataset into 2000 U.S. dollars with purchasing power parity (PPP) adjustments, we use GDP price deflators for 2000 from the Penn World Table version 6.2 (Heston *et al.*, 2006) and average nominal exchange rates from the *International Financial Statistics 2006* (IMF). For Australia, Greece, Ireland, Israel, New Zealand, Portugal, and Turkey, nominal values in the I-O tables are uniformly multiplied by the growth rates of total nominal GDP to adjust data from earlier years to the year 2000.

A.2. *Factor Endowments Data*

A.2.1. *Physical Capital Stock*

Capital stock is developed by the discounted sum of real gross fixed capital formation in 2000 PPP U.S. dollars $(GFCF_{cit})$ from 1980 to 2000:

$$K_{ci} = \sum_{t=1980}^{2000} (1 - 0.1333)^{2000-t} \, GFCF_{cit}.$$

Because industry-level values for gross fixed capital formation are not stable over the period, we do not use the perpetual inventory method (e.g., Keller, 2000). For OECD countries, values for gross fixed capital formation (GFCF) are derived from the OECD structural analysis (STAN) database (2004) and unreported data are estimated from the ISIC Rev.2 version of the OECD STAN database (1998). As many GFCF data as possible are derived from these databases but there are still some unavailable values. The following procedure is

[20]In the case of two sectors, the input usage matrix can be obtained as following:

$$\mathbf{B}_c = \begin{bmatrix} b_{11} & b_{12} \\ b_{21} & b_{22} \end{bmatrix} = \begin{bmatrix} x_{11}/Q_1 & x_{12}/Q_2 \\ x_{21}/Q_1 & x_{22}/Q_2 \end{bmatrix}$$

$$\mathbf{B}_c\mathbf{Q}_c = \begin{bmatrix} b_{11} & b_{12} \\ b_{21} & b_{22} \end{bmatrix} \begin{bmatrix} Q_1 \\ Q_2 \end{bmatrix} = \begin{bmatrix} x_{11}/Q_1 & x_{12}/Q_2 \\ x_{21}/Q_1 & x_{22}/Q_2 \end{bmatrix} \begin{bmatrix} Q_1 \\ Q_2 \end{bmatrix} = \begin{bmatrix} x_{11} + x_{12} \\ x_{21} + x_{22} \end{bmatrix}.$$

taken to interpolate these data. First, some detailed sectors (e.g., 18 "Motor vehicles, trailers and semi-trailers" and 19 "Other transport equipment") are unavailable for certain years but data for their aggregated value (18 + 19 "Transport equipment") exist for all years. We use the share of the nearest year to allocate those totals to each detailed sector. Second, if a country reports only aggregated industry sub-totals (18 + 19 "Transport equipment"), we first develop capital stocks of these sub-totals and allocate these values to each industry according to the compensation for capital (gross operating surplus) obtained from the I-O tables. This procedure is based on the idea that industry capital compensation flows are proportional to industry capital stocks (Lai and Zhu, 2007). In particular, for non-OECD countries (Brazil, Indonesia and Israel) and for Greece and Turkey, we develop country-level capital stock (gross fixed capital formation is derived from World Bank World Development Indicators and OECD STAN database) and allocate country totals to each industry according to capital compensation data from the I-O tables. Finally, unreported years of Czech Republic (1980–1994), Hungary (1980–1990), Poland (1980–1991), Portugal (1980–1994), and Slovak Republic (1980–1992) are interpolated with industry-level growth rates of available years.

One major problem with using GFCF data from the OECD STAN database (2004) is that most countries include residential investments but three (Canada, the United Kingdom and the United States) do not. In particular, agriculture and real estate are the main sources of errors from residential investments. To avoid serious errors, we obtain the difference between values for gross fixed capital formation from the OECD STAN database and those from *International Financial Statistics* (IMF) for these three countries and add these differences into the real estate sector. Unfortunately, it is impossible to adjust agriculture for residential investment and caution must be exercised when data from that sector are used in the analysis.

To convert GFCF figures into real series, we convert values into nominal U.S. dollars, divide by the price of investment from the Penn World Table 6.2, and deflate by U.S. industry-level prices for gross fixed capital formation, which is obtained from the OECD STAN database (2004). After this conversion into 2000 PPP U.S. dollars, we compute real capital stock data, using a depreciation rate of 0.1333 (e.g., Leamer, 1984; Bowen, Leamer, and Sveikauskas, 1987; and Davis and Weinstein, 2001). For Japan, industrial GFCF data are unavailable from the STAN database. Therefore, we take the total GFCF series from the World Development Indicators (World Bank) and Japan's sectoral shares are obtained from the nominal investment matrix tables of the ESRI-Histat database (ISIC Rev.3).

A.2.2. *Labor*

Sectoral labor inputs (total employment)for the year 2000 are derived from the OECD STAN databases (1998 and 2004) and the ILO LABORSTA Internet Yearly Statistics (http://laborsta.ilo.org/). Most data are available from these sources. If a country reports only aggregated industry sub-totals, we allocate these to each industry according to the compensation of employees obtained from the I-O tables. For most OECD countries, country-level annual working hours are available from the OECD Employment and Labor Market Statistics (2006) and the OECD STAN database (2004) but it is not possible to obtain industry-level annual working hours. Even though the data on weekly and/or monthly working hours for all countries in our dataset except Indonesia are available from ILO LABORSTA Internet Yearly Statistics, the definition, classification, and coverage differ from country to country. Therefore, we do not adjust total number of employees by working hours.

Appendix B: Technology Specifications in Davis and Weinstein (2001)

Specification (P4) is based on the Dornbusch–Fischer–Samuelson (1980) model in which unit factor requirements for tradables are systematically correlated with country-level capital abundance:

$$\ln a_{cif} = \theta_c + \ln \bar{a}_{if} + \gamma_f^T \ln(K_c/L_c)T_i + \varepsilon_{cif} \qquad \text{(B.1)}$$

where T_i is a dummy variable that takes on a value of one if the sector is tradable (agriculture, mining, and manufacturing sectors) and zero if the sector is non-traded goods. Here, we impose the restriction

$(\Sigma_f \gamma_f^T = 0)$ so that capital intensity does not affect the Hicks-neutral productivity (θ_c).

Specification (P5) derives from Helpman (1999), in which FPE fails and countries are in different production cones, affecting coefficients in both traded and non-traded goods:

$$\ln a_{cif} = \theta_c + \ln \bar{a}_{if} + \gamma_f^T \ln(K_c/L_c)T_i + \gamma_f^N \ln(K_c/L_c)(1 - T_i) + \varepsilon_{cif} \tag{B.2}$$

where we restrict parameters with $\Sigma_f \ (\gamma_f^T + \gamma_f^N) = 0$.

Davis and Weinstein also consider a more general specification in which they do not force country capital–labor ratios to have the same effects across sectors. This is the unrestricted technology differences of Helpman (P5′):

$$\ln a_{cif} = \theta_c + \ln \bar{a}_{if} + \gamma_{if} \ln(K_c/L_c) + \varepsilon_{cif}. \tag{B.3}$$

Appendix C: Relative Abundance and Factor-Productivity Adjustment

Here we introduce factor-augmenting productivity to the right-hand side of Equation (17). First, in the following equation we show that the inequality in relative factor abundance for the factor-productivity model does not coincide with that for the strict (or Hicks-neutral) model. Thus, the empirical prediction of Debaere's model with factor-productivity parameters differs from Debaere's original specification:

$$\frac{V_{c'f'}^*}{V_{c'f}^*} > \frac{V_{cf'}^*}{V_{cf}^*} \Leftrightarrow \frac{\pi_{c'} V_{c'f'}^*}{\pi_{c'} V_{c'f}^*} > \frac{\pi_c V_{cf'}^*}{\pi_c V_{cf}^*} \neq \frac{\pi_{c'f'} V_{c'f'}^*}{\pi_{c'f} V_{c'f}^*} > \frac{\pi_{cf'} V_{cf'}^*}{\pi_{cf} V_{cf}^*}.$$

Proof of the relative factor-abundance model with factor-augmenting productivities follows:

$$\begin{aligned}\frac{\pi_{c'f'} V_{c'f'}^*}{\pi_{c'f} V_{c'f}^*} > \frac{\pi_{cf'} V_{cf'}^*}{\pi_{cf} V_{cf}^*} &\Leftrightarrow \pi_{c'f'} V_{c'f'}^* \pi_{cf} V_{cf}^* > \pi_{cf'} V_{cf'}^* \pi_{c'f} V_{c'f}^* \\ &\Leftrightarrow \pi_{c'f'} V_{c'f'}^* \pi_{cf} V_{cf}^* + \pi_{c'f'} V_{c'f'}^* \pi_{c'f} V_{c'f}^* \\ &\quad > \pi_{cf'} V_{cf'}^* \pi_{c'f} V_{c'f}^* + \pi_{c'f'} V_{c'f'}^* \pi_{c'f} V_{c'f}^*\end{aligned}$$

$$\Leftrightarrow \frac{\pi_{c'f'}V^*_{c'f'}(\pi_{cf}V^*_{cf} + \pi_{c'f}V^*_{c'f})}{\pi_{c'f}V^*_{c'f}(\pi_{cf'}V^*_{cf'} + \pi_{c'f'}V^*_{c'f'})} > 1$$

$$\Leftrightarrow \frac{\pi_{c'f'}V^*_{c'f'}}{\pi_{c'f}V^*_{c'f}} > \frac{\pi_{cf'}V^*_{cf'} + \pi_{c'f'}V^*_{c'f'}}{\pi_{cf}V^*_{cf} + \pi_{c'f}V^*_{c'f}}$$

where π_{cf} is the factor-productivity parameter for factor f of country c.

Chapter 13

Estimating the Knowledge-Capital Model of the Multinational Enterprise*

David L. Carr, James R. Markusen and Keith E. Maskus†

The industrial-organization (IO) approach to international trade ("new trade theory") has incorporated features of increasing returns to scale, imperfect competition, and product differentiation into traditional general-equilibrium trade models. These new models offer rich predictions about the direction and volume of trade between two countries as functions of industry characteristics (factor intensities, scale economies, product differentiation) interacting with country characteristics (relative size differences, relative endowment differences, and trade costs).

However, an awkward empirical problem is that most of the firms and industries motivated by the IO approach to trade are multinational firms, while most of the theory has been about single-plant national firms. More recent theoretical developments have incorporated multinational firms, which maintain facilities in more than one country. These multinationals are often broken down into "horizontal" firms, which produce the same goods and services

*Reprinted from *American Economic Review*, Vol. 91, No. 3, pp. 693–708. Copyright 2001, with permission from The American Economic Association.
†This research was funded in part by a grant from the National Science Foundation through the National Bureau of Economic Research. We are grateful to Bruce Blonigen, James Harrigan, James Levinsohn, Marcelo Olarreaga and an anonymous referee for helpful comments and to Rebecca Neumann for research assistance.

in multiple countries, and "vertical" firms, which geographically fragment production by stages. An early example of a model with vertical multinationals is in Elhanan Helpman (1984) while an early model of horizontal multinationals is in James R. Markusen (1984).

Subsequent theoretical work has focused on horizontal firms because they seem to be more prevalent in the world. Examples include Ignatius J. Horstmann and Markusen (1987, 1992), S. Lael Brainard (1993a), and Markusen and Anthony J. Venables (1996, 1997, 1998). These models have since been subjected to empirical tests, especially by Brainard (1993b, 1997) and Karolina Ekholm (1995, 1997, 1998a,b). Results give good support to the theoretical predictions of the "horizontal" models: multinational activity should be concentrated among countries that are relatively similar in both size and in relative endowments (or per capita incomes).

Theoretical models combining both horizontal and vertical motives for direct investment are analytically difficult. Helpman's original model of vertical multinationals used the assumption of no trade costs, but in that case there is no motive for horizontal multinationals (given plant-level scale economies). For analytical tractability, the early models of horizontal firms assumed that different activities (e.g., headquarters services and plant production) use factors in the same proportion or that there is only one factor of production. However, this permits no factor-price motive for vertical fragmentation across countries.

Two recent models exist in which both vertical and horizontal firms can arise endogenously due to the simultaneous existence of trade costs and different factor intensities across activities. Analytical difficulties, however, imply that most results are derived from numerical simulations, as demonstrated in Markusen *et al.* (1996) and Markusen (1997). These simulations generate a number of testable implications, relating direct investment to country characteristics.

The purpose of this paper is therefore to take the theoretical predictions of recent theory and subject them to econometric test. We begin by reviewing this newer theory and discussing its theoretical predictions. The theory explains the volume of production of foreign affiliates of one country's firms in another country as a function of the characteristics of both countries. We then translate these and other predictions into a tractable empirical specification. Many relationships are interactive or non-linear. Results of our estimations are closely consistent with the theory. The volume of affiliate sales follows the theoretical predictions based on characteristics of both parent and host countries. Both vertical and horizontal investments are important and related to country characteristics as the model predicts.

1. The Knowledge-Capital Model

We now present a theoretical model drawn from Markusen *et al.* (1996) and Markusen (1997). We will refer to it as the "knowledge-capital model" of the multinational enterprise. This approach includes three principal assumptions. First, services of knowledge-based and knowledge-generating activities, such as R&D, can be geographically separated from production and supplied to production facilities at low cost. Second, these knowledge-intensive activities are skilled-labor-intensive relative to production. Third, knowledge-based services have a (partial) joint-input characteristic, in that they can be utilized simultaneously by multiple production facilities. The first two assumptions create a motive for the vertical fragmentation of production, locating R&D activities where skilled labor is cheap and production where unskilled labor is cheap. There will also be a market-size motive for locating production if there are plant-level scale economies. The third assumption creates firm-level scale economies and motivates horizontal investments that replicate the same products or services in different locations.

The model assumes the existence of two homogeneous goods (X and Y), two countries (h and f), and two homogeneous factors, unskilled labor (L) and skilled labor (S), which are internationally immobile. Good Y is labor-intensive and produced under constant returns to scale in a competitive industry. Good X is skilled-labor-intensive overall, exhibits increasing returns to scale, and is subject to Cournot competition with free entry and exit. Within a firm, headquarters services and plant facilities may be geographically separated and a firm may have plants in one or both countries.

With this structure, there are six firm types, with free entry and exit into and out of firm types. *Regime* denotes a set of firm types active in equilibrium. Firm types are as follows:

- Type H_h (H_f) — horizontal multinationals that maintain plants in both countries with headquarters located in country h (f).
- Type N_h (N_f) — national firms that maintain a single plant and headquarters in country h (f); they may or may not export to country f (h).
- Type V_h (V_f) — vertical multinationals that maintain a single plant in country f (h) and headquarters in country h (f); they may or may not export to country h (f).

Assumptions about the size and composition of fixed costs are crucial to the predictions of the model.[1] First, we assume the existence of multi-plant economies of scale (relevant to type-H firms) due to a joint-input property of knowledge capital. Headquarters service (blueprints, manuals, formulas, procedures, etc.) can be supplied to additional plants at low marginal cost. Thus, in good X the total fixed costs of headquarters and two plants is less than double the total fixed costs of a single-plant firm (the joint-input property of knowledge capital).

Second, it is assumed that headquarters services are more skilled-labor-intensive than production. Somewhat more controversial is our assumption that plant-level production is more skilled-labor-intensive than the composite rest of the economy. Thus, we stipulate that the ranking of skilled-labor intensity of activities is [headquarters only] > [integrated X] > [plant only] > [Y].

The model further assumes that national markets for goods are segmented and transport costs use unskilled labor. The full set of equations and inequalities characterizing equilibrium in the model is

[1]The following assumptions draw empirical support from a large number of studies, including the United Nations Conference on Trade and Development (UNCTAD) (1993), Ekholm (1995, 1997, 1998a,b), Robert E. Lipsey, *et al.* (1995), Brian Aitken, *et al.* (1996), Richard E. Caves (1996), Magnus Blomstrom, *et al.* (1997), Robert C. Feenstra and Gordon H. Hanson (1997), and Matthew J. Slaughter (1999). Data drawn from some of these sources are presented in Markusen (1997).

given in Markusen (1997) and in Markusen *et al.* (1996) and we will not repeat that exercise here.

Different country characteristics favor various firm types producing or maintaining headquarters in country h. Analogous comments apply to firms in country f. Consider first factors that favor national firms being headquartered in country h and also producing there. Assumptions of the model developed above suggest that type N_h firms will be the dominant type active in h if: (1) h is both large and skilled-labor abundant; (2) h and f are similar in size and relative endowments and transport costs are low (type N_f will sell in h); or (3) foreign investment barriers in f are high (type N_f may sell in h).[2]

Country h being large supports production there while skilled-labor abundance favors locating headquarters in h as well. Thus, an integrated type-N_h firm has a cost advantage over a type-V_h or V_f firm. A type-N_h firm also has an advantage over a type-H_h firm, which must locate costly capacity in the small f market unless trade costs are high. Type-N firms should also be dominant when the countries are similar and trade costs are small. If countries are perfectly symmetric, for example, there is no motive for type-V firms. Small trade costs favor type-N firms over two-plant type-H firms.

Type-H_h firms will be the dominant type active in country h if the nations are similar in size and relative endowments and transport costs are high (type H_f will also produce in h). Thus, horizontal multinationals firms should be associated with similarities between countries in both size and in relative factor endowments. The intuition is that if countries are dissimilar in either size or relative endowments, one country will be "favored" as a site of both production and headquarters or of one of these two activities. For example, if the countries are similar in relative endowments but of different sizes, then national firms located in the large country will be favored because they avoid costly capacity in the smaller market. If the countries are different in relative endowments but of similar size, then there is an incentive to concentrate headquarters in the skilled-labor-abundant country and production

[2]The word "dominant" means that the number of firms of this type is larger than the number of firms of any other type in the associated equilibrium.

in the skilled-labor-scarce country. Thus vertical firms headquartered in the skilled-labor-abundant countries are favored unless trade costs are high.

From this analysis, a prediction about vertical multinationals follows directly. First, type-V_h firms will be dominant in h if country h is small, skilled-labor-abundant, and trade costs from the host country back to the parent country are not excessive.

2. Simulation Results

Data exist on the volume of production in host countries by affiliates of firms in parent countries, but not on the number of firms of various types. Accordingly, we develop predictions about affiliate production, rather than the numbers of firms of various types. In this section, we solve the model numerically in order to generate such predictions on the relationship between affiliate sales and country characteristics.[3]

A preliminary issue is to define "affiliate production" in the model in a way that relates sensibly to data on affiliate sales. Parents and affiliates in the data are essentially defined in terms of ownership location. Thus, in our model we assume that the country in which a firm's headquarters is located is the parent country. Given that assumption, the production of affiliates of country-h firms in country f is the output of plants in country f "owned" by type-H_h and type-V_h firms. Similarly, the volume of production by country-h affiliates of country f firms is the production in country h of plants owned by type-H_f and type-V_f firms.

Representative simulation results are demonstrated with a series of world Edgeworth box diagrams in Figures 1–3, with the total world endowment of skilled labor on the "Y" axis and the total world endowment of unskilled labor on the "X" axis. The origin for country h is at the southwest (SW) corner of the box and the origin for country f is at the northeast (NE) corner. Along the SW–NE

[3]The full model, consisting of 41 nonlinear inequalities, is solved as a complementarity problem using Thomas F. Rutherford's (1995, 1999) solver Mathematical Programming System for General Equilibrium (MPS/GE), a subsystem of the program General Algebraic Modeling System (GAMS).

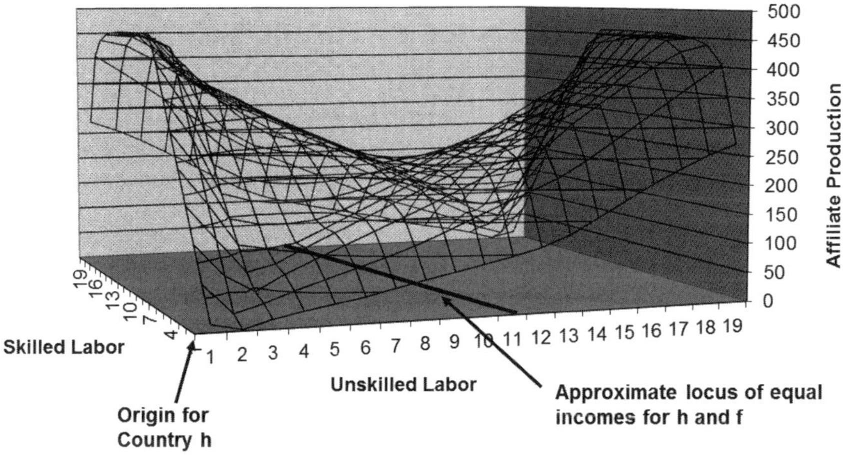

Figure 1. Volume of affiliate production: 25-percent trade costs.

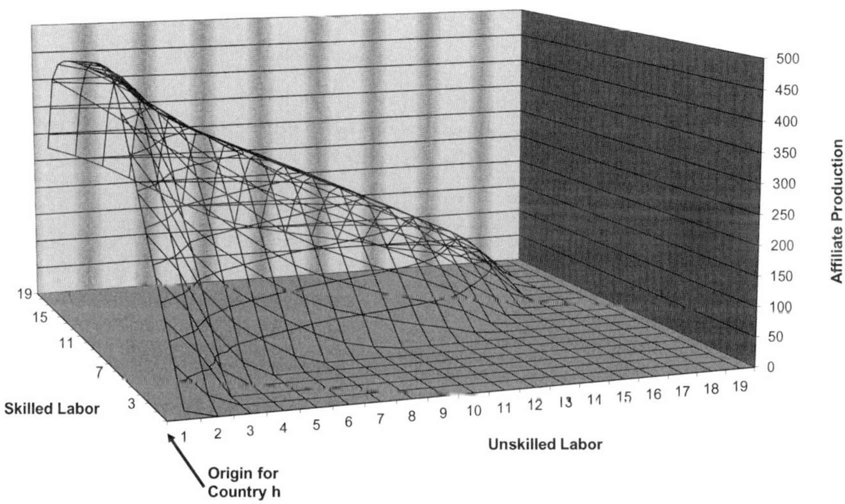

Figure 2. Volume of affiliate production: By h-owned plants in country f.

diagonal, the countries have identical relative endowments but differ in size. The approximate locus, which is not quite linear, along which the countries have equal incomes is given by the line drawn on the floor of the box in Figure 1. Country h is smaller than country f to the left of this locus and is larger to the right.

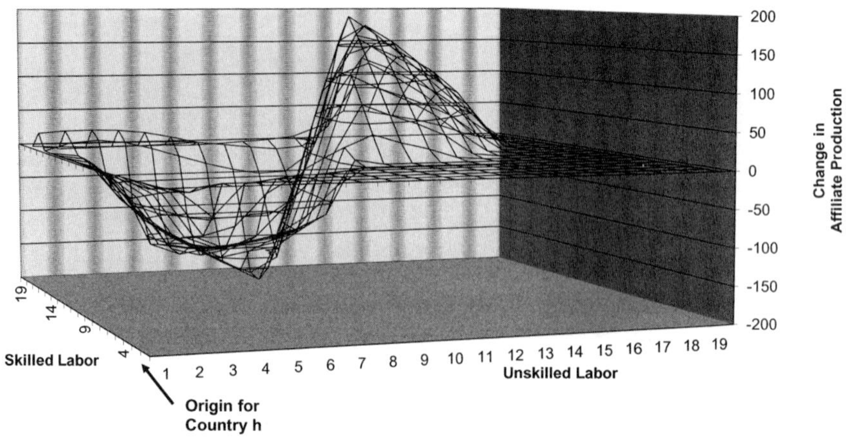

Figure 3. Change in affiliate production, h plants in f: 25-percent trade costs minus 5-percent trade costs in both directions.

Figure 1 shows simulation outcomes with high (25-percent) ad valorem trade costs in both countries. Affiliate production is the sum of the outputs of both countries' affiliated plants. Note that Figure 1 contains a classic saddle. Affiliate sales are at a minimum when the two countries are similar in relative endowments but different in size, in which case national firms headquartered in the large country dominate X production. Moving along the SW–NE diagonal (relative endowments identical), total affiliate sales reach a maximum at the mid-point where the countries are identical. At this point, all firms are horizontal multinationals. Exactly half the world output of good X is produced by affiliate plants and the other half by domestic plants of type-H firms.

A surprising result in Figure 1 is that total affiliate production is highest when one country is both moderately small and skilled-labor abundant. In such a situation, type-V firms located in that country are the dominant firm-type. Note that if only type-V firms were active in equilibrium, then *all* of the world production of X is performed by affiliates. Conversely, affiliate activity is lowest when the skilled-labor-abundant country is large, in which case all production of X is by national firms headquartered in that country.

The nonlinearities in Figure 1 present a challenge for testing. For example, the effect of differences in country size on affiliate sales depends on whether the countries are similar in relative endowments and, if they are different in size, on whether the small country is the skilled-labor-abundant country.

For further clarity, Figure 2 plots only the production by h-owned plants in country f. Here again, we see the inverted U-shaped curve of production along the SW–NE diagonal, but affiliate production is highest when country h is moderately small and highly skilled-labor abundant. The latter situation is reminiscent of Sweden, Switzerland, and The Netherlands, which are small, skilled-labor-abundant countries and important parent countries for multinationals.

Figure 3 presents results concerning the effects of trade costs, assumed here to be symmetric in both directions, on production by h-owned plants in country f. On the vertical axes of these diagrams is affiliate sales with 25-percent trade costs minus affiliate sales with 5-percent trade costs. Again, we see that the results are highly nonlinear. Figure 3 demonstrates that higher trade costs increase total affiliate sales if the countries are relatively similar in size and in relative endowments. Similarity favors horizontal multinationals and, as we noted earlier, horizontal production is encouraged by higher trade costs. This region of positive change is associated solely with *host-country* trade costs.

Higher trade costs, however, reduce total affiliate sales when there is a moderate difference in relative endowments and the skilled-labor-abundant country is somewhat smaller, as demonstrated in Figure 3. This is a region with vertical multinationals headquartered in the skilled-labor-abundant country and with a correspondingly large volume of X trade. This area of negative change in Figure 3 is associated solely with *parent-country* trade costs.

3. Econometric Specification

The results just discussed lead us to specify a central equation for estimation purposes, which we describe here. Variables are listed and defined in Table 1. The dependent variable is the real volume of production (sales) by manufacturing affiliates in each host country that are majority-owned by parents in source countries. The first independent variable is joint market size, captured by the bilateral

Table 1. Variable definitions and summary statistics.

Panel A: Variable Definitions

Real Sales is the volume of affiliate sales in millions of 1990 U.S. dollars. **GDP Sum** is the sum of parent-country and host-country gross domestic products in billions of 1990 U.S. dollars. **GDP Difference (Squared)** is the (squared) difference between parent-country and host-country GDP billions of 1990 U.S. dollars. **Skill Difference (Squared)** is the (squared) difference between the ratio of skilled labor to total labor force in the parent country and that in the host country, with skilled labor measured as managerial and professional, technical, and kindred workers. **Investment Cost Host** is an index from zero to 100 of investment impediments in the host country. **Trade Cost Host (Parent)** is an index from zero to 100 of protectionist restraints in the host (parent) country. **Distance** is the distance in kilometers of each country's capital city from Washington, DC.

Panel B: Summary Statistics for Basic Sample ($n = 509$)

Variable	Mean	Std. Dev.	Minimum	Maximum
Real Sales	15,670	24,316	0	120,070
GDP Sum	6,125	675	5,210	9,328
GDP Difference	1,146	5,219	−6,145	6,145
GDP Difference Squared	2.8e7	0.6e7	0.7e7	3.8e7
Skill Difference	0.034	0.012	−0.277	0.277
Skill Difference Squared	0.016	0.017	5.7e-7	0.077
Investment Cost Host	34.00	10.59	15.30	79.43
Trade Cost Host	33.62	12.05	6.00	85.08
Trade Cost Parent	31.74	8.61	6.00	74.34
Distance	8266	3875	734	16370

sum of real GDP levels in the parent and host countries. We expect its coefficient to have a positive sign. Indeed, a stronger hypothesis is that the elasticity of affiliate sales with respect to this sum is greater than one. The second variable is the squared difference in real gross domestic product (GDP) between parent country and host country. We expect its coefficient to be negative because our theory says that affiliate sales volume has an inverted U-shaped relationship to differences in country size, with a maximum at zero difference.

The third independent variable is the difference in a measure of skilled-labor abundance in the headquarters country relative to that in the host country. This measure in each country has a potential range from zero to one and the difference in skill endowments ranges in principle from minus one to plus one. We expect its coefficient to be positive because firms tend to be headquartered in the skilled-labor-abundant country. The fourth variable is the product of differences in economic size and skill endowments. We anticipate the sign of this interaction term to be negative for reasons shown in Figure 2. In particular, affiliates sales are highest when the country is small and skilled-labor abundant.

The fifth and sixth variables measure perceived costs of investing in, and exporting to, the host country. These variables are indices ranging from zero to 100, as discussed below. We expect the investment-cost coefficient to be negative and the trade-cost coefficient to be positive. The following variable is an interaction term between host-country trade costs and squared endowment differences. This variable is designed to capture the ideas that trade costs may encourage horizontal investment, but not vertical investment, and that horizontal investment is most important when relative endowments are similar. The coefficient should therefore be negative, weakening the direct effect of host-country trade costs. The results in Figure 3, however, show that the effect of the host-country trade costs (the area of increased affiliate production) is not symmetric around the SW–NE diagonal and is actually highest when the parent country is moderately skilled-labor abundant. Thus, this is not a theoretically sharp hypothesis and, indeed, empirical support for this term is weak, as we shall see.

The next regressor is perceived trade costs in exporting to the parent country. This variable is also an index ranging from zero to 100. Its coefficient should be negative because trade costs diminish the incentive to locate plants abroad for shipment back to the home market, as shown in Figure 3 (the area of negative change in affiliate production). Figure 3 also indicates that parent-country trade costs should be interacted with skilled-labor-endowment differences. However, the resulting variable is highly collinear with the measure of skill differences because skilled-labor-scarce countries have high trade-cost indices.[4] Thus, we exclude this interaction variable in

[4]The partial correlation between the difference in skills and the interaction between trade costs and skill differences is –0.96.

the estimates provided below. Finally, we add geographic distance in kilometers between parent and host as an independent variable. The sign of this variable is ambiguous in theory, because distance is an element in both export costs and investment and monitoring costs.

We specify the regression as linear in levels, with quadratic and interaction terms included. Collecting ideas, the regression equation is listed in Panel A of Table 2. That table also permits us to consider the interactive terms in more detail by writing the implied partial derivatives. For example, in Panel B, trade costs in the host country affect real sales volume both directly and indirectly through an interaction with squared skill differences. The direct effect is measured by coefficient $B6$. Because this coefficient is greater than zero, the derivative is expected to be positive when relative endowments are similar, reflecting the fact that host-country trade costs encourage horizontal direct investment. But it should be smaller when relative endowments differ, in which case horizontal investment is less important. This implies that the expected sign of coefficient $B7$ is negative. Discussion of Panel C is postponed.

Similarly, as noted in Panel D, the derivative of real sales with respect to GDP differences has two terms. The relationship should be an inverted U as noted above, reaching a maximum when the countries are similar in relative endowments, which is captured by the first term and the negative coefficient $B2$. However, our theory predicts that investment could fall with increases in host-country size if the host is skilled-labor-abundant. This outcome is reflected in the second term and the expected negative sign on $B4$.

Finally, in Panel E, the partial derivative of sales with respect to skill-endowment differences has three terms. The first term is a direct effect that should be positive, capturing both vertical direct investment and headquarters of horizontal firms. The direct effect is weakened as the parent country gets larger, which causes vertical firms to be replaced by national firms, headquartered in the parent nation, that serve the destination country by exports. This effect is also weakened if skill differences are large (see Figure 2). Note that both coefficients $B4$ and $B7$ appear twice in the three derivatives and are predicted to be negative in each case.

Table 2. Theoretical and empirical impacts of investment determinants on real sales volume.

Panel A: Central Regression Equation

$$\begin{aligned}\text{Real Sales} = {} & B0 + B1^*(\text{GDP Sum}) + B2^*(\text{GDP Difference Squared})\\ & + B3^*(\text{Skill Difference}) + B4^*([\text{GDP Difference}]^*[\text{Skill Difference}])\\ & + B5^*(\text{Investment Cost Host}) + B6^*(\text{Trade Cost Host})\\ & + B7^*([\text{Trade Cost Host}]^*[\text{Skill Difference Squared}])\\ & + B8^*(\text{Trade Cost Parent}) + B9^*\text{Distance}\end{aligned}$$

Panel B: Impact of Host-Country Trade Costs

$$\begin{aligned}&\text{Partial Derivative: } (\partial\text{Sales}/\partial\text{Trade Cost Host})\\ &\quad = B6 + B7^*(\text{Skill Difference Squared})\\ &\quad = 69.4 - 811.6^*(\text{Skill Difference Squared})\\ &\qquad > 0 \text{ iff } (\text{Skill Difference}) < 0.293 \text{ (WLS)}\\ &\quad = 114 - 2{,}273^*(\text{Skill Difference Squared})\\ &\qquad > 0 \text{ iff } (\text{Skill Difference}) < 0.252 \text{ (Tobit)}\end{aligned}$$

Panel C: Impact of Bilateral Trade Costs

$$\begin{aligned}&\text{Partial Derivative: } (\partial\text{Sales}/\partial\text{Trade Cost})\\ &\quad = B6 + B7^*(\text{Skill Difference Squared}) + B8\\ &\quad = 69.4 - 811.6^*(\text{Skill Difference Squared}) - 75.5\\ &\qquad < 0 \text{ all } (\text{Skill Difference}) \text{ (WLS)}\\ &\quad = 114 - 2{,}273^*(\text{Skill Difference Squared}) - 112.6\\ &\qquad > 0 \text{ iff } (\text{Skill Difference}) < 0.118 \text{ (Tobit)}\end{aligned}$$

Panel D: Impact of Difference in GDP

$$\begin{aligned}&\text{Partial Derivative: } (\partial\text{Sales}/\partial\text{GDP Difference})\\ &\quad = 2^*B2^*(\text{GDP Difference}) + B4^*(\text{Skill Difference})\\ &\quad = -0.0011^*2^*(\text{GDP Difference}) - 4.4^*(\text{Skill Difference}) \text{ (WLS)}\\ &\quad = -0.0009^*2^*(\text{GDP Difference}) - 7.7^*(\text{Skill Difference}) \text{ (Tobit)}\end{aligned}$$

(*Continued*)

4. Data Sources and Estimation Results

Data for the estimation form a panel of cross-country observations over the period 1986–1994. We take real sales volume of non-bank manufacturing affiliates in each country to indicate production

Table 2. (*Continued*)

Panel E: Impact of Difference in Skill Endowments

Partial Derivative: $(\partial\text{Sales}/\partial\text{Skill Difference})$

$$= B3 + B4^{*}(\text{GDP Difference}) + 2^{*}B7^{*}([\text{Trade Cost Host}]^{*}[\text{Skill Difference}])$$

$$= 15{,}042 - 4.4^{*}(\text{GDP Difference}) - 811.6^{*}2^{*}(\text{Trade Cost Host}^{*}\text{Skill Difference})\ (\text{WLS})$$

$$= 29{,}366 - 7.7^{*}(\text{GDP Difference}) - 2{,}273^{*}2^{*}(\text{Trade Cost Host}^{*}\text{Skill Difference})\ (\text{Tobit})$$

activity. The U.S. Department of Commerce provides annual data on sales of foreign affiliates of American parent firms and on sales of U.S. affiliates of foreign parent firms. Thus, the data are bilateral with the United States, which is either the parent country or the host country in every observation. There are 36 countries in addition to the United States for which we have at least one year of complete data. Annual sales values abroad are converted into millions of 1990 U.S. dollars using an exchange-rate adjusted local wholesale price index, with exchange rates and price indexes taken from the *International Financial Statistics* (IFS) of the International Monetary Fund.

Real gross domestic product is measured in billions of 1990 U.S. dollars for each country. For this purpose, annual real GDP figures in local currencies were converted into dollars using the market exchange rate. These data are also from the IFS.

Skilled labor abundance is defined as the sum of occupational categories 0/1 (professional, technical, and kindred workers) and 2 (administrative workers) in employment in each country, divided by total employment. These figures are compiled from annual surveys reported in the *Yearbook of Labor Statistics* published by the International Labor Organization. In cases where some annual figures were missing, the skilled-labor ratios were taken to equal the period averages for each country. The variable Skill Difference is the relative skill endowment of the parent country less that of the affiliate country.

The cost of investing in the affiliate country is a simple average of several indices of perceived impediments to investment, reported in the *World Competitiveness Report* of the World Economic Forum.[5] The investment barriers include restrictions on the ability to acquire control in a domestic company, limitations on the ability to employ foreign skilled labor, restraints on negotiating joint ventures, strict controls on hiring and firing practices, market dominance by a small number of enterprises, an absence of fair administration of justice, difficulties in acquiring local bank credit, restrictions on access to local and foreign capital markets, and inadequate protection of intellectual property. The resulting indices are computed on a scale from zero to 100, with a higher number indicating higher investment costs.

A trade-cost index is taken from the same source and is defined as a measure of national protectionism, or efforts to prevent importation of competitive products. It also runs from zero to 100, with 100 being the highest trade costs. All of these indices are based on extensive surveys of multinational enterprises. It should be noted that both the investment-cost and trade-cost indices are ordinal and qualitative in nature, without "natural units". Thus, regression coefficients represent the partial effects of a change in the average perceived costs of investing and trading.

We also incorporate a measure of distance, which is simply the number of kilometers of each country's capital city from Washington, DC. It is unclear whether this variable captures trade costs or investment costs, since both should rise with distance.

Panel B in Table 1 lists summary statistics for these variables. The final data set, after eliminating any row with missing variables, contains 509 observations.[6] An additional 119 observations are complete except that no foreign affiliate sales are listed in the Commerce Department data. On examination, these countries in all cases are relatively poor and generally small. Thus, we conjecture that the missing observations are in fact zeros. We then perform alternative

[5]Some of these data were kindly provided by the staff of the United States International Trade Commission.

[6]Correlation coefficients and a list of countries are available on request.

estimations using a Tobit procedure, adding these cases to the data set for a total of 628 observations.

Results for the central-case regressions are shown in Tables 3 and 4. The regression in the first column of Table 3 is estimated with ordinary least squares (OLS). However, a Breusch–Pagan test indicated the presence of heteroskedastic errors in the OLS specification. Consequently we present in the second column a weighted least-squares (WLS) estimation.[7] The third column is the Tobit equation, adding 119 observations as noted above. The first four variables capture the relationships shown in Figures 1–2. All of the coefficients on these variables have the hypothesized signs and are highly significant. The two variables involving skill differences have much larger magnitudes in the Tobit regression. This is intuitively sensible because the zero-sales observations added in the Tobit regression are overwhelmingly cases where the potential parent nation is skilled-labor-scarce and smaller than the potential affiliate nation (the United States). Excluding them from the OLS and WLS estimation in Table 3 likely biases downward the role of skilled labor.

The next four variables involve the trade and investment cost measures. Virtually all signs are consistent with the theory, with the sole exception being the positive coefficient on the interaction between host-country trade costs and squared skill differences in the OLS case. However, that coefficient is not significant in any regression. Host-country trade costs have a significantly positive impact on affiliate sales in the WLS case and a nearly significant coefficient in the Tobit case. Parent-country trade costs are not significant in the OLS and WLS regressions, but attain near-significance in the Tobit equation. Investment costs in the host country have signficantly negative coefficients in all specifications. Thus, controlling for distance, the decisions of multinational enterprises in setting output levels of affiliates are responsive to perceived costs of investing in the country and the strength of import protection. This outcome is sensible given our measures of investment costs and trade costs, which are

[7]The test indicated that the residual variances could best be estimated as a linear function of the squared sum of GDP. This estimation was used to compute weights for the WLS estimation.

Table 3. Basic results for model of real sales volume of affiliates: OLS, WLS and Tobit.

Variable	OLS	Sign as Predicted?	WLS	Sign as Predicted?	Tobit	Sign as Predicted?
GDP Sum	10.80 (7.01) (0.0001)	Y	13.92 (9.80) (0.0001)	Y	15.04 (10.27) (0.0001)	Y
GDP Difference Squared	−0.0012 (−6.89) (0.0001)	Y	−0.0014 (−8.90) (0.0001)	Y	−0.0010 (5.89) (0.0001)	Y
Skill Difference	33,743 (3.77) (0.0002)	Y	31,044 (4.01) (0.0001)	Y	61,700 (7.28) (0.0001)	Y
GDP Difference * Skill Difference	−6.34 (−2.62) (0.09)	Y	−4.27 (−2.12) (0.04)	Y	−10.20 (4.34) (0.0001)	Y
Investment Cost Host	−516.6 (−3.79) (0.0001)	Y	−455.6 (−3.92) (0.0001)	Y	−378.6 (2.82) (0.005)	Y
Trade Cost Host	119.2 (1.16) (0.25)	Y	190.6 (2.20) (0.03)	Y	156.2 (1.51) (0.13)	Y
Trade Cost Host*Squared Skill Difference	605.2 (0.36) (0.72)	N	−569.9 (−0.41) (0.68)	Y	−1264 (0.75) (0.57)	Y
Trade Cost Parent	93.7 (−0.99) (0.32)	Y	−93.3 (−1.14) (0.26)	Y	−122.0 (1.46) (0.14)	Y
Distance	1.82 (−7.75) (0.0001)		−1.34 (−6.63) (0.0001)		−1.48 (6.47) (0.0001)	
Intercept	16,630 (1.08) (0.28)		−5,381 (−0.42) (0.68)		−23,282 (1.61) (0.11)	
Observations	509		509		628	
Adjusted R^2	0.47		0.60			
Log-likelihood					−5,755	

Notes: Figures in first row of parentheses below coefficients are t-statistics in OLS and WLS. In Tobit, figures in first row of parentheses below coefficients are square roots of chi-square statistics. Marginal significance levels of the coefficients are indicated in the second row of parentheses. Y indicates "Yes" and N indicates "No".

Table 4. Fixed-effects estimation of basic model: OLS, WLS and Tobit.

Variable	OLS	Sign as Predicted?	WLS	Sign as Predicted?	Tobit	Sign as Predicted?
GDP Sum	13.41	Y	13.72	Y	16.57	Y
	(12.81)		(13.62)		(17.44)	
	(0.0001)		(0.0001)		(0.0001)	
GDP Difference Squared	−0.0010	Y	−0.0011	Y	−0.0009	Y
	(−8.07)		(−9.81)		(8.01)	
	(0.0001)		(0.0001)		(0.0001)	
Skill Difference	20,084	Y	15,042	Y	29,366	Y
	(1.57)		(1.34)		(2.34)	
	(0.12)		(0.18)		(0.02)	
GDP Difference* Skill Difference	−5.91	Y	−4.44	Y	−7.71	Y
	(−2.42)		(−2.09)		(3.22)	
	(0.02)		(0.04)		(0.001)	
Investment Cost Host	−198.8	Y	−173.2	Y	−41.3	Y
	(−1.49)		(−1.52)		(0.32)	
	(0.14)		(0.13)		(0.75)	
Trade Cost Host	74.9	Y	69.4	Y	144.0	Y
	(0.96)		(1.02)		(1.93)	
	(0.34)		(0.31)		(0.05)	
Trade Cost Host* Squared Skill Difference	−388.2	Y	−811.6	Y	−2,273	Y
	(−0.24)		(−0.57)		(1.49)	
	(0.81)		(0.57)		(0.14)	
Trade Cost Parent	−87.7	Y	−75.5	Y	−112.6	Y
	(−1.63)		(−1.60)		(2.43)	
	(0.10)		(0.11)		(0.02)	
Distance	−1.08		−0.87		−0.77	
	(−5.45)		(−4.95)		(4.28)	
	(0.0001)		(0.0001)		(0.0001)	
Intercept	−22,492		−24,552		−53,341	
	(−2.00)		(−2.57)		(5.24)	
	(0.05)		(0.01)		(0.0001)	
Observations	509		509		628	
Adjusted R^2	0.83		0.87			
Log-likelihood					−5,436	

Notes: Figures in first row of parentheses below coefficients are t-statistics in OLS and WLS. In Tobit, figures in first row of parentheses below coefficients are square roots of chi-square statistics. Marginal significance levels of the coefficients are indicated in the second row of parentheses. Y indicates "Yes" and N indicates "No".

indexes of perceived costs and protectionism developed from surveys of multinational managers. The survey questions do not ask about geographical distance, implying that the respondents do not factor it into their answers. Thus, we have conceptually distinctive measures of distance and costs. Indeed, the distance measure is essentially uncorrelated in the data with investment costs and trade costs.

The results presented above are from a panel data set and it is of interest to isolate cross-section and time-series effects. Before discussing this task, we emphasize that the theoretical results apply equally well to time-series and cross-section processes. That is, the theory should correctly characterize both the time-path of the interactions between two countries and the interactions among countries in a single year. For example, as two countries grow in total GDP and become more similar in size over time, direct investment between them should grow in the manner suggested by the theory. Among a set of countries in a given year, the same bilateral relationships should apply.

One way to isolate the cross-section contribution to the results is to run regressions on data for particular years or on data averaged over all years. However, we encounter severe collinearity in attempting this task because the United States constitutes one half of each observation pair. Thus, for example, [GDP Sum] and [GDP Difference Squared] have a correlation coefficient of 0.995 in the 1990 cross-section, despite the fact that they have a correlation of –0.60 in the full panel. Thus, the time-series variation in U.S. GDP is vital to identifying the separate contributions of these two variables and this information is discarded in the averaging procedure or in the use of a single year. Furthermore, the variable [Trade Cost Parent] has the same value for all U.S. outward investments and [Trade Cost Host] and [Investment Cost Host] have the same values for all investments in the United States. In conjunction with the fact that there are only 63 cross-section observations in the most complete year, it is not possible to obtain precise estimates of our basic specification. Despite this problem, in our cross-section regression using data averages, all coefficient signs are correct except for [Trade Cost Host * Squared Skill Difference], which was not significant in the panel regressions of Table 3 in any case. We do not report these results here, but they are available from the authors or may be found in Carr *et al.* (1998).

In order to distinguish the time-series contributions to the results we employ country fixed effects. Table 4 lists results where each equation contains a dummy variable for the host country. That is, the fixed effects pertain only to recipient countries. We do not include a dummy variable for the United States because such inclusion would generate perfect collinearity. As in the case of Table 3, OLS, WLS, and Tobit results are presented. We do not report the coefficients on the dummy variables, but most are significant.[8]

The results are qualitatively similar to those in Table 3 for the first group of four variables in the OLS and WLS regressions, except that the coefficient on [Skill Difference] is reduced by half. The magnitudes of the coefficients on [Investment Cost Host] and [Trade Cost Host] are considerably smaller in Table 4 as well and both have lower significance levels. Thus, although the sign pattern is robust to the inclusion of country fixed effects, it is difficult to identify confidently the contribution of trade costs and investment costs to multinational production.

The Tobit results in Table 4 also show a smaller coefficient for [Skill Difference] with the country dummies added. The variables measuring perceived trade costs are significant in this specification in the hypothesized directions, but the investment-cost variable is not. Overall, it appears that the addition of country fixed effects does not change the results qualitatively, but a smaller role for endowment differences is predicted.[9] It is noteworthy that in the Tobit specification, which incorporates many more developing countries with zero reported affiliate sales, the magnitudes and significance levels of trade costs in both host and parent countries are

[8]Results are available on request. We also tried country-pair dummies but with the United States as a partner in each case this procedure could not well distinguish individual country effects. We should note that the variable [Distance] is a perfect linear combination of the country dummies, so one country dummy is dropped in the fixed-effects regressions. Because [Distance] is time-invariant, its interpretation in this specification is somewhat unclear.

[9]Most of the countries in the sample are less skilled-labor-abundant than the United States. It may be that the country dummies are capturing some of this effect that should be correctly attributed to endowment differences, as it is in the panel estimation.

expanded in comparison to the WLS case, as are those of relative endowment differences. This result provides some support for the notion that horizontal FDI and vertical FDI respond differently to host-country and parent-country trade protection.[10]

Overall, we believe the results in Tables 3 and 4 provide strong support to the knowledge-capital model of foreign direct investment. Indeed, the strong statistical fit of the model suggests that bilateral variables explain much of the variation in affiliate sales, despite the fact that the world has many countries that could complicate bilateral investment relations. In the regression, affiliate sales are strongly sensitive to bilateral aggregate economic activity, squared differences in GDP, differences in skilled labor endowments, and the interaction between size and endowment differences. The evidence suggests more weakly that affiliate activity depends on investment costs and trade costs in the hypothesized directions. We wish to use these results to characterize the various direct and indirect impacts more fully, which is the next task.

5. Interpreting the Coefficients

In this section, we interpret the magnitude of the coefficients and interpret the partial derivatives discussed above. For this purpose, we employ the coefficients from the model in Table 4 (fixed effects included) and apply them to average data values from the year 1991.

First, consider increases in perceived trade costs in the host country, as measured by the index [Trade Cost Host]. It is clear from the estimation that trade costs increase affiliate production when countries have identical relative endowments of skilled labor ([Skill Difference] = 0). This is consistent with horizontal investment. This effect is weakened when the countries differ in relative endowments, but theory suggests it should not be reversed: vertical investments should be discouraged by *parent-country* trade costs, but not so much by *host-country* trade costs. Results from Table 4 are

[10]In a final specification including both country dummies and year dummies, we find that all coefficients retain their anticipated signs and neither their magnitudes nor significance levels change much from those reported in Table 4. These results are available on request.

expressed as a partial derivative in Panel B of Table 2. In the data, the WLS derivative is always positive for all parent-to-host observations, and the Tobit derivative is positive for all but one case (United States to Indonesia). We can therefore state the following empirical conclusion from Panel B of Table 2:

RESULT 1: *An increase in the host-country's trade costs will raise production by affiliates of parent-country firms.*

While we do not attempt any measure of trade versus investment, this result suggests that inward trade costs induce a substitution of local production for imports.

Second, consider an equal bilateral increase in trade costs in both parent and host. This should encourage horizontal investments but discourage vertical investments. Results from Table 4 are expressed as partial derivatives in Panel C of Table 2. Using terminology somewhat tautologically, suppose we define trade and investment as "complements" if higher bilateral trade costs discourage investment and "substitutes" if higher trade costs encourage investment. The result in Panel C, Table 2 can then be interpreted as follows:

RESULT 2: *A bilateral increase in parent and host-country trade costs*:

(*A*) *decreases affiliate production, so trade and investment are "complements"* (*WLS*);

(*B*) *generally decreases affiliate production when the non-U.S. country is a developing country ("complements") but increases affiliate production when the non-U.S. country is another high-income country ("substitutes")* (*Tobit*).

The Tobit result in Part (B) accords well with the intuition from the theory model. Investment between two developed countries (small [Skill Difference]) is generally horizontal, and therefore encouraged by trade costs. Investment between countries of quite different income levels (large [Skill Difference]) is generally vertical, which is discouraged by trade costs.

Third, consider an increase in parent-country GDP, holding total world GDP constant (that is, the host's GDP change is the negative of the parent's change). When countries have identical relative endowments ([Skill Difference] = 0), this derivative is positive with [GDP Difference] < 0, zero at [GDP Difference] = 0, and negative with [GDP Difference] > 0. With the parent country more skilled-labor-abundant than the host country, the theory and simulations predict that this derivative switches sign, from positive to negative, at a lower value of [GDP Difference] (see Figure 2). Econometric results from Table 4 are expressed as partial derivatives in Panel D of Table 2. The latter imply that an increase in a parent country's GDP will increase its affiliate sales abroad only if it is small ([GDP Difference] < 0) and/or skilled-labor scarce ([Skill Difference] < 0).[11]

One interesting interpretation of these results involves the convergence in income between the United States and its trading partners, holding total two-country income constant (GDP Sum is constant). Using values of [Skill Difference] from the data, it turns out that the contribution of the second term in the derivative is small and is always dominated by the first term. Note that [GDP Difference] is always positive if the United States is the parent and negative if the United States is the host country.

RESULT 3: *A convergence in income (GDP) between the United States and any host country (holding the sum of their incomes constant) increases affiliate sales in both directions.*

This finding is connected to results in Figures 1 and 2. The volume of affiliate sales along the SW–NE diagonal of Figure 1 is the sum of the values in Figure 2 (h to f) and corresponding values for the other direction, f to h (not shown). Both are inverted U-shaped relationships and while they are not identical, there are regions where convergence toward the center of the box raises affiliate sales in both

[11]There are 66 country pairs (i, j observations) with positive affiliate sales from parent to host in 1991. Fifty-nine of these have complete data for this exercise. Thirty-six of the 59 are affiliate sales of U.S. firms abroad and 23 are foreign-firm affiliate sales in the United States.

directions. Moving toward the center, national firms headquartered in the large country (i.e., the United States) are replaced by horizontal multinational firms headquartered in both countries.

Fourth, consider an increase in the skilled-labor abundance of the parent country relative to the host country. Our results in Table 4 are expressed as partial derivatives in Panel E of Table 2. Results indicate that this derivative is generally positive for similar countries, but its (absolute) value is reduced by a higher relative endowment difference or a larger GDP difference. Large values of [Skill Difference] and [GDP Difference] weaken the effects of an increase in parent-country skilled-labor abundance on outward affiliate sales. To put it the other way around, an increase in the *host-country* skilled-labor abundance (Δ Skill Difference < 0) may increase inward investment if that host country is small relative to the parent.

Inserting values for [Skill Difference], [Trade Cost Host], and [GDP Difference] for the 1991 data, these results imply the following:

RESULT 4: (*A*) *When the United States is parent, an increase in host-country skilled-labor abundance increases U.S. affiliate production in the host country (production by U.S. affiliates is attracted to skilled-labor-abundant host countries).*

(*B*) *When the United States is host, an increase in parent-country skilled-labor abundance increases the parent-country's affiliate production in the United States.*

Result (A) is consistent with the well-known stylized fact that the poorest countries in the world receive a much smaller share of world direct investment than their share of world income (Kevin Honglin Zhang and Markusen, 1999).

As a final point, we note that the theory suggests a sharper hypothesis on the coefficient of [GDP Sum] than that it is simply positive. Higher total income should lead to some shifting from national firms, which are high marginal-cost suppliers to foreign markets, to horizontal multinationals, which are high fixed-cost suppliers (Markusen and Venables, 1998). In regions of parameter space in which regime shifting does not occur, affiliate production should rise in proportion to total world income. Overall, this suggests that affiliate sales should be elastic with respect to world income. We therefore used the results in Table 4 to calculate the implied elasticity of total affiliate sales [Real Sales] with respect to total income [GDP Sum] for 1991 in the data. Our results are that this elasticity is 1.35 at the mean values for [GDP Difference Squared] and [Skill Difference], using the WLS estimates with country fixed effects.[12] This estimate is significantly larger than one. We can therefore state the following result:

RESULT 5: *Affiliate production is income elastic, in that a bilateral increase in parent- and host-country incomes increases affiliate production by a greater proportion.*

This result is consistent with the well-known stylized fact that direct investment, whether measured by stocks or affiliates sales, has risen much faster than world income and trade since the mid-1970s (Markusen and Venables, 1996, 1998; Markusen, 1997). It also bolsters the finding in Jonathon Eaton and Akiko Tamura (1994) that the elasticity of U.S. foreign direct investment (FDI) abroad with respect to increases in host-country per capita income lies between 1.2 and 1.6, though their estimates of the elasticity of inward FDI in the United States are larger, lying between 2.0 and 2.2. Our approach is distinctive, however, in that we compute the elasticity of affiliate sales with respect to the bilateral sum of real GDP.

6. Concluding Remarks

The knowledge-capital approach to the multinational enterprise as outlined in this paper is operational and yields clear, testable hypotheses. It this sense, it is more useful than some other theories

[12]We do not report an elasticity based on the Tobit regression because it would be misleading. In particular, the cases included in that regression with zero values of [Real Sales] may be country pairs in which a large increase in [GDP Sum] is required to generate positive sales, rather than a marginal increase, suggesting that local elasticity estimates are questionable.

of FDI, such as the "transactions cost" approach to multinational enterprises.

In this paper, we test hypotheses regarding the importance of multinational activity between countries as a function of certain characteristics of those countries, particularly size, size differences, relative endowment differences, trade and investment costs, and certain interactions among these variables as predicted by the theory. In our view, the data fit the model well, lending considerable support to the theory. The panel estimates in Tables 3 and 4 yield correct signs and strong statistical significance for the central variables [GDP Sum], [GDP Difference Squared], [Skill Difference], [GDP Difference * Skill Difference], [Investment Cost Host], and [Trade Cost Host]. Other variables [Trade Cost Parent] and [Trade Cost Host * Skill Difference Squared] have correct signs but display weak statistical significance. Our efforts to separate the panel results into cross-section and time-series impacts are made problematic by multicollinearity in the cross-section data. Because of the bilateral nature of these data, the time-series variation in the U.S. observations is critical to identification of the contributions of several variables. Estimation with country fixed effects produced results consistent with the panel approach.[13]

According to our findings, outward investment from a source country to affiliates in a host country is increasing in the sum of their economic sizes, their similarity in size, the relative skilled-labor abundance of the parent nation, and the interaction between size and relative endowment differences. Some of these findings are consistent with earlier results, particularly those of Brainard (1997) and Ekholm (1997). But the precise formulations here are different and closely tied to one particular model. This model allows for simultaneous horizontal and vertical motives for direct investment and emphasizes certain interactions, such as that between size and endowment differences. We should also note that the theoretical model fully endogenizes trade flows in its calculations, allowing direct predictions on

[13]For comparison purposes we ran a simple gravity equation on the panel data set, with log real affiliate sales regressed on log real GDP in host and parent countries and log distance. The gravity equation displayed a considerably lower adjusted R^2 (0.46) than did the panel equation based on the theoretical model (0.60).

affiliate sales without requiring us to worry about questions of trade versus investment. Trade, like factor prices and commodity prices, is endogenous in generating the predictions of the model.

Subsequent to the estimation, we interpreted the estimates in the language of comparative-static questions about the world economy. Results indicate first that increases in host-country trade costs raise inward affiliate production. Bilateral increases in trade costs produce results that suggest that trade and investment are "complements" but may be "substitutes" (Tobit regression) for similar countries, with the latter result being consistent with horizontal investment predicted between similar countries. A convergence in country size between the United States and any host country will increase affiliate sales in *both* directions. An increase in any host country's skilled-labor abundance will increase U.S. outward affiliate sales, demonstrating that U.S. outward investment is attracted to more skilled-labor-abundant countries. Finally, affiliate production is elastic in total two-country GDP as predicted by theory.

In summary, we are enthusiastic about the results, and believe that they fit well with theory. We hope that the model will therefore prove useful in future policy analysis.

References

Aitken, Brian; Harrison, Ann and Lipsey, Robert E. "Wages and Foreign Ownership: A Comparative Study of Mexico, Venezuela, and the United States." *Journal of International Economics*, May 1996, *40* (3–4), pp. 345–371.

Blomstrom, Magnus; Fors, Gunnar and Lipsey, Robert E. "Foreign Direct Investment and Employment: Home Country Experience in the United States and Sweden." National Bureau of Economic Research (Cambridge, MA) Working Paper No. 6205, October 1997.

Brainard, S. Lael. "A Simple Theory of Multinational Corporations and Trade with a Trade-off between Proximity and Concentration." National Bureau of Economic Research (Cambridge, MA) Working Paper No. 4269, February 1993a.

Brainard, S. Lael. "An Empirical Assessment of the Factor Proportions Explanation of Multinationals Sales." National Bureau of Economic Research (Cambridge, MA) Working Paper No. 4583, December 1993b.

Brainard, S. Lael. "An Empirical Assessment of the Proximity-Concentration Trade-off Between Multinational Sales and Trade." *American Economic Review*, September 1997, *87* (4), pp. 520–544.

Carr, David L.; Markusen, James R. and Maskus, Keith E. "Estimating the Knowledge-Capital Model of the Multinational Enterprise." National Bureau of Economic Research (Cambridge, MA) Working Paper No. 6773, October 1998.

Caves, Richard E. *Multinational Enterprise and Economic Analysis*, 2nd Ed. London: Cambridge University Press, 1996.

Eaton, Jonathan and Tamura, Akiko. "Bilateralism and Regionalism in Japanese and U.S. Trade and Direct Foreign Investment Patterns." *Journal of the Japanese and International Economies*, December 1994, *8* (4), pp. 478–510.

Ekholm, Karolina. *Multinational Production and Trade in Technological Knowledge*. Lund Economic Studies (Lund, Sweden) No. 58, Lund: Lund University, March 1995.

Ekholm, Karolina. "Factor Endowments and the Pattern of Affiliate Production by Multinational Enterprises." CREDIT (University of Nottingham) Working Paper No. 97/19, 1997.

Ekholm, Karolina. "Headquarter Services and Revealed Factor Abundance." *Review of International Economics*, November 1998a, *6* (4), pp. 545–553.

Ekholm, Karolina. "Proximity Advantages, Scale Economies, and the Location of Production," in Pontus Braunerhjelm and Karolina Ekholm, eds., *The Geography of Multinational Firms*. Boston: Kluwer Academic Publishers, 1998b, pp. 59–76.

Feenstra, Robert C. and Hanson, Gordon H. "Foreign Direct Investment and Relative Wages: Evidence from Mexico's Maquiladoras." *Journal of International Economics*, May 1997, *42* (3–4), pp. 371–393.

Helpman, Elhanan. "A Simple Theory of Trade with Multinational Corporations." *Journal of Political Economy*, June 1984, *92* (3), pp. 451–471.

Horstmann, Ignatius J. and Markusen, James R. "Strategic Investments and the Development of Multinationals." *International Economic Review*, February 1987, *28* (1), pp. 109–121.

Horstmann, Ignatius J. and Markusen, James R. "Endogenous Market Structures in International Trade (Natura Facit Saltum)." *Journal of International Economics*, February 1992, *32* (1–2), pp. 109–129.

Lipsey, Robert E.; Blomstrom, Magnus and Ramstetter, Eric. "Internationalized Production in World Output." National Bureau of Economic Research (Cambridge, MA) Working Paper No. 5385, December 1995.

Markusen, James R. "Multinationals, Multi-Plant Economies, and the Gains from Trade." *Journal of International Economics*, May 1984, *16* (3–4), pp. 205–226.

Markusen, James R. "The Boundaries of Multinational Firms and the Theory of International Trade." *Journal of Economic Perspectives*, Spring 1995, *9* (2), pp. 169–189.

Markusen, James R. "Trade versus Investment Liberalization." National Bureau of Economic Research (Cambridge, MA) Working Paper No. 6231, October 1997.

Markusen, James R. and Venables, Anthony J. "The Increased Importance of Multinationals in North American Economic Relationships: A Convergence Hypothesis." In Matthew W. Canzoneri, Wilfred J. Ethier, and Vitoria Grilli, eds., *The New Transatlantic Economy*. London: Cambridge University Press, 1996, pp. 169–188.

Markusen, James R. and Venables, Anthony J. "The Role of Multinational Firms in the Wage-Gap Debate." *Review of International Economics*, November 1997, *5* (4), pp. 435–451.

Markusen, James R. and Venables, Anthony J. "Multinational Firms and the New Trade Theory." *Journal of International Economics*, December 1998, *46* (2), pp. 183–203.

Markusen, James R.; Venables, Anthony J.; Eby-Konan, Denise; and Zhang, Kevin Honglin. "A Unified Treatment of Horizontal Direct Investment, Vertical Direct Investment, and the Pattern of Trade in Goods and Services." National Bureau of Economic Research (Cambridge, MA) Working Paper No. 5696, August 1996.

Rutherford, Thomas F. "Extensions of GAMS for Complementarity Problems Arising in Applied Economics." *Journal of Economic Dynamics and Control*, November 1995, *19* (8), pp. 1299–1324.

Rutherford, Thomas F. "Applied General Equilibrium Modeling with MPSGE as a GAMS Subsystem: An Overview of the Modeling Framework and Syntax." *Computational Economics*, October 1999, *14* (1–2), pp. 1–46.

Slaughter, Matthew J. "Production Transfer within Multinational Enterprises and American Wages." *Journal of International Economics*, April 2000, *50* (2), pp. 449–472.

United Nations Conference on Trade and Development. *World Investment Report*. New York: United Nations, 1993.

Zhang, Kevin Honglin and Markusen, James R. "Vertical Multinationals and Host-Country Characteristics." *Journal of Development Economics*, August 1999, *59* (2), pp. 233–252.

Chapter 14

Quantifying the Impact of Services Liberalization in a Developing Country*

Denise Eby Konan and Keith E. Maskus[†]

1. Introduction

Many developing countries sharply lowered manufacturing tariffs and other impediments to goods trade in the 1990s, both through unilateral trade liberalization and agreements reached in the World Trade Organization (WTO) and regional trade accords. Nevertheless, as Rodrik (2001) has pointed out, traditional trade liberalization seems not to have fully delivered on its promise for prosperity. In this paper we suggest that part of the explanation rests in the continued insulation, and resulting poor condition, of the domestic services sectors in many developing countries. These service sectors tend to deliver high-cost and unproductive input services, thereby limiting economic efficiency gains from trade reform. We explore this claim in a computational general equilibrium model of the Tunisian economy.

The mechanisms by which services trade liberalization might improve welfare differ from those of goods trade. Producers respond to a lowering of border barriers in products by reorienting production towards goods in which an economy possesses a comparative

*Reprinted from *Journal of Development Economics*, Vol. 81, No. 1, pp. 141–162. Copyright 2006, with permission from Elsevier.
[†]We gratefully acknowledge Sherry Stephenson, Bernard Hoekman, Mohamed Goaied, Ari Van Assche, and Dominique van der Mensbrugghe for valuable discussions throughout the writing of this paper. We also thank Robert Lipsey, participants at the European Trade Study Group in Nottingham, UK, and two anonymous referees for constructive comments.

advantage. Tariff cuts thus involve a restructuring of the economy. Liberalization of merchandise trade also tends to redistribute income, with abundant factors benefiting disproportionately in the long run.

In contrast, trade in services is not solely a cross-border phenomenon. As the paper by Markusen, Rutherford, and Tarr (2005) points out, many foreign services are best transferred through foreign direct investment (FDI) due to the personal contact required between the service provider and client. Restrictions on FDI or the movement of professional personnel may reduce services trade far more than tariffs or other border barriers limit trade in goods. Indeed, in many developing countries, Tunisia included, laws and regulatory agencies erect entry barriers that essentially rule out FDI in key service sectors.

Foreign direct investment involves inflows of capital and personnel, but more importantly for developing countries it also tends to embody transfers of technology that can upgrade productivity in the domestic economy. Because financial, communications, and professional services are key intermediate inputs into production in all sectors, cost reductions in these sectors could have the effect of upgrading overall productivity (Lipsey, 2001; Markusen, 1989). Thus, whereas liberalizing goods trade moves an economy toward specialization, liberalizing services trade through permitting foreign establishments could lead to more balanced output expansion.[1]

In Tunisia there is a large state presence in many key service sectors. Regulatory agencies limit competition by restricting markets available to producers. Foreign participation is highly restricted and often limited to the services of non-resident Tunisian suppliers. As we discuss below, anecdotal evidence indicates that Tunisia's communication, financial, insurance, distribution, and professional services are expensive, of poor quality, and often inaccessible. Thus, the economy might particularly gain from more openness in the service sector.

[1]Similar comments pertain to trade in intermediate and capital goods. One interesting difference, which we find in our simulations, is that liberalization of goods trade may have sharply stronger impacts on relative factor prices than liberalization of services, because the former generates stronger production reallocation effects favoring the abundant factor.

In some respects, Tunisia has pursued a strategy of liberalization and alignment with the global economy (World Bank, 2004). For example, Tunisia concluded an Association Agreement with the European Union in 1995. Under this Agreement, border restrictions on industrial-goods imports from the EU were to be phased out over 12 years and the participants agreed to work toward greater financial and technical cooperation. However, the trade opening was confined largely to merchandise, excluding services.

Indeed, Tunisia's 1995 commitments under the General Agreement on Trade in Services (GATS) were few and generally represented a preservation of status quo policies.[2] Reluctance to open services markets to foreign suppliers, which would also involve privatization and deregulation, reflects concerns of policymakers about whether the economic benefits would justify the adjustment costs involved. Moreover, Tunisian services are supplied by important domestic interests and this fact makes liberalization difficult to undertake without some indication of the potential for gains and their likely distribution. Hoekman and Djankov (1996) in particular note the potentially large gains in well-being for Tunisian citizens of liberalization in services.

Unfortunately, as discussed in Section 2, empirical studies of the effects through FDI of services liberalization are scarce, in part due to a variety of measurement problems. It is against this backdrop that we conduct the present study. We develop a computable general equilibrium (CGE) model to compare the impacts of services liberalization to that of trade liberalization. A significant innovation is to implement GATS definitions and consider, separately, liberalization of both cross-border services trade and the establishment of domestic presence by FDI. For this purpose we develop estimates of the distortions that the present, relatively closed, environment in Tunisia imposes on the price of services. By simulating the removal of these distortions, we are able to quantify the impacts of liberalization.

---

[2]Of the twelve service sectors defined under the GATS classification scheme Tunisia made commitments in only three: financial services, tourism, and telecommunications. These commitments reflected little more than a codification of Tunisia's existing legal system and have not been expanded.

The potential efficiency gains would be large, on the order of four to five percent of initial consumption, while output in all merchandise sectors would rise. Furthermore, the real prices of both capital and labor would increase.

2. Conceptual Issues

Trade in services is distinguished from trade in goods by the intangible nature of many transactions. Unlike physical goods, which must cross borders and thus are subject to customs procedures and tariffs, services often involve direct transactions between the consumer and producer. This fact complicates measurement of both service flows and their corresponding impediments. The founders of GATS recognized the importance of the various channels by which services are transacted. GATS identifies four modes of supply: cross-border supply (mode 1), consumption abroad (mode 2), commercial presence (mode 3), and the presence of natural persons (mode 4). Thus GATS extends traditional trade law to include both foreign direct investment (mode 3) and movements of labor (mode 4).

Unfortunately, these legal definitions do not accord well with present international balance of payments accounting practices, as discussed in detail by Karsenty (2000), making difficult the measurement of services trade and its components. Equally problematic is the measurement of impediments to services trade.[3] For example, Hoekman (1995) developed a frequency indicator as an initial attempt to quantify the presence of barriers based on the GATS schedule of commitments by country. While this index provides some indication of the extent of commitments, it is not designed to measure the level of service barriers. Francois (1999) estimated a gravity model of services trade, adopting Singapore and Hong Kong as free-trade benchmarks. Discrepancies in predicted trade patterns were used to indicate the severity of policy barriers. Warren and Findlay (2000) described an ongoing Australian services research project attempting to measure the wedge between price and marginal

[3]Warren and Findlay (2000) and Hoekman (2000) provide excellent surveys of these measurement issues. See also Brown and Stern (2001) and Stern (2002) for discussion of measurement and modeling in services.

cost in service sectors in order to get an indication of the impact that impediments might have on prices. The project also seeks to develop quantity impact measures by comparing domestic output to international standards. Unfortunately, the study is incomplete and may not provide information that could be applied to developing economies.

Given severe data limitations, econometric studies on services trade and their barriers are rare. In contrast, a CGE model can provide quantifiable insights with relatively sparse datasets.[4] An early model by Brown *et al.* (1996) converted Hoekman's frequency indices into an approximation of cross-border barriers to services trade and simulated liberalization in the multi-country Michigan model. Several studies of services barriers exist using the multi-country database established in the Global Trade Analysis Project (GTAP).[5] Hertel (2000) approximated cross-border barriers with the gravity-equation estimates of Francois and treated liberalization as an elimination of resource-using barriers. Robinson, Wang, and Martin (2002) focused on enhancements in total factor productivity arising from imported producer services in a ten-region, eleven-sector CGE model.

A significant limitation of these models is that they do not capture the potential benefit of FDI liberalization in services. Recently, however, FDI has been incorporated into multi-country CGE models. Two examples are Dee and Hanslow (2001) and Brown and Stern (2001). Their approaches to measuring services were somewhat different, with the latter authors using price-cost margins for a number of countries. Both studies assumed imperfect competition in manufactures and services. Typically, FDI was treated as a capital flow and barriers to the right of establishment were modeled as a tax on both the existing investment flow and local sales. Liberalization (removal of the tax) resulted in a global reallocation of capital as sector-level rates of return adjust to a new equilibrium. This approach represents a significant improvement over earlier work, which focused only on cross-border services trade (mode 1). These models are informative about the potential global impacts of the GATS approach.

[4]Whalley (2003) and Dee *et al.* (2000) survey the computational literature in services.
[5]GTAP is explained in Hertel (1997).

A contribution of particular note is Brown, Deardorff, and Stern (1997), who analyzed the potential impacts of the Tunisia–EU Association Agreement in a model with monopolistic competition in non-agricultural sectors. They permitted an exogenous increase in FDI equal to 10 percent of the initial Tunisian capital stock in response to the Agreement, with the incoming capital allocated according to sectoral returns with industry-specific capital. They do not contemplate liberalization of services, however. Their findings suggest that the EU Agreement would have a modest impact on Tunisian GDP, ranging from −0.1 percent to +3.3 percent depending on the scenario.

Unfortunately, the models available to date fall short of capturing domestic effects of services liberalization in such highly regulated developing economies as Tunisia for two reasons. First, service sectors that experienced no FDI in the benchmark economy were assumed to be non-tradable and could not, therefore, be recipients of FDI in counterfactual scenarios. Yet in Tunisia, we might reasonably expect that FDI would occur in several important sectors (such as telecommunications or insurance) in which, under current policy, foreign suppliers are simply not permitted to serve Tunisian residents. Prior methodologies do not offer a reasonable way to predict how responsive FDI flows would be in such sectors in a liberalized environment. Second, the market structure in most service sectors in Tunisia is highly regulated and imperfectly competitive, perhaps best characterized as cartels. As Francois and Wooten (2001) discussed, shallow liberalization without deregulation could essentially be an invitation to foreign firms to join the cartel. Deeper liberalization, involving also deregulation permitting new entry, should bring about a more competitive market structure.

An additional point of departure for our work is to consider the impact of services liberalization on the structure of Tunisia's economy relative to that of commodities trade liberalization. The Tunisian input–output table allows us to disaggregate services into 15 sectors. We consider the regulatory environment at a disaggregated level and are able to compare the relative impact of liberalization in individual service sectors. With data on services as intermediate inputs, our simulations indicate how service liberalization may restructure the domestic economy.

As discussed in the following section, our model is designed to capture several static effects of services and goods trade liberalization, including efficiencies from production reallocation, pro-competitive gains from reducing cartel power, and efficiencies from adopting best-practice technologies. Analysis of pro-competitive impacts in the CGE context stems from the pioneering work of Harris (1984). However, we do not compute gains from increasing returns to scale, as emphasized by Harris (1984), Brown, Deardorff, and Stern (1997), and others. Nor do we consider the potential for product variety gains from liberalizing services trade, an element introduced by Romer (1994) and introduced into dynamic CGE models by Rutherford and Tarr (2002).

3. Tunisia: Services and Impediments

Services play a significant role in the Tunisian economy, as is evident from Table 1. Nearly half of all output (excluding tourism) was comprised of services in 1995, according to the input–output table for that year (Institute National de la Statistique, 1998). Services made up one-third of household consumption and 18 percent of intermediate demand. Tourism was the largest of the tradable services, representing over 15 percent of imports and more than 19 percent of all exports. Tourism export services are considered to be rather open.[6]

There is no statistical information on the importance of FDI in Tunisian services trade, as the earnings and sales of foreign-owned subsidiaries are not reported separately from those of domestic firms. However, foreign participation clearly is minimal in the main service activities due to a variety of investment measures, exchange controls and limits on movement of foreign personnel. The most important restriction is the Foreign Commercial Activities Law of 1961 (Decree Law Number 61-14), by which non-Tunisians are required to obtain a trader's permit, *carte de commercant*, to engage in any form of

[6]Tourism may be exported without being produced because tourists are treated as "foreign consumers" in the input–output table and enter as a component of final demand. The production of tourism services exists in the other service sectors, such as transportation, hotels, and restaurants.

Table 1. Sectoral output and factor shares (%).

	Production	Imports	Household consumption	Intermediate consumption	Exports
Aggregate Sectors (% of total)					
Agriculture and Fishing	17.5	10.2	32.4	20.1	6.6
Manufacturing	30.0	63.2	29.7	51.3	55.3
Utilities, Mining, Petroleum	5.8	5.8	4.0	10.9	6.8
Services	46.7	20.8	33.8	17.7	31.2
Service Sectors (% of total)					
Construction	8.2	0.0	0.3	0.4	0.0
Distribution/ Commerce	6.9	0.0	0.0	0.0	0.0
Transportation	5.6	2.7	5.7	4.3	8.7
Communication	1.0	0.1	0.3	1.7	0.4
Hotel	1.5	0.0	3.9	0.1	0.0
Restaurant	4.1	0.0	10.9	0.0	0.0
Finance	2.5	0.2	0.1	4.8	0.3
Insurance	0.3	0.2	0.3	0.6	0.0
Business	1.4	2.1	0.1	2.5	2.5
Real Estate	2.6	0.0	5.0	1.3	0.0
Repair	1.3	0.0	1.1	1.8	0.0
Health and Education	2.0	0.0	4.9	0.2	0.0
Public	9.0	0.0	0.6	0.0	0.0
Other Services	0.2	0.0	0.6	0.0	0.0
Tourism	—	15.4	—	—	19.3

Source: Institute National de la Statistique, 1998, *Les Comptes de la Nation Base 1983, agregats et tableaux d'ensemble 1993–1997.*

commercial activity. This permit, granted on a discretionary basis, has effectively precluded foreign participation in most wholesale and retail markets. The Investment Code of 1993 is also an important barrier to majority foreign equity projects, with foreign entry requiring approval by regulatory councils. Such approvals are rare and FDI is strictly controlled.[7]

[7]Despite some movements toward liberalization of FDI in 2001, foreign equity participation in Tunisian services other than hotels remains "negligible" according to the World

Domestically, the level of competition varies across services sectors. Certain sectors, including computer services, real estate, construction, restaurants, and some distribution activities, are characterized by many participants, even though licensing requirements and other regulations impose burdensome costs on domestic entry and production. In several important sectors, such as automobiles and electronics, exclusive-distributor laws require that foreign suppliers appoint only Tunisian citizens as local agents and also ban parallel imports (Maskus and Lahouel, 2000). This situation grants effective monopoly power to local agents and substantially raises consumer prices.

In other areas, such as financial and banking services, insurance, and many professional services, entry is tightly controlled by regulatory boards that limit the ability of providers to offer a full range of services and also limit market access. That such restrictions increase costs and limit market size is particularly evident in the insurance industry. Tunisia's 1997 expenditures on insurance premiums were only 1.58 percent of GDP, a penetration rate far behind those of nearly all middle-income countries and even behind that of several African nations, with an average rate of 5.7 percent (Swiss Re, 1999). The insurance market is highly concentrated and dominated by state-owned enterprises, which accounted for roughly half of total insurance premiums (Vittas, 1995). Foreign involvement remains strictly limited. According to the Insurance Code (Article 44) Tunisian residents, and risks situated solely in Tunisia, may only be insured directly by policies from Tunisian-owned companies. Foreign subsidiaries are permitted as a public limited or mutual company and may only serve non-residents.

As Bahlous and Nabli (2000) discussed, the government is also heavily involved in the banking system. Five of the 12 commercial banks in the mid-1990s were public.[8] They initiated 68 percent of loans and tended to favor state-owned enterprises, which are

Bank (2004). The essential reasons include reservations from FDI of 225 activities, prior approval requirements in other activities, and cumbersome entry requirements. It is evident that the market was not more open to FDI in 1995, our benchmark year.

[8]This situation is still evident in that in the year 2000 there were only 14 commercial banks and these held 2/3 of the economy's financial assets (World Bank, 2004). The government controls the three largest banks.

heavily indebted. Many small and medium-sized enterprises reported difficulties in gaining access to credit (World Bank, 2000, 2004). The ratio of non-performing loans to total assets was 36 percent and 72 percent in 1993 for the two largest public banks. Goaied (1999) estimated the cost frontier of Tunisian deposit commercial banks with panel data for 1980 to 1995 and found that the mean cost-inefficiency of private banks was 28.5 percent and of public banks was 50.5 percent.

By standard measures the Tunisian banking system also seems to engage in less monetary intermediation than would be anticipated for a country of its development level. In general, the ratio of domestic credit to GDP rises as countries become richer. The ratio in Tunisia in 1995 was 0.57, which was markedly lower then those in Egypt (0.77) and Jordan (0.88), Arab League countries with even lower income levels.[9] Thus, as a rough indication, monetary intermediation through the Tunisian banking sector is perhaps 30 percent less than might be expected.

Finally, several service sectors have traditionally been provided by the government and are only now being slowly liberalized to allow limited private participation. These industries include telecommunications, postal services, air transport, health, and education. For example, Tunisie Telecom remains the sole provider of nearly all telecommunications services and regulation is the domain of the Ministry of Telecommunications. As shown in Table 1, communications services comprise only one percent of production in the economy, whereas they make up four percent of GDP in Morocco (Varoudakis and Rossotto, 2001). Tunisian prices are likewise four times those of Morocco (World Bank, 2001). Recent investments in infrastructure have reduced the waiting time for fixed-line installation and the rate of service disruption. The telephone network is fully digitized. Yet Tunisian teledensity falls short of that of other middle-income countries, particularly in rural areas. Internet service accounts are expanding rapidly in response to the granting of two licenses to provide residential and commercial internet services but monthly

[9]The source for these data is International Monetary Fund, *International Financial Statistics Yearbook 1997*.

access rates are double those in Morocco and much higher than those observed in most liberalized markets.[10]

4. The Model and Benchmark Data

In this section we present the model structure and describe the database used, the details of which are in a technical appendix. We employ what is, in most respects, a standard CGE model of a small open economy. Our contribution is in the explicit treatment of services production and investment. In the benchmark, production decisions in the services sector are distorted by regulations that raise entry costs and limit the rights of foreign enterprises to establish facilities in Tunisia. Counterfactual experiments involve the removal of regulatory investment barriers. As new firms are now permitted to enter, the market structure becomes competitive (eliminating rents) and more efficient (introducing lower cost firms).

Benchmark regulation distorts prices and quantities through two primary channels. One is the *cartel effect* whereby barriers to FDI and excessive regulation limit both domestic and foreign participants in certain service sectors, thus hampering competition and supporting market power on behalf of local firms. This market power creates the opportunity in sector i to charge price markups v_i over marginal cost c_i:

$$c_i(1 + v_i) = p_i. \tag{1}$$

The markup depends on the number of firms, price elasticity of demand, and conjectures about reactions of rival firms. We assume that the rents generated by markups accrue to the representative agent.

Second is a *cost inefficiency effect*. Marginal costs in a regulated environment may be excessively high as low-cost foreign suppliers are excluded from the market. Additionally, domestic suppliers may be forced to absorb into their costs various regulations on provision and

[10]Modernization of this sector seems to have improved technical efficiency considerably by 2003 (World Bank, 2004), though entry remains tightly controlled.

bureaucratic procedures. These activities do not contribute to output and generate pure economic waste (Hoekman and Konan, 2000). Thus, resource-using service barriers λ_i raise marginal costs above 'best practice' marginal costs c_i^* that would prevail in a liberalized environment:

$$c_i^*(1+\lambda_i) = c_i. \tag{2}$$

Combining equations (1) and (2) shows that the wedge between price and true marginal costs depends on the product of an ad valorem markup and a proportionate waste factor.

Ultimately it would be preferable and interesting to locate empirical information on these two wedges in each sector in order to simulate the effects of their separate and joint removal. Unfortunately, we only have estimates of the entire price–cost wedge and cannot perform this decomposition except on assumed shares. Thus, our counterfactual scenarios make two assumptions. First, we assume that barriers to FDI generate wedges consisting of half rents and half waste. Second, based on the discussion in the prior section we distinguish three types of service sectors: (i) competitive with inefficient technologies, or "low rent, high cost" (construction, distribution, hotels, restaurants, real estate, and repair); (ii) monopolized with efficient technologies, or "high rent, low cost" (communication); and (iii) monopolized with inefficient technologies, or "high rent, high cost" (transport, finance, insurance, and business services).

Services liberalization involves confronting a domestic cartel with the possibility of competitive markets with foreign entry. It is important to note that liberalization does not, in our model, generate endogenous changes in FDI flows. Rather, the scenarios involve changes in ownership and market structure in ways that improve efficiency and alter the distribution of rents. This assumption reflects the fact that in several service sectors there is no foreign participation (that is, FDI) in the benchmark equilibrium. In that context it is impossible to determine what the impact of liberalization would be on "marginal" FDI flows, which would not be meaningful. It also permits us to retain a fixed aggregate capital stock in the model, rather than engaging in dynamic simulations of endogenous investment and capital allocation. In this context, the estimates of welfare changes

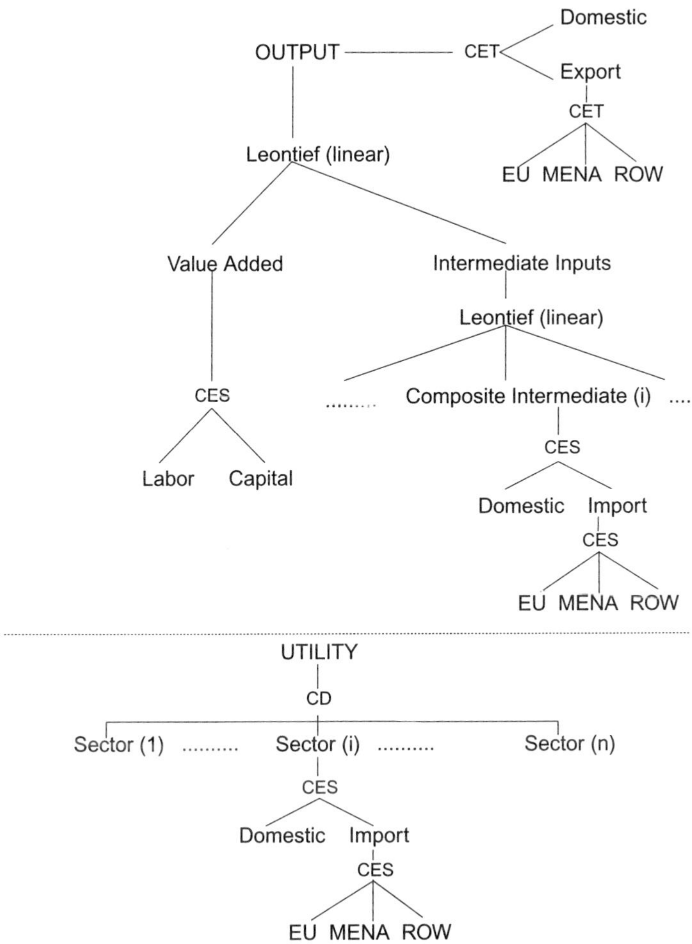

Figure 1. Model structure.

from liberalization of establishment rules are likely to be understated relative to full long-run gains.

Model equations are presented in the Appendix and the model is depicted in Figure 1. We assume that production of agricultural, mineral, and manufacturing goods is characterized by constant returns to scale and perfect competition, implying that prices equal marginal costs of output.[11] Services production is subject to constant

[11]See, for example, Konan and Maskus (2000).

returns as well, though in the initial benchmark it is monopolized as discussed above. It should be stressed that a number of services, such as telecommunications, transport, and finance, likely are subject to increasing returns and, to the extent that trade liberalization expands output in those sectors, additional gains would accrue to the economy that we do not capture here. Note also that we engage solely in static simulations, without permitting endogenous increases in product differentiation and variety. For these reasons, in addition to our conservative estimates of price–cost wedges, computed welfare gains are liable to be underestimated.

In all sectors, production functions are approximated with Leontief technologies using composite intermediate inputs and real value added. A constant elasticity of substitution (CES) production function describes the substitutability between labor and capital inputs in producing real value added. Intermediate inputs and final goods are differentiated by country of origin according to the Armington assumption, so that export and import prices differ across regions.[12] The three trading regions are the European Union (EU), the Arab League countries,[13] and the rest of the world (ROW).

In each sector, demand for domestically produced and imported goods is represented by a CES function, and intermediate imports are also differentiated across regional sources of supply in a CES structure. Similarly, Tunisian industries supply regionally differentiated goods to both domestic and foreign markets (exports). Production follows a nested two-stage constant elasticity of transformation (CET) function. Total output is first calculated as the sum of domestic supply and total exports, with the latter then being allocated across the same destination regions according to a sub-CET function. Capital and labor are assumed to be freely mobile across

[12]In some contexts the Armington assumption can be problematic for it may generate large terms-of-trade effects that dominate welfare calculations. However, De Melo and Robinson (1989) showed that models allowing product differentiation are well behaved under a small open economy assumption; in effect the economy is a price taker at the level of aggregate trade flows and each region's aggregation is sufficiently distinctive to support the Armington assumption. In any event, we consider here only Tunisian import (FDI) liberalization, which would be expected to worsen the country's terms of trade. The fact that we find large welfare increases suggests that the issue is not important.

[13]These countries are designated MENA, for Middle-East North-African nations.

sectors, whereas the stocks of factor endowments are exogenous, implying that our simulations pertain to long-run outcomes of liberalization.

A representative consumer maximizes a nested CES utility function with a corresponding multi-staged budget constraint. In the first stage, the consumer decides how much to spend on goods from each sector, given the budget constraint. Income elasticities across sectors are set at unity as given by a Cobb–Douglas (CD) utility nest. In the second nest, the consumer determines domestic and aggregate import expenditures in each sector according to a CES function. Then given a budget for imports, the consumer selects purchases of imports from each region. These latter functions also characterize the split between government consumption and investment spending on domestic and imported goods and services. The representative consumer receives income from primary factors (labor and capital), net transfers from the government, the current-account deficit, and any net economic rents from the operation of restrictions on services trade.

Two standard closure rules are imposed: the savings–investment balance and a fixed current-account balance. The savings–investment balance is based on the assumption that the capital stock is exogenously fixed at the benchmark level. This stock is financed through forced consumer savings that act as a direct (lump-sum) tax. The interest rate (an index price of the composite capital stock) is endogenous and determined by factor-demand conditions. The current-account balance is the sum of the merchandise trade balance, the services balance, net foreign worker remittances, and (negative) net payments on foreign capital.[14] We assume that foreign reserves will be held constant so that the current account will be just offset by (the negative of) the capital account. The current-account balance itself is held constant in real terms throughout the simulations. Income from foreign remittances less foreign capital payments enters as an exogenous addition to the representative agent's income. To hold the current-account balance fixed while international prices are constant

[14]In the 1995 benchmark year, foreign remittances were approximately 650 million Dinars while net capital income totaled negative 680 million Dinars according to the IMF *Balance of Payments Statistics Yearbook*.

requires a balancing item. This is accomplished by means of a change in the home "real exchange rate," which refers implicitly to a change in the home price index (generated by changes in price of home-produced goods) sufficient to sustain a constant current-account balance as import and export volumes change.[15]

The government budget deficit is a deduction in available income for the representative agent, constituting a transfer to government consumption. The deficit is held fixed during our simulations. Thus, if a policy reform causes prices to fall, thereby reducing the tax revenues required to finance government expenditures, this tax saving is transferred to the representative agent. At the same time, if trade liberalization results in lost tariff revenues, the revenues are recouped by means of allowing household lump-sum taxes to vary endogenously.

The data for the model consist of a Social Accounting Matrix (SAM) and other parameters, such as import and export trade flows by region, sectoral tax and tariff rates, and elasticities of substitution and transformation. Because there is little empirical evidence on relevant elasticities for the Tunisian market, we make standard assumptions about their values. In particular, labor–capital substitution is set at unity in a Cobb–Douglas value-added production function. Benchmark trade elasticities are drawn from Rutherford *et al.* (1995) and Konan and Maskus (2000). The various trade elasticities are 2.0 for substitution between domestic and imported goods, 5.0 for substitution among regional imports and for transformation between domestic output and exports, and 8.0 for transformation among regional export destinations. These data are assembled into a consistent set of relationships between intermediate demand, final demand, and value-added transactions using the 1995 input–output table for Tunisia on a diskette provided by the Institut National De La Statistique (INS) along with the 1998 INS *Les Comptes de la Nation* report.

Trade and tariff data, provided in Table 2, were aggregated to the input–output sectoral basis using import weights consistent

[15]Tunisia has maintained a managed floating exchange rate system. The system has become more flexible, and the real exchange rate has depreciated somewhat leading up to implementation of the EU free trade agreement, World Bank (2004).

Table 2. Benchmark trade shares and tariffs.

	EU trade share (%)		Trade-weighted import tariff	
Sectors	Imports	Exports	EU	Rest of World
Agriculture	38.7	68.7	13.5	13.0
Process Foods	55.3	71.9	17.6	18.5
Chemicals and Glass	48.8	24.0	21.1	23.6
Non-Ferrous Metals	57.7	70.4	15.3	21.2
Metalwork	67.8	58.8	15.3	17.5
Machinery	77.0	86.0	6.5	8.5
Automobiles & Trucks	87.6	40.8	6.3	10.8
Automobile Parts	57.1	71.0	0.1	1.7
Electrical Parts	66.4	50.4	4.3	7.8
Electronics	66.4	50.4	4.3	7.8
Household Appliances	66.4	50.4	4.3	7.8
Chemicals	75.7	39.2	9.0	10.3
Clothing & Textiles	92.3	94.8	17.7	21.6
Leather	93.3	96.6	25.1	28.3
Wood	41.1	66.7	16.4	16.6
Paper	74.4	20.6	11.1	5.3
Plastics	72.9	28.0	14.5	18.7
Other Manufacturing	72.9	76.2	8.1	15.8
Mining	30.6	86.0	17.4	2.5
Petroleum	63.3	38.7	12.2	20.2
Services	70.0	76.0		

with the concordance between the input–output table and the tariff classification. Tariff rates were determined by collections data for 1995 and vary across regions due to duty drawback provisions as well as preferential treatment of the EU and the Arab League. There are no data on tariff collections on services, reflecting the absence of formal trade taxes, and we take their tariff rates to be zero. This treatment is the same as that in Chatti (2000), while Brown *et al.* (1997) assume there are no barriers to trade in Tunisian services.

Table 3 lists our estimated price wedges due to service barriers, which are the key parameters to be altered in the simulations. Mode 1 restrictions on cross-border trade are treated as ad valorem tariff equivalent non-tariff barriers. These are simply set to zero in the liberalization exercises. Sectors listed as non-traded (indicated by *nt*) remain non-traded post-liberalization. In terms of mode 3,

Table 3. Barriers to trade in services (NTB ad valorem price equivalent %).

	Mode 1 Cross-border trade	Mode 3 Foreign presence
Construction	*nt*	3
Distribution	*nt*	5
Transportation	50	3
Communications	200	30
Hotel	*nt*	3
Restaurant	*nt*	3
Finance	30	30
Insurance	50	50
Business, Insurance & Leasing	10	10
Real Estate	10	10
Repair	*nt*	3
Health and Education	*nt*	*nt*
Public	*nt*	*nt*
Other Services	*nt*	*nt*

Note: *nt*, non-traded modes of supply.

we would ideally like to estimate the impact that services barriers have on both price markups and on resource costs to distinguish between the pro-competitive effects and the cost-reduction effects. Warren and Findlay (2000) suggested computing the pro-competitive impacts using price–cost margins (or "net interest margins"). We would also like to capture the cost-reduction effects by comparing actual costs to a constructed estimate of costs if services were provided according to a 'world's best-practice' cost function. Unfortunately, none of these measurements are attainable for Tunisia, as is the case for most countries.

The services barriers given in Table 3 are based on industry studies in Tunisia, extensive discussions with Tunisian industry experts, country economists and government officials, and Zarrouk (2000). The financial services barriers estimates are taken from our observation that the level of monetary intermediation in the banking system is about 30 percent lower than in comparable countries (Bahlous and Nabli, 2000) and on Goaied's (1999) estimation of the cost inefficiencies in the financial sector. This is in line with the estimates of Kalirajan *et al.* (2000) for the banking sectors in Chile, Singapore, South Korea, and Thailand. The price wedges in insurance, communications, and transportation reflect the high level of benchmark regulation in those sectors and comparisons with markets in similar countries (Vittas, 1995; World Bank, 2000). The distribution and retail sectors show large inefficiencies and are very fragmented, making our five-percent inefficiency measure conservative. Many professional services are subject to a nationality requirement, thus restricting foreign participation, and it is likely that our ten-percent estimated price wedge is low. While the construction, hotel, and restaurant sectors are viewed here as already substantially liberalized and subject to a small price cost wedge, foreign participation remains subject to the investment code and projects are granted upon approval of a *Cahier des Charges.* Remaining sectors (health and education, public services, and other services) are taken to be exempt from potential liberalization.

It is important to note that these price wedges, while not constituting formal tariffs, do serve to restrict cross-border trade (mode 1) in services. Further, they increase costs of producing and trading other goods through the input–output structure. Thus, for example, the high costs in telecommunications make it relatively difficult for Tunisian exporters to sell textiles and carpets abroad through internet services and telephone contacts. In this regard, our approach captures the cost impacts that would arise from modeling service tariffs.

5. Results

The liberalization scenarios considered can be broken down in several ways. First, liberalization of investment barriers in services is considered. Sensitivity analysis demonstrates the importance of the decomposition of the price wedge into resource-using versus rent-generating barriers. The second step is to distinguish between liberalization of border barriers and investment barriers, broadly. Next, the impacts of liberalization are broken down on a sectoral basis. The final set of scenarios compares the impact of services liberalization to that of traditional liberalization of trade in goods.

Note that we do not consider any reciprocal liberalization in exporting regions. It is possible that Tunisia will experience efficiency gains from the EU Association Agreement, its membership in the WTO, and any FDI preferences that may exist in the Arab League. In particular, we could expect large benefits from a decision by the European Union to permit greater temporary movement of service providers (mode 4), given that in 1995 approximately 480,000 Tunisian migrant workers lived there and contributed remittances constituting 34 percent of Tunisian GDP (Jbili and Enders, 1996). However, our focus is to isolate the potential gains from unilateral services liberalization associated with competition and efficiency.

Thus, Table 4 presents potential impacts of lifting barriers to foreign investment in services (mode 3 delivery). As discussed in the previous section, investment barriers are assumed to drive an observable wedge between price and marginal cost. Unobserved is the decomposition of the wedge into two sources of distortion: rent-generating (cartel effect) and resource-using (inefficiency effect). Four benchmark possibilities are considered in Table 4. First (column 1), assume that barriers to foreign investment solely preserve a domestic cartel and generate pure economic rents for Tunisian interests.

Table 4. Liberalization of foreign direct investment in services under alternative assumptions on benchmark investment barriers.

	Rent Scenario (1)	Baseline Scenario (2)	Mixed Scenario (3)	Inefficiency Scenario (4)
Macroeconomic Variables (% change)				
Welfare, household income (EV)	0.33	4.00	4.31	7.68
Consumer price index (CPI)	−7.09	−7.11	−7.11	−7.13
Real returns to labor	4.39	3.20	3.24	2.04
Real returns to capital	6.55	6.99	6.97	7.43
Labor turnover	3.44	3.42	3.40	3.79
Capital turnover	4.87	4.90	4.92	5.06
Production (share of GDP)				
Agriculture (bench = 19.5%)	21.1	21.2	21.2	21.3
Manufacturing (32.8%)	28.5	28.7	28.7	28.9
Mining and utilities (7.0%)	6.8	6.8	6.8	6.7
Services (40.7%)	43.6	43.3	43.3	43.1

That is, Tunisian service producers use world-class technologies and face costs equivalent to those of the low-cost world producer. Upon liberalization, foreign entry is assumed to eliminate the price-wedge as markets become competitive. Domestic rents are completely dissipated to the benefit of Tunisian services consumers. The result is a modest gain in welfare of the representative household of one-third of one percent, measured as equivalent variation in the representative agent's real income.

At the other extreme (column 4), assume services markets are perfectly competitive in the benchmark, but that Tunisian firms do not employ world-class production techniques and are thus inefficient. Entry by foreign firms introduces cost-saving innovations and services prices fall. Welfare increases dramatically, by nearly eight percent. This large difference from the initial benchmark case reflects the fact that elimination of a pure rent wedge generates a small net gain in efficiency from resource reallocation, but most of the gross gains are transfers from the prior rent stream earned by the representative agent.[16] In both cases consumer prices fall about seven percent. Note that while both scenarios increase returns to both labor and capital (non-labor value added), labor gains relatively more in the pure rent-wedge case. This suggests that one effect of the protected cartel is to restrain wages in relation to what they would be under full efficiency. Capital gains are relatively larger when the pure cost wedge is removed.

Two additional scenarios involve combinations of rent-generating distortions and benchmark inefficiencies within services sectors. In the "baseline scenario," it is assumed that the price-wedge is an equal combination of the rent-generating and resource-using distortions (column 2 of Table 4). Welfare impacts mix efficiency gains with rent losses for the representative agent. The result is a four-percent increase in welfare and a seven-percent fall in the consumer price index. Gains from liberalization are skewed toward capital, the price of which increases seven percent. Five percent of the capital

[16] As discussed in Anderson and van Wincoop (2001), liberalization of trade costs that generate rents for home agents have smaller welfare impacts than liberalization of "real" trade costs that absorb resources.

stock changes sector of employment. Real returns to labor increase by three percent and three percent of the labor force changes sector of employment.

The 'mixed' scenario considers an environment where the composition of the price–cost wedge varies across sectors (column 3 of Table 4). The margin attributed to rents is assumed to be 33 percent in relatively competitive sectors, including construction, distribution, hotels, restaurants, real estate, and repair services. The remaining 67 percent is consumed in cost inefficiencies. Rents account for 67 percent of the wedge in communications. The margin in remaining sectors, transportation, finance, insurance and business, is assumed to be evenly distributed between rents and inefficiency costs. Foreign ownership liberalization improves welfare by 4.3 percent increase in equivalent variation and lowers the CPI by 7.11 percent. Real returns to labor and capital improve by 3.24 percent and 6.99 percent, respectively. Interestingly, the macroeconomic outcome closely resembles that of the 'baseline scenario,' which provides support for taking baseline conditions as a reasonable approximation for further analysis.

The bottom part of Table 4 indicates impacts on major sectoral production shares. The figures in parentheses in the left-most part indicate benchmark output shares before any liberalization. Liberalization itself tends to favor relative output expansion in agriculture and services, while restraining the manufacturing and mining shares. Thus, the initial structure of protection for services actually restrains domestic output in that sector. The remaining columns compare these shares across FDI-liberalization cases. While these shares are essentially stable across benchmark assumptions, moving from the rent-wedge case to the efficiency-wedge case slightly favors agriculture and manufacturing, while reducing the share of services.

Table 5 provides liberalization scenarios by the two modes of service delivery considered, assuming an equal split of price markups between rents and costs (baseline scenario) in the benchmark. Assume that Tunisia eliminates all border barriers (mode 1 barriers) on tradable services, and does so on a non-discriminatory or most-favored-nation (MFN) basis. This would raise welfare, as measured by equivalent variation in the representative agent's real income,

Table 5. Baseline services liberalization scenarios.

	Border liberalization (Mode 1)	Investment liberalization (Mode 3)	Full service liberalization (Modes 1 & 3)
Macroeconomic Variables (% change)			
Welfare, household income (EV)	1.22	4.00	5.30
Consumer price index	−1.02	−7.11	−8.04
Real returns to labor	0.57	3.19	4.23
Real returns to capital	1.32	6.99	8.23
Labor turnover	0.78	3.42	3.73
Capital turnover	0.98	4.90	5.35
Production (share of GDP)			
Agriculture (bench = 19.5%)	19.9	21.2	21.4
Manufacturing (32.8%)	31.7	28.7	28.2
Mining and utilities (7.0%)	7.1	6.8	6.8
Services (40.7%)	41.3	43.3	43.6

by about 1.2 percent. While both factors would gain in terms of real returns, the benefits would be larger for capital (a 1.32 percent increase in price), with the real wage increasing by less than one percent. In terms of factor adjustment, 0.8 percent of the labor force and one percent of capital would turn over industry of employment.

The investment liberalization (mode 3) scenario shown in column 2 of Table 5 is the same as that discussed above (column 2 of Table 4). Finally, the aggregate impact of liberalizing both investment barriers and border barriers (modes 1 and 3) is slightly more than additive as shown in column 3, raising welfare by 5.3 percent. Note that roughly 75.5 percent of estimated welfare gains may be attributed to investment liberalization, while 23.0 percent are due to border liberalization. This leaves a small positive residual gain of 1.5 percent reflecting the fact that both types of barriers together interact to generate a larger income loss than they would separately. It is interesting that this interaction process affects labor and capital differently. The individual liberalization impacts on the real wage sum to less than the joint impact of full liberalization, suggesting that the barriers interact to restrain wages. The opposite is true for the real price of capital. Finally, note in the bottom panel that any policy reform expands the share of services compared to the initial benchmark. However, mode-1 liberalization favors manufacturing and mining over full liberalization in mode 1 and mode 3, which favors agriculture and services. Indeed, full services liberalization would increase the services share of GDP from 40.7 percent to 43.6 percent, a large shift in relative output.

It is possible also to consider the gains from liberalizing individual service sectors, as reported in Table 6.[17] For this purpose we compute the impacts of liberalizing both mode 1 and mode 3 delivery options in six of the 11 tradable service sectors. About 41 percent of the welfare gains of full liberalization may be attributed to reforms in financial and transportation sectors. The impact of liberalization in business services, distribution, and communications are also substantial. Note that these individual trade reforms are not neutral with respect to factor prices. Opening up the construction sector would actually reduce the real wage moderately. Capital income would gain significantly from liberalization of transportation, business services, and finance.

The final set of scenarios compares liberalization of goods and services. Goods-trade liberalization is modeled as a non-discriminatory and unilateral elimination of the commodity tariffs given in Table 2. Results are shown in Table 7. Household welfare is estimated to increase by 1.5 percent, in contrast to the 5.3 percent gain from full services liberalization. As Stolper–Samuelson theory would predict, the gains are largely experienced by the abundant factor, labor, while there is relatively little change in the real returns to other value added. The real wage increases nearly 19 percent while returns to capital increase 3.5 percent. Goods trade involves a shift of production to sectors in which Tunisia demonstrates a comparative advantage. Thus, factor turnover is relatively high at 10 percent for both workers and capital. There is a significant increase in

[17]Note that sector-level reforms are assumed to occur in isolation while full liberalization entails the simultaneous liberalization of all modeled service sectors. The aggregate impact does not equal the sum of the sectoral impacts due to interactions between sectors with joint liberalization.

Table 6. Baseline service liberalization (mode 1 and mode 3), by sector.

	Communications	Construction	Transportation	Business, insurance, & leasing	Distribution	Finance	All service sectors
Welfare (EV)	0.46	0.32	1.09	0.60	0.44	1.08	5.30
Real returns to labor	0.90	−0.13	0.62	0.66	1.23	0.60	4.23
Real returns to capital	0.46	0.10	1.44	1.69	0.54	2.29	8.23
Labor turnover	0.35	0.15	1.03	1.34	0.68	1.84	3.73
Capital turnover	0.40	0.11	1.23	1.89	0.89	2.62	5.35

Table 7. Comparing goods and services liberalization (baseline case).

	Goods liberalization (Eliminate tariffs)	Services liberalization (Modes 1 & 3)	Goods and services liberalization
Macroeconomic Variables (% change)			
Welfare, household income (EV)	1.52	5.30	6.67
Consumer price index	−1.07	−8.04	−9.07
Real returns to labor	18.93	4.23	17.67
Real returns to capital	3.46	8.23	13.61
Labor turnover	9.83	3.73	3.40
Capital turnover	10.38	5.35	5.89
Production (share of GDP)			
Agriculture (bench = 19.5%)	15.3	21.4	19.1
Manufacturing (32.8%)	42.8	28.2	33.6
Mining and utilities (7.0%)	5.2	6.8	5.6
Services (40.7%)	36.7	43.6	41.7

the manufacturing share of total output from 32.8 percent in the benchmark to 42.8 percent in the free goods-trade environment. Agriculture declines substantially as a relative component of output, as do services.

Liberalizing services barriers and goods tariffs simultaneously yield a potential gain in welfare of 6.7 percent. These gains are less than the sum available under the isolated liberalization packages, indicating that there is a slightly offsetting interaction between the two commitments. Gains from joint liberalization, however, are more evenly distributed across labor and capital than they are in either individual reform. Thus, removal of goods tariffs strongly favors labor while removal of services barriers favors capital. Both factors experience significant real price increases with joint liberalization.

That the blended goods and services liberalization package does not generate gains for the abundant factor (labor) at the expense of the scarce factor (capital), as Stolper–Samuelson trade theory would predict, is remarkable. It is also notable that the structure of output is closer to that with an isolated service liberalization than to one with solely goods tariff elimination, as noted in the bottom panel. The productivity gains associated with lower service costs raise real returns for both factors and make service-intensive sectors relatively more globally competitive. Thus, in an important sense, removing commodity tariffs without services liberalization tends to "over-adjust" the economy toward manufacturing and away from agriculture and services. Freeing up services moves the structure back toward the fully efficient outcome in the final column. Indeed, a pure goods-trade liberalization results in a much greater movement of labor and more dramatic changes in the composition of production.

6. Conclusions

This paper considers how services liberalization differs from that of goods liberalization using an applied general equilibrium model of the Tunisian economy. Not surprisingly, goods-trade liberalization reorients production towards manufacturing (especially in textile and electronics) in line with Tunisia's benchmark comparative advantage. In contrast, the overall composition of production with service liberalization remains fairly stable, with only slight increases in manufacturing and service provision and a small decrease in agriculture and mining. The gains under service liberalization are more evenly distributed across factors than are those under goods liberalization, where gains are strongly concentrated in the hands of workers (the abundant factor). Fewer workers are required to change positions to accommodate service liberalization, implying less impact on frictional unemployment. Thus, we might expect less political resistance to services liberalization relative to goods liberalization.

Interestingly, combining goods and services liberalization appears to offer the best of both worlds. The gains from jointly freeing up goods and services are nearly additive in the two independent reforms. Yet the overall structure of the economy remains similar to that of the benchmark, with relatively small changes in the employment location of workers. Services liberalization therefore eases the adjustment costs involved in the liberalization of goods trade as the expansion of economic activity would no longer be restrained by the lagging competitiveness of Tunisia's service sector.

The potential welfare implications of services liberalization are clearly positive and substantial. Even given modest assumptions about the current environment for trade in services, welfare (measured as equivalent variation) and GDP are both estimated to increase more than seven percent. These gains are more than three times the magnitude of the estimated gains from goods-trade liberalization alone. Further, perhaps 75 percent of services liberalization gains may be achieved from the liberalization of foreign investment barriers that impede mode-3 delivery of services. Granting the right of establishment to foreign firms stands to increase real household income by four percent. In contrast, liberalization of cross-border trade in services (mode 1) results in a roughly one-percent gain in household income.

The finding that Tunisia would gain from unilateral services liberalization, with balanced gains to factors, but has failed to do so clearly suggests that political-economy constraints are important. It is likely that significant rents accrue to incumbent service providers, which are able to block market deregulation. In this context, the government may be faced with the prospect of compensating these incumbents in some fashion to effect true change. One means might be to pursue reciprocal services liberalization in major export markets while supporting efficiency-enhancing investments in Tunisian services. The GATS approach has yet to effect significant opening of important services in the EU or the Arab League countries that would benefit Tunisian exporters. One obvious approach would be to encourage the EU to provide greater opportunities for temporary workers from Tunisia through mode 4 negotiations. While this change would benefit Tunisian laborers it would not directly compensate existing domestic service providers. Thus, some of the income gains to workers might need to be transferred to service incumbents to buy out their resistance to reforms.

We stress that the uncertainties surrounding estimated price wedges and the allocation of those wedges between rent and production inefficiency deserve further exploration. Improved estimates of services barriers are critical for a more accurate determination of the magnitude of the potential impacts of liberalization. Still, recall that conservative assumptions were made here about the existing distortions and the potential competitive effects and efficiency improvements with liberalization. It is likely that a liberalizing reform would bring even greater gains than those reported here.

References

Anderson, James E. and Eric van Wincoop, 2001, "Borders, Trade, and Welfare," in Susan Collins and Dani Rodrik, editors, *Brookings Trade Forum 2001*, 207–243.

Bahlous, Mejda and Mustapha K. Nabli, 2000, "Financial Liberalization and Financing Constraints on the Corporate Sector in Tunisia," Economic Research Forum for the Arab Countries, Working Paper No. 2005.

Brown, Drusilla K., Alan V. Deardorff, and Robert M. Stern, 1997, "Some Economic Effects of the Free Trade Agreement between Tunisia and the European Union," in Ahmed Galal and Bernard Hoekman, editors, *Regional Partners in Global Markets: Limits and Possibilities of the Euro-Med Agreements*, London: Centre for Economic Policy Research.

Brown, Drusilla K. and Robert M. Stern, 2001, "Measurement and Modeling of the Economic Effects of Trade and Investment Barriers in Services," *Review of International Economics*, 9: 262–286.

Chatti, Rim, 2000, "General Equilibrium Assessment of Trade Liberalization Effects under Cournot Oligopoly Market Structures: The Case of Tunisia," Economic Research Forum for the Arab Countries, Iran, and Turkey, Working Paper No. 2009.

Dee, Philippa and Kevin Hanslow, 2001, "Multilateral Liberalization of Services Trade," in Robert M. Stern, editor, *Services in the International Economy*, Ann Arbor: University of Michigan Press.

Dee, Philippa, Alexis Hardin, and Leanne Holmes, 2000, "Issues in the Application of CGE Models to Services Trade Liberalization," in Christopher Findlay and Tony Warren, editors, *Impediments to Trade in Services: Measurement and Policy Implications*, London and New York: Routledge.

De Melo, Jaime and Sherman Robinson, 1989, "Product Differentiation and the Treatment of Foreign Trade in Computable General Equilibrium Models of Small Economies," *Journal of International Economics*, 27: 47–67.

Francois, Joseph, 1999, "A Gravity Approach to Measuring Services Protection," Erasmus University, manuscript.

Francois, Joseph and Ian Wooton, 2001, "Market Structure, Trade Liberalization, and the GATS," *European Journal of Political Economics*, 17: 389–402.

Goaied, Mohamed, 1999, "Cost-Frontier Analysis of Tunisian Commercial Banking Sectors," manuscript.

Harris, Richard G., 1984, "Applied General Equilibrium Analysis of Small Open Economies with Scale Economies and Imperfect Competition," *American Economic Review*, 74: 1016–1032.

Hertel, Thomas W., 1997, *Global Trade Analysis: Modeling and Applications*, New York: Cambridge University Press.

Hertel, Thomas W., 2000, "Potential Gains from Reducing Trade Barriers in Manufacturing, Services and Agriculture," in *Multilateral Trade Negotiations: Issues for the Millenium Round*, St. Louis: Federal Reserve Bank of St. Louis.

Hoekman, Bernard, 1995, "Tentative First Steps: An Assessment of the Uruguay Round Agreement on Services," World Bank Policy Research Working Paper 1455, Washington DC.

Hoekman, Bernard, 2000, "The Next Round of Services Negotiations: Identifying Priorities and Options," in *Multilateral Trade Negotiations: Issues for the Millenium Round*, St. Louis: Federal Reserve Bank of St. Louis.

Hoekman, Bernard and Simeon Djankov, 1996, "Catching Up with Eastern Europe? The European Union's Mediterranean Free Trade Initiative," World Bank, Policy Research Working Paper 1562, Washington DC.

Hoekman, Bernard and Denise Eby Konan, 2000, "Rents, Red Tape, and Regionalism: Economic Effects of Deeper Integration," in Bernard Hoekman and Jamel Zarrouk, editors, *Catching Up with the Competition: Trade Policy Challenges and Options for the Middle East and North Africa*, Ann Arbor: University of Michigan Press.

Institute National De La Statistique (INS), 1998, *Les Comptes de la Nation, agregats et tableaux d'ensemble 1993–1997*, Tunis: INS Press.

Jbili, Abdelali and Klaus Enders, 1996, "The Association Agreement between Tunisia and the European Union," available at www.worldbank.org/fandd/english/0996/articles/040996.htm.

Kalirajan, Kaleeswaran, Greg McGuire, Duc Nguyen-Hong, and Michael Schuele, 2000, "The Price Impact of Restrictions on Banking Services," in Christopher Findlay and Tony Warren, editors, *Impediments to Trade in Services: Measurement and Policy Implications*, London and New York: Routledge.

Karsenty, Guy, 2000, "Assessing Trade in Services by Mode of Supply," in Pierre Sauve and Robert M. Stern, editors, *GATS 2000: New Directions in Services Trade Liberalization*, Washington DC: Brookings Institution Press.

Konan, Denise Eby and Keith E. Maskus, 2000, "Joint Trade Liberalization and Tax Reform in a Small Open Economy: The Case of Egypt," *Journal of Development Economics*, 61: 365–392.

Lipsey, Robert, 2001, "The New Economy: Theory and Measurement," paper for 27th Annual PAFTAD Conference: The New Economy: Challenges for East Asia and the Pacific, Canberra, Australia.

Markusen, James R., 1989, "Trade in Producer Services and in Other Specialized Intermediate Inputs," *American Economic Review*, 79: 85–95.

Markusen, James R., Thomas F. Rutherford, and David Tarr, 2005, "Trade and Direct Investment in Producer Services and the Domestic Market for Expertise," *Canadian Journal of Economics*, 38: 758–777.

Maskus, Keith E. and Mohamed Lahouel, 2000, "Competition Policy and Intellectual Property Rights in Developing Countries," *The World Economy*, 23: 595–611.

Robinson, Sherman, Zhi Wang, and Will Martin, 2002, "Capturing the Implications of Services Trade Liberalization," *Economic Systems Research*, 14: 3–33.

Rodrik, Dani, 2001, "Trading in Illusions," *Foreign Policy*, March/April: 54–62.

Romer, Paul M., 2004, "New Goods, Old Theory, and the Welfare Costs of Trade Restrictions," *Journal of Development Economics*, 43: 5–38.

Rutherford, Thomas, E.E. Rutstrom, and David Tarr, 1995, "The Free Trade Agreement Between Tunisia and the European Union," World Bank, manuscript.

Rutherford, Thomas and David Tarr, 2002, "Trade Liberalization, Product Variety and Growth in a Small Open Economy: A Quantitative Assessment," *Journal of International Economics*, 56: 247–272.

Stern, Robert M., 2002, "Quantifying Barriers to Trade in Services," in Bernard Hoekman, Aaditya Mattoo, and Philip English, editors, *Development, Trade and the WTO: A Handbook*, Washington DC: World Bank.

Swiss Re, 1999, *World Insurance in 1997: Booming Life Business, but Stagnant Non-Life Business*, Sigma No. 3/99.

Varoudakis, Aristomene and Carlo Maria Rossotto, 2001, "Regulatory Reform and Performance in Telecommunications: Unrealized Potential in the MENA Countries," World Bank, manuscript.

Vittas, Dimitri, 1995, "The Insurance Sector in Tunisia," Financial Sector Development Department, World Bank Working Paper, March 1995.

Warren, Tony and Christopher Findlay, 2000, "Measuring Impediments to Trade in Services," in Pierre Sauve and Robert M. Stern, editors, *GATS 2000: New Directions in Services Trade Liberalization*, Washington DC: Brookings Institution Press.

Whalley, John, 2003, "Liberalization of China's Key Service Sectors Following WTO Accession: Some Scenarios and Issues of Measurement," Industry Canada, manuscript.

World Bank, 2000, *Tunisia: Social and Structural Review 2000: Integrating into the World Economy and Sustaining Economic and Social Progress*, World Bank, Washington, DC.

World Bank, 2001, *Republic of Tunisia: Information and Communications Technology Strategy Report*, Washington DC.

World Bank, 2004, *Republic of Tunisia: Development Policy Review: Making Deeper Trade Integration Work for Growth and Jobs*, Washington DC.

Zarrouk, Jamel, 2000, "Regulatory Regimes and Trade Costs," in Bernard Hoekman and Jamel Zarrouk, editors, *Catching Up with the Competition: Trade Opportunities and Challenges for Arab Countries*, Ann Arbor: University of Michigan Press.

Appendix: Model Equations and Notation

A. Production

1. Value Added Function $$V_i = [a_{Li} L_i^{(\sigma i-1)/\sigma i} + a_{Ki} K_i^{(\sigma i-1)/\sigma i}]^{\sigma i/(\sigma i-1)}$$
2. Imported Intermediates $$M_{iN} = [\Sigma_r \delta_{ri} m_{riN}^{(\eta i-1)/\eta i}]^{\eta i/(\eta i-1)}$$

3. Composite Intermediate	$z_{ji} = [\gamma_{di} d_{ji}^{(\eta j-1)/\eta j} + \gamma_{mi} m_{ji}^{(\eta j-1)/\eta j}]^{\eta j/(\eta j-1)}$
4. Final Goods Technology	$Y_i = \min[z_{1i}/a_{1i}, \ldots, z_{ni}/a_{ni}, V_i/a_{VA}]$
5. Domestic & Foreign Sales	$Y_i = [\alpha_{Di} D_i^{(\varepsilon i-1)/\varepsilon i} + \alpha_{Xi} X_i^{(\varepsilon i-1)/\varepsilon i}]^{\varepsilon i/(\varepsilon i-1)}$
6. Export Allocation	$X_i = [\Sigma_r \beta_{ri} X_{ri}^{(ei-1)/ei}]^{ei/(ei-1)}$
7. Marginal Cost Condition	$(1+\lambda_i)c_i Y_i = \Sigma_j(1+v_j)p_j d_{ji} + \Sigma_j\Sigma_r(1+ u_j + t_{rj})p_{rj}^m m_{rji} + (w_K K_i + w_L L_{1i})$

B. Utility

8. Utility Function	$U = \Pi_i C_i^{bi};\ \Sigma_i b_i = 1$
9. Domestic & Import Consumption (applies also to G_i and I_i^F)	$C_i = [\phi_{Di} D_{iC}^{(\psi i-1)/\psi i} + \phi_{MiC} M_{iC}^{(\psi i-1)/\psi i}]^{\psi i/\psi i-i}$
10. Import Allocation (applies also to M_{iG} and M_{iI}^F)	$M_{iC} = [\Sigma_r \delta_{ri} M_{riC}^{(\eta i-1)/\eta i}]^{\eta i/\eta i-1}$

C. Constraints and Balancing Items

11. Agent's Budget Constraint	$\Sigma_i \tilde{p}_i^C C_i = w_K \bar{E}_K + w_L \Sigma_i L_i + -\Sigma_i \tilde{p}_i^{IF} I_i^F - \Sigma_i p_i I_i^I - r^F K^F - D + \Sigma_i v_i Y_i$
12. Government Budget Constraint	$\Sigma_i \tilde{p}_i^G G_i = D + \Sigma_i \tau_{Vi} \tilde{p}_i^C V_i + \Sigma_i \Sigma_r t_{ri} p_{ri}^m (M_{riC} + M_{riI}^F)$
13. Current Account Balance	$0 = \Sigma_r \Sigma_i (1/e)(p_{ri}^m M_{ri} - p_{ri}^x X_{ri} - w_L^F L^F + r^F K^F)$
14. Product Market Clearance	$S_i = \Sigma_j a_{ij} Y_j + G_i + I_i^F + I_i^I + C_i$
15. Factor Market Clearance	$\Sigma_i K_i = \bar{E}_K; \Sigma_i L_i = \bar{E}_{1L}$
16. Zero Profits	$p_i D_i + \Sigma_r p_{ri}^x X_{ri} = c_i Y_i$
17. Supply Value Balance	$\tilde{p}_i S_i = \tilde{p}_i^Z \Sigma_j a_{ij}(1+v_i)Y_j + \tilde{p}_i^C D_{iC} + \tilde{p}_i^{IF} D_{iI}^F + \tilde{p}_i^G D_{iG} + \tilde{p}_i^{IF} I_i^I + \Sigma_r(1+u_i + t_{ri})p_{ri}^m (M_{riC} + M_{riG} + M_{riI}^F)$

D. Price Relationships and Identities

18. Components of Domestic Sales	$D_i = D_{iC} + D_{iI}^F + I_i^I + D_{iG}$
19. Components of Import Sales	$M_i = M_{iN} + M_{iC} + M_{iI}^F + M_{iG}$
20. Domestic Price of Intermediate Imports (holds also for imports for G)	$p_{ri}^N = (1 + u_i + t_{ri})p_{ri}^m$
21. Domestic Price of Imports for C (holds also for imports for I^F)	$p_{ri}^C = (1 + u_i + t_{ri})p_{ri}^m$
22. Consumer Price of Domestic Goods (holds also for purchases for I^F)	$p_i^C = (1 + v_i)p_i$
23. Capital-Market Equilibrium	$\tau_{K1} + w_{K1} = \cdots = \tau_{Kn} + w_{Kn}$ (mobile capital sectors)

List of Variables

L_i	Domestic labor inputs, sector $i (i = 1, \ldots, 34)$
K_i	Capital (other value added) inputs, both mobile and immobile
V_i	Value added
M_i	Total imports
M_{ri}	Imports from region $r (r =$ EU, MENA, ROW)
M_{iN}	Imports of commodity i for intermediate use
m_{riN}	Imports for intermediate use from region $r (r =$ EU, MENA, ROW)
z_{ji}	Composite intermediate input of j into $i (j = 1, \ldots, 34)$
d_{ji}, m_{ji}	Intermediate usages of domestic and imported goods
Y_i	Output of good i

D_i, X_i	Output for domestic sales and exports
D_{iC}, D_{iG}, D_{iI}^F	Domestic sales: private and public consumption, capital formation
X_{ri}	Exports of good i to region r
c_i	Index of marginal cost of production
p_i	Domestic producer price index
$\tilde{p}_i^Z, \tilde{p}_i^C, \tilde{p}_i^{IF}, \tilde{p}_i^G$	Domestic price indexes (home and imported prices)
w_K, w_L	Factor price indexes
U	Utility
$\tilde{p}_i$	Composite price index for total domestic supply
C_i, G_i	Private and public consumption
I_i^F, I_i^I	Fixed capital formation and inventory investment
M_{iC}, M_{iG}	Imports for private and public consumption
M_{iI}^F	Imports for fixed capital formation
M_{riC}, M_{riG}	Imports for private and public consumption from region r
M_{riI}^F	Imports for fixed capital formation from region r
K^F	Net payments on foreign capital holdings
e	Real exchange rate (price index for foreign exchange)
B	Current-account balance
D	Government budget deficit (held fixed)
S_i	Supply on domestic market $(D_i + M_i)$
p_{ri}^N	Domestic price index for intermediate imports
p_{ri}^C, p_{ri}^G	Domestic price indexes for imports of private and public consumption
p_{riI}^F	Domestic price index for imports for gross capital formation
p_i^C, p_{iI}^F	Price index for private consumption/fixed capital of domestic goods
p_{ri}	Producer price index for goods exported to region r

List of Parameters

σ_i	Substitution elasticity between capital and labor
η_a	Substitution elasticity between intermediates and value added

η_i	Armington elasticity on imports between regions
η_j	Substitution elasticity between domestic and imported intermediates
ε_i	Transformation elasticity between domestic and exported output
e_i	Transformation elasticity on exports between regions
ψ_i	Substitution elasticity between domestic and imported consumption
t_{ri}	Tariff rate on imports from region $r(t_{ri} = 0$ for service sectors)
u_i	Resource-using services border barriers ($u_i = 0$ for non-service sectors)
v_i	Service rents on output ($v_i = 0$ for non-service sectors)
λ_i	Service resource-using barriers on output ($\lambda_i = 0$ for non-service sectors)
$\bar{E}_K, \bar{E}_{1L}$	Endowment of capital and labor
p^m_{ri}	Price of imports from region r
p^x_{ri}	Price of exports in region r
r^F	Price of foreign capital payments

PART IV

POLICY ANALYSIS OF GLOBAL PUBLIC GOODS AND TECHNOLOGY TRANSFER

Chapter 15

Regulatory Standards in the WTO: Comparing Intellectual Property Rights with Competition Policy, Environmental Protection and Core Labor Standards*

Keith E. Maskus

1. Introduction

With the implementation of the Agreement on Trade-Related Aspects of Intellectual Property Rights (TRIPS), intellectual property rights (IPRs) became, on the part of WTO member states, obligations of commercial policy that cannot be escaped. Intellectual property rights are thus enforceable rules governing establishment and treatment of the rights and terms of competition. Adoption and enforcement of at least the minimum standards required will procure considerably stronger global protection of intellectual assets (Maskus, 2000).

The protection of IPRs raises the question of whether other regulatory regimes may be introduced sensibly into the WTO framework. Critics of TRIPS wonder why, if IPRs are included in the WTO to protect intellectual capital, labor standards are not also needed to protect workers, environmental regulations to protect natural resources, and competition policy to protect consumers. This issue was addressed in the Doha Declaration. It calls for negotiations on competition policy after the Ministerial meeting in Mexico in 2003, with the Working Group on the Interaction between Trade

*Reprinted from *World Trade Review*, Vol. 1, No. 2, pp. 135–152. Copyright 2002, with permission from Cambridge University Press.

and Competition Policy to focus on defining core principles and modalities before that time. It also sets the stage for negotiations on issues of environmental regulation, such as the relationship between WTO rules and trade obligations in multilateral environmental agreements (MEAs), reductions in barriers to trade in environmental goods and services, and labeling requirements. No mention is made of labor standards.

The Doha Declaration thus provides a partial answer to the basic question. However, the decisions taken at Doha in this regard themselves need analytical scrutiny. In this paper I address the question by comparing IPRs, competition policy, environmental standards, and labor rights in terms of the logic of including each within a system of international trade rules. Specifically, economists recognize four general principles when considering the need for a negotiated set of multilateral disciplines to restrain government action or mandate certain standards. First, are the existing regulations, such as variable IPRs or environmental standards, "trade-related" in the sense of significantly distorting trade flows? Second, do those regulations impose externalities on other countries that limit the attainment of global optimality and can multilateral rules internalize those costs? Third, if left to unilateral action, do countries choose policies that result in globally inadequate regulation, requiring policy coordination? Fourth, is it possible to compute and assign damages from failing to comply with international rules, thereby making dispute settlement a feasible task?

Note that only the first of these principles directly implicates the WTO as the mechanism for achieving more efficient global regulation. The second and third issues could, in principle, be managed by coordination outside the trading system, perhaps through the establishment of additional agreements and organizations. However, for internalization and coordination to be meaningful, there must be some form of binding enforcement, as suggested by the fourth principle. For many reasons the global community has, to date, assigned only to the WTO this ability to enforce negotiated obligations. Thus, to the extent that internalization and coordination are important and dispute resolution is feasible, the WTO perforce becomes the locus of action under current international institutional arrangements.

Recognition of this point, however, raises a further danger, which is simply that the WTO systemically may be incapable of managing a broad set of global regulatory disciplines. The institution may be stretched beyond its ability and resources, a situation that arguably exists already given the broad scope of WTO agreements, including TRIPS. This observation argues against further extensions of its regulatory purview. At the same time, a failure to incorporate rules on issues of international social protection may reduce public support for the trading system, arguing in the opposite direction. Thus, a final question I pose is whether there are systemic reasons for including additional regulatory language into the WTO, given that TRIPS already exists.

2. Why Did IPRs Make It into the WTO?

The TRIPS Agreement makes protection of intellectual property rights a foundation block of the World Trade Organization. It is natural to ask why IPRs attained this status, while other major forms of business regulation did not. To a great extent, the answer relies on considerations from political economy (Ryan, 1998). Three powerful and easily organized industries (pharmaceuticals, recorded entertainment, and software) in the United States recognized the opportunity afforded by the Uruguay Round to protect their intellectual property in the future and made IPRs a core issue for the United States Trade Representative. This approach complemented their efforts to publicize the damages they faced from weak international IPRs and to push for aggressive unilateral trade actions by the United States under Section 301. These efforts were joined by intellectual property-producing interests in the European Union and Japan early in the Uruguay Round negotiations. As intellectual property became better recognized as a trade policy issue, more American, European, and Japanese export-oriented industries signed on to the effort, making TRIPS a required condition for success in the negotiations. In this context, IPRs were simply an issue ripe for inclusion in the WTO, at least in terms of economic interests. The other broad issues of competition, environmental protection, and labor rights were not similarly positioned, largely because of difficulties in organizing the relevant interests.

It should be recognized that the political economy interests in global protection were complemented by certain technical factors underlying the rising demand for, and supply of, global intellectual property protection (Maskus, 2000). These factors may be summarized as follows. First, in today's globalizing economy the creation of knowledge and its adaptation to product designs and production techniques are increasingly essential for market success. In this environment firms wish to exploit their technical advantages on an international scale and to limit appropriation of those advantages by competitive rivals. Both tasks are made easier with stronger and more harmonized IPRs. Second, the costs of copying and imitating products and technologies in key sectors continue to fall, making infringement easier and more prevalent. This is evident not only in computer software, video disks, and pharmaceuticals, but also in industrial designs and such new areas as biotechnology, broadcasts, and internet transmissions. Thus, even though many observers associate TRIPS with simple counterfeiting and piracy, the economic interests go well beyond those issues. Third, the dynamics of competition in information-based technologies and products tend to outstrip the ability of the traditional intellectual property regime to accommodate needs for protection, causing the system to move over time into new and stronger forms of exclusive rights.

For their part, developing countries were able to use the Uruguay Round to secure other trade advantages from the major developed economies, in terms of increased market access in textiles and apparel and commitments to future liberalization in agriculture, in compensation for agreeing to stronger standards in the intellectual property area (Maskus, 2000; Watal, 2001). However, given that many of these promises are either back-loaded (textiles and apparel) or only agreements to consider future negotiations (agriculture), many observers argue that the main effect of TRIPS will be simply to redistribute economic rents from poor countries to rich countries (McCalman, 2001; World Bank, 2001). While there is considerable legitimacy to this claim, it should be noted that many middle-income developing economies in recent years have come increasingly to the view that greater linkages to globalization processes through access to technology and information is critical for growth and that stronger IPRs can play an important role in providing that access. Business interests inside rapidly developing economies also are mounting effective campaigns for stronger protection, recognizing that their own innovative efforts are disadvantaged by weak systems (Maskus, 1998a; Sherwood, 1997).

Advocates of including intellectual property rights in the WTO argued that such rights are integral to the trading system on grounds that weak standards invite unauthorized use of creative materials, which use constitutes unfair competition at best and cross-border theft at worst (United States Chamber of Commerce, 1987). Labeling such competition unfair places IPRs into a category akin to anti-dumping rules, though the former standards go well beyond border enforcement against unfair trade. This perspective, while debatable on utilitarian grounds, does raise the question of whether weak and variable IPRs have distortionary impacts on trade flows.

As shown in recent empirical studies, IPRs are strongly trade-related, meaning they meet the first criterion for inclusion (Maskus, 2000). These studies do not focus on counterfeiting and piracy as deterrents to trade, which is the issue many commentators have in mind when they mention the trade impacts of IPRs. Rather, they consider carefully the role of differential protection of technology, in the form of variable legal patent rights, on trade, foreign direct investment, and licensing. For example, Maskus and Penubarti (1995, 1997) estimated changes in imports of manufacturing goods and high-technology manufactures that could be induced by stronger patent rights. A patent index from Rapp and Rozek (1990) was increased by various amounts for different countries to reflect roughly the commitments required by TRIPS. The anticipated impacts on trade volumes depended on the extent of patent revisions, market size, and reductions in the imitation threats from complying with TRIPS. Estimated effects on trade ranged from small impacts in the United States and Switzerland, which were not required to undertake much legal revision, to substantial increases in imports in China, Thailand, Indonesia, and Mexico, which were required to adopt stronger rights. It should be noted that most of the largest predicted impacts were in nations with strong imitation capacities, such as Argentina and Brazil. In contrast, India and Bangladesh would experience relatively weak, though positive, trade impacts. Smith (1999) found similar impacts.

Patent laws also influence FDI flows, which are a primary channel of technology transfer. Firms with easily copied products and technologies, such as pharmaceuticals and software, are quite concerned about the ability of the local IPRs system to deter imitation. Firms considering investing in a local R&D facility pay particular attention to protection of patents and trade secrets (Mansfield, 1994, 1995). Thus, the strength of IPRs and the ability to enforce contracts could have important effects on decisions by multinational firms in certain sectors on where to invest and whether to transfer advanced technologies. Maskus (1998b) provided an econometric estimation of a model of FDI and patent rights. Using an index of patent rights from Ginarte and Park (1997), the study found a negative elasticity of FDI assets with respect to patents in high-income economies, but a strongly positive elasticity among developing economies. These result suggest that the strengthening of patents within TRIPS should increase significantly U.S.-owned FDI assets in Brazil, Mexico, Chile, and Indonesia. Indeed, the increase in the Mexican FDI assets would be 2.6 percent of the 1994 stock of U.S.-owned assets in that country and that in Brazil would be 7.4 percent. Other studies bear complementary results. Lee and Mansfield (1996) statistically related the investment decisions of U.S. multinational enterprises to their perceptions of the weaknesses of IPRs in a sample of developing countries. They found that FDI is negatively affected by weak protection. Smarzynska (2001) discovered with firm-level data that foreign investors considering operations in the countries of Eastern Europe and the Former Soviet Union pay attention to patent rights. In particular, investment in technology-intensive sectors is deterred by weak protection and in all sectors weak protection discourages investment in production facilities but does not deter investment in distribution. Smith (2001) also found that international FDI flows are positively related to IP protection. Moreover, patent rights significantly affect investment and licensing decisions, which have been shown to be reduced by weak and variable IPRs (Yang and Maskus, 2001).

Taken together, such findings suggest that a well-designed system of IPRs should support the efficient functioning of international markets. Conversely, by agreeing to establish and respect standards for intellectual property protection in the WTO, governments recognize that their existing, separate regimes may be sub-optimal in some dimensions. Surrendering some discretion to international rules forcing stronger standards may promote both collective and national welfare, though, as suggested above, beneficial outcomes may not emerge for all nations without other compensation being paid, presumably in the form of market access, by major IPR exporters.

3. Other Standards under Consideration

The logical grounds for including IPRs, or any other set of regulatory standards, rest on the severity of inherent trade impediments posed by varying standards, the ability of stronger standards to support the trading system, and the role of multilateral regulation in overcoming international market and policy coordination failures. In this context, it is interesting to compare IPRs to the other types of standards for which Doha set the stage (competition policy and environmental regulation) or that remain current in global debates about trade policy (labor rights). In these areas, proposals for the WTO exist at various stages of complexity and inclusiveness. Thus, it is difficult to capture what may be intended in policy terms except in broad brush.

Competition policies refer to laws and regulations designed to maintain market contestability, in both static and dynamic terms. Indeed, there is a strong relationship between IPRs and competition regulation, though the latter area covers broader elements, such as merger control, market dominance, cartels, tied sales, and other forms of behavior that might restrict competition. As for their inclusion in the trading system, proposals range from minimalist concepts of consultation to extensive harmonization and proscriptions of anti-competitive activity (Graham and Richardson, 1997a). The language in the Doha Declaration is sufficiently broad that it could encompass a wide range of approaches. The definition I take here is fairly comprehensive and focuses on trade-related anti-monopoly concepts. Specifically, countries would be expected to prevent export cartels, to relax distribution monopolies that deter imports, and to discipline licensing practices that anti-competitively restrain marketing and product development. Some policies may require blanket proscriptions while others may be based on a rule-of-reason approach.

I define environmental standards here as national regulations aimed primarily at influencing the use of environmental resources by producers and consumers. A distinction should be made between those standards with largely domestic impacts and those having cross-border effects generating international externalities. Presumably an international trade agreement would be aimed solely at the latter regulations, or their absence (Esty, 1994). What form a WTO agreement might take is unclear though the Doha Declaration suggests that it might begin by clarifying the linkages between WTO rules and requirements under MEAs. Those requirements involve outright bans on production or trade of particular pollutants, such as chloroflourocarbons in the Montreal Protocol. Or they set out targeted reductions in emissions, such as those for carbon dioxide in the Kyoto Agreement. They also call for national promotion of internationally agreed environmental goals, such as the preservation of biological varieties in the Biodiversity Convention.

It may be that the scope of WTO involvement will be limited to insulating trade sanctions in environmental agreements from WTO recourse. However, it is not difficult to envision a broader scenario unfolding during negotiations. The MEAs have not been ratified by all (or even most) WTO members. Thus, an important issue arises as to whether the application of MEA trade sanctions could induce economic damages in third parties that are members of the WTO and invite a counter-reaction. In this context pressure could rise to globalize the MEAs through adoption of their rules into the WTO.

A WTO agreement on core labor standards would require each country to agree to recognize and enforce five general rights that are considered by many to be fundamental human rights. These include banning exploitative use of child workers, eliminating forced labor (slavery, bonded labor, and prison labor), preventing discrimination in the workplace, allowing free association of workers, and permitting workers to undertake collective bargaining (Maskus, 1997). The role of the WTO would be to permit trade sanctions against countries that fail in these endeavors under certain circumstances. One variant under discussion would be to allow countries to ban imports of offending goods (or their equivalent value) under a consensus view of abhorrent practices (Rodrik, 1997).

4. A WTO Ranking of Standards Based on Economic Grounds

I consider several criteria that could support the inclusion of standards into the WTO, and how well the issue areas fit them, in Table 1. These criteria are divided into five general items and some more specific questions. The first four general items provide a basis for comparison of IPRs with competition policy, environmental regulation, and labor standards. They include how trade-related the areas are, the importance of international externalities that trade rules might overcome, coordination failures of countries to enforce collective interests through stronger standards, and the ability of dispute settlement to deal with them effectively. The final general area looks at systemic questions, presuming that IPRs are already in the WTO and analyzing the other issues conditioned upon that fact.

In each of these areas I assign a qualitative ranking indicating the extent to which the issue may be supported in theoretical and empirical terms. The word "absent" indicates no relationship or evidence, "murky" indicates a weak or highly ambiguous relationship, "moderate" indicates a relationship of medium strength, and "clear" indicates an identifiable and strong relationship.[1] To economists the proper resolution of complex policy questions rests on both theory and empirical evidence. The qualitative descriptions reflect both the apparent importance of the problem and the ability of multilateral coordination through enforceable trade rules to deal with it adequately. Because the rankings are qualitative only they inevitably exhibit a degree of arbitrariness. For example, some entries under "evidence" are "absent", reflecting an absence of empirical research devoted to those areas rather than implying that no such evidence could be found. Readers will differ in their own assessments of how

[1]This approach follows that in Graham and Richardson (1997).

Table 1. Qualitative analysis of arguments for various types of standards to be incorporated into the WTO.

	IPRs	Competition policy	Environmental standards	Core labor standards
I. Unconditional Rankings				
A. *Trade Impacts*				
1. Directly trade-related	Theory Clear Evidence Clear	Theory Clear Evidence Murky	Theory Clear Evidence Moderate	Theory Clear Evidence Murky
2. FDI-related and technology transfer	Theory Clear Evidence Moderate	Theory Clear Evidence Moderate	Theory Clear Evidence Murky	Theory Murky Evidence Absent
B. *International Externalities*				
1. Static	Theory Clear (unpriced diffusion) Evidence Moderate	Theory Moderate (anti-competitive spillovers) Evidence Murky	Theory Clear (environmental spillovers) Evidence Moderate	Theory Moderate (utility spillovers; wage suppression) Evidence Murky
2. Dynamic	Theory Moderate (over- or under-investment in global or targeted R&D) Evidence Murky	Theory Moderate (over-investment in global entry deterrence) Evidence Absent	Theory Clear (under-investment in global abatement) Evidence Murky	Theory Murky (under-investment in global human capital) Evidence Absent

(Continued)

Table 1. (*Continued*)

	IPRs	Competition policy	Environmental standards	Core labor standards
C. *Policy Coordination Failures*				
1. Inadequate standards	Theory Moderate (too-weak rights) Evidence Murky	Theory Moderate (insufficient competition maintenance) Evidence Moderate	Theory Moderate (race to bottom) Evidence Murky	Theory Murky (race to bottom) Evidence Murky
2. Standards build market infrastructures	Theory Clear (property rights) Evidence Moderate	Theory Clear (entry conditions) Evidence Moderate	Theory Murky (raise environmental demand) Evidence Absent	Theory Clear (labor unions) Evidence Murky
D. *Meaningful Dispute Resolution*	Theory Clear Evidence Moderate	Theory Clear Evidence Moderate	Theory Murky Evidence Murky	Theory Murky Evidence Murky
Unweighted Rank	Clear/Moderate	Moderate	Moderate/Murky	Murky
Favoring Trade Weights*	Clear	Moderate	Moderate	Murky
Favoring Externality Weights*	Moderate	Moderate/Murky	Clear/Moderate	Murky/Absent

(*Continued*)

Table 1. ***(Continued)***

	IPRs	Competition policy	Environmental standards	Core labor standards
Favoring Coordination Weights*	Moderate	Moderate	Murky	Murky
Favoring Dispute Resolution Weights*	Moderate	Moderate	Murky	Murky
(*assigns weights of 1.0 to designated issues and 0.5 to others)				
II. Rankings Conditional upon IPRs in WTO				
E. *Systemic Issues*				
1. WTO linkages	na	Clear (IPRs, AD, subsidies)	Moderate (IPRs, agriculture, subsidies)	Murky (subsidies)
2. Institutional fit	na	Moderate	Absent	Absent
3. Support for WTO system	na	Moderate	Clear	Clear
Conditional Rank	na	Clear/Moderate	Moderate	Moderate
Grand Rank	na	Moderate	Moderate/Murky	Murky

the standards fit the underlying criteria. However, the criteria and rankings can serve as a basis for discussion about the desirability of extending the WTO to additional issue areas.

4.1. *Trade Impacts*

Inclusion of IPRs in the WTO was originally justified by their relationship to trade in goods. Economic theory makes the essential point that weak IPRs can operate as a non-tariff barrier to trade by reducing domestic demand for goods imported under patent or trademark protection. However, the theoretical result is ambiguous as a strengthening of IPRs could enhance market power at the same time that it shifts demand toward imports. Net impacts would depend on a variety of factors in each country and product. As discussed earlier, empirical evidence finds that IPRs are closely trade-related, with a strong positive relationship in large and middle-income countries. Accordingly, IPRs are strongly trade-related in both theory and fact, earning a ranking of "clear" on each scale.

Competition policy also has potentially strong impacts on trade in theory (Levinsohn, 1996), but has attracted little empirical study. In this case, the ranking of "murky" under evidence reflects simply the practical attention the issue has drawn in recent years to particular cases (Graham and Richardson, 1997a). Environmental standards (Levinson, 1996) and labor rights (OECD, 1996, 2000) also affect trade in theory. Evidence is stronger in the former case. Rodrik (1997) produced some evidence that trade in labor-intensive goods is marginally affected by core labor rights, as measured by membership in international labor conventions. Most other examinations could find no credible evidence that deficient core labor rights have any impact on trade flows (OECD, 1996; Maskus, 1997; Rodrik, 1996). Mah (1997) reported a negative correlation between a country's export share of GDP and freedom-of-association rights and rights to non-discrimination, but the results are questionable because he failed to include control variables in his regressions.

Standards might also have important indirect impacts on trade through their influence on FDI and technology licensing, as noted in the second row. This theory is well established for IPRs, competition policy, and environmental standards, though clear evidence of these impacts exists only for IPRs. That competition policies could affect investment decisions seems clear from historical and descriptive evidence but the issue has attracted little systematic study. One recent paper could find no impacts (Noland, 1998). Numerous attempts have failed to discover systematic evidence in the environmental realm (Levinson, 1997; Low and Yeats, 1992; Jaffe *et al.*, 1995), though a recent study found some sensitivity of FDI in the United States to state-level heterogeneity in the stringency of environmental regulations (List and Co, 2000). Regarding core labor rights, their absence does not necessarily attract FDI even in theory because denying such rights is tantamount to cost-increasing distortions in many cases (Maskus, 1997). Neither is there any evidence of such effects (OECD, 1996; Maskus, 1997; Drezner, 2001).

4.2. *International Externalities*

To economists a primary justification for an international policy regime is its ability to internalize cross-border externalities that are harming global welfare. Policy intervention in each of the four issue areas is advocated on these grounds by various observers. Externalities may be static or dynamic and I have listed (non-exhaustive) examples of each kind in Panel B of Table 1. In IPRs the main static concern is that weak and variable international protection promotes uncompensated international technology diffusion and product copying, which could slow down invention and innovation. This case is easily made in theoretical terms, though in fact it is not clear whether effective diffusion happens more readily under weak or strong IPRs. Moreover, the impact on global welfare is not easily determined. Evidence exists that diffusion transpires through a variety of international channels (Coe, Helpman and Hoffmaister, 1997), including the legitimate use of patented technologies (Eaton and Kortum, 1996). The evidence is not entirely clear, however, and I assign it a rank of "moderate".

The externality theory in competition policy is essentially that weak rules encourage anti-competitive and exclusionary behavior that permits abuse of monopoly power abroad or in defending home markets. Export cartels are one example, as is predatory dumping that might emanate from protected scale. So also are collusive agreements among firms in one set of jurisdictions to divide markets elsewhere. These concerns are reasonably well established in theoretical terms, though doubts expressed by industrial organization specialists about the efficiency losses associated with monopolization and vertical arrangements in closed economies apply as well at the international level. Evidence on the prevalence of anti-competitive practices spilling over borders is weak and indirect.

That there are static spillovers in the environmental area is clear theoretically and constitutes the basic argument for international intervention. There is moderate but convincing evidence on this point, at least through a wide range of case studies. In core labor rights, the spillover argument is that consumers in rich nations are made worse off by awareness of exploitative labor conditions abroad. It is also argued that weak core labor standards may operate to suppress wages of low-skilled workers in countries that import products made by exploited workers. The theoretical basis for the utility spillovers is sensible, if based on an untested assumption about preferences, but there is little basis for the wage story even in theory (Maskus, 1997). Evidence is weak in any case.

The dynamic externality in IPRs is that weak rights may generate inadequate global investment in R&D, a theoretical point set out by Grossman and Lai (2001). The evidence on this point is mixed and hardly conclusive. A subtler variant is that weak IPRs in developing countries result in too-little R&D aimed at meeting particular needs of consumers in those countries, such as tropical medicines. Theory points in this direction (Diwan and Rodrik, 1991), while evidence in the pharmaceutical sector suggests strongly that global research efforts suffer from this distortion (Lanjouw, 1997).

In competition policy, dynamic markets are distorted to the extent that weak rules induce excessive investment in entry deterrence, such as excess capacity and closed distribution networks. This theory is respectable but I am unaware of any empirical studies of the issue.

Similarly, the issue in environmental regulation is that weak standards generate insufficient global investment in abatement efforts, causing future generations to suffer excessive pollution. Again, the theory is well established but evidence is scarce and anecdotal. Finally, in labor rights the question becomes whether inadequate protection of children results in sub-optimal levels of education that spills over into lower global growth rates. In theory, low schooling rates are more closely associated with poverty than with weak proscriptions against child labor, while strengthening rules against child labor paradoxically could result in less schooling (Maskus, 1997). Evidence on this crucial point is again missing.

4.3. *Policy Coordination Failures*

Economists also justify international policy regimes on the basis of overcoming failures of countries to advance their long-term interests through collective action. Indeed, the essential purposes of GATT were to prevent countries from unilaterally raising tariff rates on behalf of domestic political interests and to establish a multilateral forum for reducing tariffs to the joint benefit of members. In IPRs, it is argued that countries may choose standards that are weaker than optimal in an effort to promote local production through copying more rapidly than neighbor countries (McCalman, 1999). If all nations (or a collection of developing nations) did this, each country might be caught in a low-level equilibrium set of standards that discourage growth and technical change. Indeed, Grossman and Lai (2001) present a model in which the Nash Equilibrium patent lengths in both the North (developed countries) and South (developing countries) are shorter than those that would be required under a globally efficient patent regime, although their model does not support harmonization. Higher standards in the TRIPS agreement could overcome such a problem. The theory supporting this claim is established in the IPRs area but there is little evidence that countries compete on these grounds.

In the competition area it may be argued similarly that countries choose individually and collectively sub-optimal regulations to compete for production and investment that respond to lax competition maintenance. The absence of competition policies in much of the developing world may be taken as indirect confirmation of this problem and I assign somewhat higher rankings in this area. It should be noted, however, that unless competition policy standards are appropriately chosen and well administered, their absence is not necessarily sub-optimal for any country.

In environmental standards and labor rights the basic argument is the "race to the bottom", in which countries choose to lower those standards (or not to raise them) despite the negative impacts on national and global environmental quality and worker protection. The theory is reasonably defensible in the case of environmental regulation, though evidence remains mixed and scarce (Levinson, 1996). The theory makes little sense in the labor rights case, though some point to the proliferation of export processing zones as evidence of standards deterioration.

A second coordination failure that trade rules might address is that forcing higher standards could help build market-supporting infrastructures, such as a professional judiciary to enforce property rights, administrative transparency in government agencies, and countervailing power among economic agents. Such institutions may be underdeveloped in poor countries because of the difficulty of organizing interests in their behalf. This argument is perhaps the *sine qua non* for IPRs, because property rights are at the core of sophisticated market systems. Similarly, competition policy is a fundamental support for efficient markets, for in its absence concentrated interests are liable to dominate economic systems. There is informal evidence from developing economies that IPRs support business development and innovation, while it seems clear from case studies of market deregulation and trade liberalization that entry can proceed rapidly when unblocked. Hence, IPRs and competition policy score highly on this criterion.

It has been argued that environmental standards help set the stage for cumulative improvements in protection by raising the demand for environmental inputs and expanding awareness of environmental problems. The logic in this claim seems somewhat circular and evidence of the incentive effects of environmental standards is weak. A similar argument does hold water in at least one area of labor rights, because the establishment of labor unions and collective bargaining can improve prospects for employee training, eliminate inefficient monopsony hiring practices, and expand workers' commitments to acquire firm-specific capital (Sengenberger, 1991; Freeman, 1993). However, evidence suggests that unions frequently do not operate in a manner that raises market efficiency and growth, especially in developing nations (Rama, 1995; Farber, 1986).

4.4. *Meaningful Dispute Resolution*

That such coordination problems exist does not mean their resolution could readily be made operational in the WTO. Dispute resolution is more manageable when the issue stems from commercial damages arising from weak standards rather than from questions of morality. This task is conceptually most straightforward in the case of IPRs, where copying is aimed at particular products and technologies that may be identified through court proceedings. It is surely less straightforward in competition policies, where business practices may exclude particular identifiable competitors but translating those impacts into consumer costs is daunting. Nonetheless, a history of enforcement in the United States and the European Union provides guidelines for calculating and assigning costs, so I assign reasonably high marks in competition policy.

Such exercises in the environmental and labor rights fields are fraught with conceptual and practical difficulties, making their inclusion in the WTO questionable. For example, the immediate losers from exploitative use of child labor are children themselves but it is difficult to assess these costs in relation to available alternatives. Moreover, trade sanctions easily could harm those agents they are designed to protect (Maskus, 1997; Freeman, 1994; Elliott, 1998). On the other side, it is difficult to assign monetary values to the disutility experienced by consumers of products made under exploitative conditions. Experimentation with product labeling suggests that consumers may be willing to pay some amount to avoid such products. However, willingness-to-pay studies have not been systematically performed. One is left with the nebulous policy of simply banning such imports on the basis of unmeasured psychic costs (Rodrik, 1997), with potentially damaging impacts on recipients of the sanctions.

Summing up these various effects analyzed, IPRs achieve an unweighted rank of "clear/moderate", suggesting there are justifiable motivations for its entry into the WTO. Competition policy ranks next as "moderate" and environmental regulation third as "moderate/murky". It must be admitted that these differences are not significant scientifically. Indeed, I believe that these areas should be grouped together as issues for which international policy coordination is sensible. The balance sheet is less favorable to core labor rights, essentially because the alleged spillovers are questionable and evidence suggests the international trade effects are minimal.

Because various analysts would emphasize different motivations for developing international trade rules, the next four rows weight the rankings by each type of argument, with implicitly a weight of 1.0 for particular designated arguments and 0.5 for the others.[2] This exercise makes the case relatively stronger for IPRs when emphasis is placed on trade-relatedness. It ranks environmental issues at the top if the focus is on international externalities. In all cases, core labor standards achieve a low ranking, suggesting it is of questionable value to consider their incorporation into the WTO.

4.5. *Conditional Rankings*

A final type of justification is that the systemic structure of the WTO invites or repels extension of its coverage to new issue areas, as shown in Panel E. I do not attempt a distinction between theory and evidence because these questions are amenable less to empirical confirmation and more to rhetoric. Because IPRs are already in, I do not rank them on this score. A first question is whether the WTO rules themselves give rise to clear linkages between existing coverage and the new issues. This is strongly so in competition policy, which is directly relevant for IPRs (indeed, the TRIPS agreement invites countries to use competition disciplines in the area) and also strongly connected to antidumping and subsidies. The linkages are perhaps less clear in environmental protection, though there are potential

[2]For this exercise, the basic rankings were assigned index values from zero (for "absent") to three (for "clear"). The resulting verbal descriptions reflect the numerical rankings that ensued.

synergies with IPRs and agriculture (in biogenetic technologies), subsidies, and the Agreement on Sanitary and Phytosanitary Standards. Again, the weakest relationship emerges for labor rights.

A second point is that the WTO may not be an appropriate institution to handle the new issues. Indeed, this argument was made about IPRs before their introduction via the Uruguay Round and it remains a contentious issue. Although the potential difficulties in managing competition regimes via the WTO are significant, the focus of competition policy on market access and its intimate linkage to IPRs suggests that the trading system could be the appropriate locus. The case for environmental and labor standards in the WTO is far shakier on these grounds. Both areas are primarily about domestic regulations with incidental (if perhaps important) impacts on trade and do not incorporate significant elements of market access. In this context, if there is a strong case for international coordination in the environmental area, as these rankings would suggest, it points more toward a separate institutional structure (Esty, 1994).

Finally, it may be argued that failure to proceed in these issue areas could significantly erode support for the international trading system, a claim that is often made on behalf of environmental and labor standards. Competition policy carries considerable importance for the international business community in this context, but likely arouses less concern among the general public than environmental policy and labor standards.

Averaging these rankings suggests that competition policy comes in highest on the conditional entry criterion, largely because of its linkages to existing areas and its potential institutional fit. Environmental and labor standards fare less well and raise real questions about how their inclusion would affect the trading system. Overall, the "grand rank" across all five general criteria provides an advantage to competition policy over environmental regulation. The criteria adopted here reject core labor rights as an appropriate area for the WTO.

5. Concluding Remarks

In this paper I have attempted to rank four fundamental areas of commercial regulation — IPRs, competition policy, environmental standards, and core labor standards — in terms of their suitability for inclusion in the multilateral trading system. By setting out several basic criteria for this purpose, including trade relatedness, international externalities, policy coordination problems, and systemic appropriateness of the WTO, I find that IPRs were a reasonable regulatory area for inclusion. Competition policy seems also to be appropriate for consideration by the WTO. However, environmental regulations fare less well, unless a high weight is placed on their cross-border externalities. There is little ground for inclusion of worker rights, either in theory or in terms of available evidence.

Perhaps a useful way to think of these results is along institutional lines. The WTO seems an appropriate and effective locus for developing coordinated standards in the areas of intellectual property rights and competition rules, primarily because of their focus on market access and competition. It is less appropriate for environmental standards because of the systemic problems that could emerge in using the WTO for this purpose. Nonetheless, the international externalities (both static and dynamic) in this area are sufficiently in evidence to provide a strong argument for coordinated international action on environmental protection. This suggests that a separate institutional structure, such as a World Environmental Organization, is worthy of consideration. In this regard, if the intention of the Doha Declaration is to limit the WTO's ambit to establishing exemptions from WTO disciplines of trade sanctions imposed within the confines of MEAs — an exemption that could be difficult to sustain — then the Declaration seems well grounded. Finally, the weak theory and limited evidence on the externality and coordination aspects of worker rights, along with potential systemic problems, do not support their inclusion into the WTO. They could provide an argument for strengthening the existing international structure of worker rights through the International Labor Organization.

References

Coe, David T., Elhanan Helpman, and Alexander W. Hoffmaister. 1997. North-South R&D Spillovers. *The Economic Journal* 107: 134–149.

Diwan, Ishak and Dani Rodrik. 1991. Patents, Appropriate Technology, and North-South Trade. *Journal of International Economics* 63: 79–90.

Drezner, Daniel W. 2001. Fears of a "Race to the Bottom" on Wages and Labor Standards: Why Aren't they Justified? Manuscript, University of Chicago.

Eaton, Jonathan and Samuel Kortum. 1996. Trade in Ideas: Patenting and Productivity in the OECD. *Journal of International Economics* 40: 251–278.

Elliott, Kimberly A. 1998. International Labor Standards and Trade: What Should Be Done? In *Launching New Global Trade Talks*, ed. by Jeffrey J. Schott. Washington DC: Institute for International Economics.

Esty, Daniel C. 1994. *Greening the GATT: Trade, Environment, and the Future.* Washington DC: Institute for International Economics.

Farber, Henry S. 1986. The Analysis of Union Behavior. In *Handbook of Labor Economics: Volume 2*, ed. by Orley Ashenfelter and Richard Layard. Amsterdam: Elsevier Science Publishers.

Freeman, Richard B. 1993. Labor Market Institutions and Policies: Help or Hindrance to Economic Development? In *Proceedings of the World Bank Annual Conference on Development Economics 1992.* Washington DC: The World Bank.

Freeman, Richard B. 1994. A Hard-Headed Look at Labor Standards. In *International Labour Standards and Economic Interdependence*, ed. by Werner Sengenberger and Duncan Campbell. Geneva: International Labour Organization.

Ginarte, Juan Carlos and Walter G. Park. 1997. Determinants of Patent Rights: A Cross-National Study. *Research Policy* 26: 283–301.

Graham, Edward M. and J. David Richardson. 1997a. *Global Competition Policy.* Washington DC: Institute for International Economics.

Graham, Edward M. and J. David Richardson. 1997b. *Competition Policies for the Global Economy.* Policy Analyses in International Economics 51. Washington DC: Institute for International Economics.

Grossman, Gene M. and Edwin Lai. 2001. International Protection of Intellectual Property. Manuscript, Princeton University.

Hoekman, Bernard M. and Michel M. Kostecki. 1995. *The Political Economy of the World Trading System: from GATT to WTO*, Second Edition. Oxford: Oxford University Press.

Jaffe, Adam B., *et al.* 1995. Environmental Regulations and the Competitiveness of U.S. Manufacturing: What Does the Evidence Tell Us? *Journal of Economic Literature* 33: 132–163.

Lanjouw, Jean O. 1998. *The Introduction of Pharmaceutical Product Patents in India: "Heartless Exploitation of the Poor and Suffering"?* NBER Working Paper 6366. Cambridge, MA: National Bureau of Economic Research.

Lee, J.-Y. and Edwin Mansfield. 1996. Intellectual Property Protection and U.S. Foreign Direct Investment. *Review of Economics and Statistics* 78: 181–186.

Levinsohn, James. 1996. Competition Policy and International Trade. In *Fair Trade and Harmonization: Prerequisites for Free Trade? Volume 1: Economic Analysis*, ed. by Jagdish N. Bhagwati and Robert E. Hudec. Cambridge, MA: MIT Press.

Levinson, Arik. 1996. Environmental Regulations and Manufacturers' Location Choices: Evidence from the Census of Manufactures. *Journal of Public Economics* 62: 5–29.

List, John A. and Catherine Y. Co. 2000. The Effects of Environmental Regulations on Foreign Direct Investment. *Journal of Environmental Economics and Economic Management* 40: 1–20.

Low, Patrick and Alexander Yeats. 1992. Do "Dirty" Industries Migrate? In International Trade and the Environment. World Bank Discussion Papers. Washington DC: The World Bank.

Mah, I.S. 1997. Core Labour Standards and Export Performance in Developing Countries. *The World Economy* 20: 773–785.

Mansfield, Edwin. 1994. *Intellectual Property Protection, Foreign Direct Investment, and Technology Transfer.* Washington DC: International Finance Corporation Discussion Paper 19.

Mansfield, Edwin. 1995. *Intellectual Property Protection, Direct Investment, and Technology Transfer: Germany, Japan, and the United States.* Washington DC: International Finance Corporation Discussion Paper 27.

Maskus, Keith E. 1997. *Should Core Labor Standards Be Imposed through International Trade Policy?* World Bank Development Research Group, Policy Research Working Paper 1817. Washington DC: The World Bank (August).

Maskus, Keith E. 1998a. The Role of Intellectual Property Rights in Encouraging Foreign Direct Investment and Technology Transfer. *Duke Journal of Comparative and International Law* 9: 109–161.

Maskus, Keith E. 1998b. The International Regulation of Intellectual Property. *Weltwirtschaftliches Archiv* 134: 186–208.

Maskus, Keith E. 2000. *Intellectual Property Rights in the Global Economy.* Washington DC: Institute for International Economics.

Maskus, Keith E. and Mohan Penubarti. 1995. How Trade-Related Are Intellectual Property Rights? *Journal of International Economics* 39: 227–248.

Maskus, Keith E. and Mohan Penubarti. 1997. Patents and International Trade: An Empirical Study. In *Quiet Pioneering: The International Economic Legacy of Robert M. Stern*, edited by Keith E. Maskus, *et al.* Ann Arbor: University of Michigan Press.

McCalman, Phillip. 1999. *National Patents, Innovation and International Agreements.* Canberra: Australian National University (Photocopy).

McCalman, Phillip. 2001. Reaping What You Sow: An Empirical Analysis of International Patent Harmonization. *Journal of International Economics* 55: 161–186.

Noland, Marcus. 1998. *Competition Policy and FDI: A Solution in Search of a Problem?* Washington DC: Institute for International Economics (Photocopy).

Organization for Economic Cooperation and Development. 1996. *Trade, Employment, and Labor Standards: a Study of Core Workers' Rights and International Trade.* Paris: OECD.

Organization for Economic Cooperation and Development. 2002. *International Trade and Core Labour Standards.* Paris: OECD.

Rama, Martin. 1995. Do Labor Market Policies and Institutions Matter? The Adjustment Experience in Latin America and the Caribbean. *Labour* XX: S243–S269.

Rapp, Richard T. and Richard P. Rozek. 1990. Benefits and Costs of Intellectual Property Protection in Developing Countries. *Journal of World Trade* 24: 75–102.

Rodrik, Dani. 1996. Labor Standards in International Trade: Do They Matter and What Do We Do About Them? In *Emerging Agenda for Global Trade: High Stakes for Developing Countries*, ed. by R. Z. Lawrence, D. Rodrik, and J. Whalley. Overseas Development Council Essay 20, Washington DC: Johns Hopkins University Press.

Rodrik, Dani. 1997. *Has Globalization Gone Too Far?* Washington DC: Institute for International Economics.

Ryan, Michael P. 1998. *Knowledge Diplomacy: Global Competition and the Politics of Intellectual Property.* Washington: Brookings Institution Press.

Sengenberger, Werner. 1991. The Role of Labour Market Regulation in Industrial Restructuring. In *Towards Social Adjustment*, ed. by G. Standing and V. Tokman. Geneva: International Labor Organization.

Sherwood, Robert M. 1997. The TRIPS Agreement: Implications for Developing Countries. *IDEA: The Journal of Law and Technology* 37: 491–544.

Smarzynska, Beata. 2001. Composition of Foreign Direct Investment and Protection of Intellectual Property Rights: Evidence from Transition Economies. Manuscript, Washington DC: The World Bank.

Smith, Pamela J. 1999. Are Weak Patent Rights a Barrier to U.S. Exports? *Journal of International Economics* 48: 151–177.

Smith, Pamela J. 2001. How Do Foreign Patent Rights Affect U.S. Exports, Affiliate Sales, and Licenses? *Journal of International Economics* 55: 411–439.

United States Chamber of Commerce. 1987. *Guidelines for Standards for the Protection and Enforcement of Intellectual Property Rights.* Washington DC: United States Chamber of Commerce.

Watal, Jayashree. 2001. *Intellectual Property Rights in the WTO and Developing Countries.* The Hague: Kluwer Law International.

World Bank. 2001. *Global Economic Prospects and the Developing Countries 2002: Making Trade Work for the World's Poor.* Washington DC: The World Bank.

Yang, Guifang and Keith E. Maskus. 2001. Intellectual Property Rights and Licensing: An Empirical Investigation. *Weltwirtschaftliches Archiv* 137: 58–79.

Chapter 16

Economic Perspectives on a Multilateral Agreement on Open Access to Basic Science and Technology*

John H. Barton and Keith E. Maskus†

1. Introduction

In this essay we set out a general proposal for a multilateral agreement, most appropriately reached at the World Trade Organization (WTO), on "free trade" (more accurately, "open access") in basic science and technology. The idea for this agreement was advanced earlier by one author (Barton, 2003b). We discuss the economic foundations for such an accord, which we call an Agreement for Access to Basic Science and Technology (ABST), and analyze the principal issues that would emerge in establishing it. The agreement could be structured around open access for inputs (coordination and movement of research projects and scientific personnel), open access to outputs (basic research results) or both, and be founded on basic WTO principles. Moreover, the agreement could include provisions for preferential treatment for developing countries. Its central purposes would be to ensure widespread access to essential scientific results and to enhance the transfer of basic technological information to the developing world at reasonable cost.

*Reprinted from *Economic Development and Multilateral Trade Cooperation*, pp. 349–368, Simon J. Evenett and Bernard M. Hoekman, editors. Copyright 2006, with permission from Palgrave/MacMillan and World Bank.
†Prepared for presentation at The World Trade Forum, held at World Trade Institute, Berne, June 16–17, 2003. The authors are grateful to Richard Baldwin, Bernard Hoekman and other participants of the Forum.

We propose such an agreement for three essential reasons, which we explain further in the following sections. First, we share the increasingly widespread concern that government restrictions on access to data and research results could harm the pace of global scientific advance and the diffusion of knowledge, particularly to the detriment of transition and developing countries (Maskus and Reichman, 2004; David, 2003). Put simply, there has been a sharp policy shift toward making knowledge a private commodity, despite its inherent character as a public good, raising fundamental questions for science, education, and the diffusion of information.

Second, issues of technology transfer are at the center of the contentious debate about how the WTO Agreement on Trade-Related Aspects of Intellectual Property Rights (TRIPS) might affect developing countries. There is a common perception that, despite the promise held out by TRIPS that stronger technology protection would expand flows of knowledge to poor countries, very little gains have emerged in this regard. At the same time, the exclusive rights offered by TRIPS bear the potential for limiting access of developing countries to even publicly generated basic research that might otherwise enable greater competition and local innovation.

Finally, the economics of knowledge creation and the non-rival nature of information implies that global investments in basic science and technology are surely under-funded in comparison with a global optimum. Knowledge is a prime example of a global public good that can be more effectively provided by cooperative multilateral actions.

Thus, in our view it is important to preserve and enhance the global commons in science and technology, while setting out a public mechanism for increasing the international flow of technical information without unduly restricting private rights in commercial technologies. The agreement we describe in this paper bears promise for meeting these objectives. There are precedents for an ABST. The United States is party to numerous bilateral scientific exchange accords, some as treaties and some as arrangements between counterpart institutions. However, they apply solely to designated cooperative public sector programs. Recently scholars have proposed an agreement among universities and research institutions on free exchange in scientific data and in coordinating licensing arrangements (Reichman and Uhlir, 2003).

In the following section we set out our concerns underlying the need for the proposed agreement. In the third section we discuss the economic principles justifying coordinated intervention in the diffusion of basic research. In the fourth section we describe the essentials of the proposal for an ABST and analyze both its structure and political viability. We offer concluding remarks in the fifth section.

2. The Proliferation of Proprietary Rights in Basic Research

As noted above, an essential reason for proposing an ABST is to offset the imbalances in access to knowledge arising from two fundamental policy changes. First, governments are increasingly imposing restraints on the use of knowledge generated by public research or through public funding. Such restrictions arise in part from regulations designed to protect national competitiveness, as in the following examples. Under the U.S. Bayh–Dole Act, a license to use technology developed from federally funded research must be awarded preferentially to national firms. In 2002 the European Union established a European Research Area, with several reservations for EU firms as regards use of publicly funded research results.

Other restrictive regulations emerge from the increasing scope of intellectual property protection for basic scientific results in the United States and the EU, which remain the primary generators of knowledge. This protection is expanding beyond products and applied technologies to basic ideas, procedures and materials, which had not been the subjects of proprietary rights under traditional intellectual property norms. For example, the United States grants patents on basic research tools and scientific methods of discovery in genetic sciences. In doing so the classical utility standard, under which an invention had to be reduced to a commercially useful product or process (as opposed to a scientific method), has been diminished significantly as a bar to gaining exclusive proprietary rights to basic knowledge. At the same time, the United States permits patents on research tools with broad reach-through claims, extending property rights on basic inventions to applied innovation. This policy raises fundamental concerns about "anti-commons" effects in biotechnology and materials sciences as the transactions costs of acquiring licenses rise (Heller and Eisenberg, 1998; David, 2003). Furthermore, a recent U.S. court decision greatly narrowed the research exemption in patent law that was long permitted as a means of legitimate science and competition.[1]

In its 1996 Directive on the Legal Protection of Databases, the European Commission established a *sui generis* regime for the protection of data compilations (David, 2004; Reichman and Uhlir, 2003). Exclusive rights are available for the developers of commercially useful databases, even where those data are compiled from public research results and information already in the public domain. Thus, the regime offers proprietary rights on creations of limited inventiveness and, because the protection is renewable, it effectively is stronger than limited-duration patent grants. These restrictions on data use in the EU offer little scope for fair use procedures that hitherto permitted reasonable access by researchers and educational institutions. Certainly those rights are considerably stronger than the global standard (as established by the WTO Agreement on Trade-Related Aspects of Intellectual Property Rights, or TRIPS) of protecting databases solely with copyrights. For our purposes the primary concern is that such protection may reduce access to publicly generated, basic research results. For example, meteorological data generated by government research stations are now sold frequently under private rights. The European Commission's Directive may soon be emulated in U.S. law, as legislation to protect databases on a similar standard has repeatedly been brought forward in Congress.

Finally, we are concerned that excessively restrictive standards for granting proprietary rights to basic knowledge may be extended globally through various negotiated initiatives. For example, under the auspices of the World Intellectual Property Organization (WIPO), deliberations are underway regarding a global Substantive Patent Law Treaty. The essential objective of these negotiations

[1]Madey v. Duke University, 307 F. 3d 1351 (CAFC 2002).

is to harmonize standards for patent eligibility and examination procedures in order to minimize transactions costs. It is possible that recently evolved standards in the United States, including minimal utility, low inventiveness and novelty, and broad claim coverage, could be established in this treaty, thereby significantly limiting access of researchers and competing innovators on a global scale.[2]

A second concern relates to the effects of IPRs on international technology transfer. Governments in many countries place great importance on the ability of local firms and researchers to acquire technological information on reasonable terms. Economic evidence suggests that, in the long run, the stronger technology protection standards required by TRIPS should provide firms in developing economies significantly larger private technology inflows, mediated through imports, foreign direct investment (FDI), and licensing.[3] However, the likelihood of these increases depends on such national characteristics as size, proximity to markets, infrastructure quality, local imitative capacity, and the ability to absorb, modify, and deploy new technologies. Indeed, much of the additional technology transfer on offer may be from parents to affiliates within a multinational enterprise, with diminished capacities for spillovers to local firms. More directly, strengthened proprietary rights to knowledge raise the likelihood that firms may act more monopolistically in setting prices for information. Thus, increases in technology flows are not guaranteed and may not emerge at all in the poorest economies (Smith, 2001).

While there is considerable small-scale innovation in developing countries, they remain net importers of new technological information, which is the primary source of technical change and an important determinant of productivity growth (Hoekman, Maskus, and Saggi, 2004). In this context, access to general knowledge emanating from basic research in the technologically advanced economies, no less for developing countries than for the developed nations themselves,

[2]Barton (2004) analyzes prospects for this treaty and argues for more competitive and development-regarding standards.

[3]Maskus (2000) provides a comprehensive review of such evidence and Smith (2001) offers recent empirical evidence.

is central to the processes of experimentation, adaptive innovation, and product development. Put differently, fundamental science generates the enabling technologies from which dynamic competition springs (David, 2003). Significant restraints on access to such scientific results could pose difficulties for competition, particularly in developing countries.

Such concerns about technology transfer surfaced in TRIPS itself. Thus, Article 7 lists as a primary objective of the Agreement that intellectual property rights should contribute "... to the transfer and dissemination of technology, to the mutual advantage of producers and users of technological knowledge..." To give content to this objective and to promote technology transfer to poor countries, negotiators agreed in Article 66.2 that "Developed country Members shall provide incentives to enterprises and institutions in their territories for the purpose of promoting and encouraging technology transfer to least-developed country Members in order to enable them to create a sound and viable technological base."

Authorities in many developing countries are disappointed in the evidently meager response of technology flows to these indirect and direct incentives.[4] For this reason, WTO Members affirmed in Paragraph 11.2 of the Doha Ministerial Declaration that the provisions of Article 66.2 are mandatory. The Declaration calls for a mechanism for ensuring full implementation of that obligation and also set up a Working Group on Trade and Technology Transfer. To date the Working Group has collected submissions from developed countries describing the incentives they have in place to encourage technology transfer to poor countries. While these submissions are many and detailed, the programs described tend to be small, focused on specific projects, and not specific to the needs of the least developed countries. Moreover, the Article is restricted to the "least-developed countries," which is a subset of the developing countries and, indeed, a

[4]The assertion that technology transfer has not increased much since TRIPS is not based on systematic evidence, a study of which is needed. There is some evidence, reviewed in Barton (2003a), that the world's move toward stronger intellectual property rights contributes to greater foreign direct investment in some sectors, and may favor direct foreign investment over development through indigenous technology-based enterprises.

subset in which the private sector technology transfer envisioned by the Article is particularly difficult.

That supply responses in technology transfer have been meager to date is not surprising in economic terms. "Technology transfer" refers to a host of complex transactions that depend on a variety of factors, only one of which is the intellectual property system. New intellectual property regimes are only recently in place in many countries and have yet to achieve strong enforcement support. Most significantly, international technology transactions are undertaken primarily in private markets. While governments can provide some subsidies, tax advantages, research subventions, and credit guarantees, ultimately "... they cannot force the private sector to transfer its technologies" in the words of the EU submission to the Working Group.[5]

Though the issue surfaces most directly in TRIPS, the need to enhance technology transfer appears in several other WTO Agreements, suggesting that it is a broader concern than may be reflected in intellectual property rights. Such language exists also in GATS, the TRIMS Agreement, the Agreement on Technical Barriers to Trade, and the Agreement on Sanitary and Phytosanitary Measures. Further, any prospective agreement on competition policy presumably would address various aspects of technology markets. Despite this coverage, many developing countries believe that the WTO Agreements do not adequately address means for facilitating effective technology transfers to their markets (Roffe, 2002). In some degree this reflects a preference that the transfer of private commercial technologies be subsidized or made available without private property rights attached (Becker, 2002). Neither possibility is likely, for fiscal subsidies to transfer commercial technologies would run into political opposition in home countries, while eliminating intellectual property rights in commercial information (through patent buyouts, compulsory licensing, and the like) would run counter to the nature of TRIPS itself.

[5] "Implementation of Article 66.2: Information from Developed Country Members: European Communities and Their Member States," 2002.

3. Economic Perspectives

To advocate a multilateral policy intervention on behalf of open access to basic science and technology, we must demonstrate that the approach would help resolve market failures in the private generation of knowledge and its use, would provide advantages over a system based extensively on proprietary rights to basic science and technology, and would improve multilateral policy coordination in this area. To this end we begin by considering the economic nature of basic technological knowledge.

3.1. *Economics of Knowledge Creation and Diffusion*

Generations of writers have explained the fundamental characteristics of knowledge as a public good. Thomas Jefferson, for example, compared an idea to the flame of a candle, which could be used to light other candles without diminishing the original light.[6] Thus, the defining feature of knowledge is its non-rivalry: access distributes the gains from its use widely without reducing the ability of developers to use it. In turn, the social value to multiple uses is the sum of all the individual valuations, a sum potentially far larger than its value to an individual user. However, this same characteristic immediately generates a free-rider problem because anyone who can gain access to an existing idea would rationally refuse to offer to bear any of the initial costs of developing it. Non-rivalry is a technical characteristic suggesting non-convexities in the use of ideas and optimal social policy would call for the widest possible use of existing knowledge, assuming the marginal cost of additional provision is small.

A second feature of knowledge is that it may be difficult to maintain it in exclusive possession while putting it to some useful purpose. Attempts at secrecy often fail and, in the extreme, some technologies, such as medicines and software, bear their technical properties "on their face" and are easily learned through simple imitation. There may be technical and legal solutions that aim at generating exclusivity, but the stronger these are the more costly is access to use of information.

[6]David (1993) describes Jefferson's writings in this context.

Taken together, these characteristics imply that the returns to investing in ideas often may not be captured by an original creator. This non-appropriability problem necessarily implies that knowledge is a public good; indeed, in Arrow's (1962) classic conception it is the purest form of public good. Accordingly, market actors would not undertake costly investments in developing new information and society would suffer from a diminished rate of technical progress.

This essential market failure calls for public intervention, which has by tradition taken place through a mix of two canonical policies. Numerous governments provide substantial direct funding to research in order to solve the underinvestment problem. Thus, in the United States annual federal government spending on research grants to universities, government laboratories, and private research centers amounts to tens of billions of dollars. There are also subsidies through tax incentives for private firms to undertake research and development. Systems in Europe and Japan are similar, if not as large. These supports have directly accounted for the development of massive amounts of basic technology that supported extensive commercial invention.

The second approach is to safeguard the ability to earn returns to investment in research by providing exclusive intellectual property rights. Patents, copyrights, trademarks, trade secrets and related devices protect different forms of innovation and operate in different fashions but all offer exclusivity in the production, use, sale, and licensing of designated subject matter. These are essentially market-based inducements to R&D, for firms are free to invest in whatever programs they think will achieve market success. They are also incentives for placing new products and ideas on the market, the ultimate source of welfare gains from innovation. In contrast, government direction or funding of commercial research is generally ineffective for the public sector may have limited information about dynamic market prospects and is liable to make politically motivated and inefficient allocations of research funds. To be sure, IPRs can in some circumstances support substantial market power and induce wasteful duplication of R&D spending. Nevertheless, they seem to be an integral support for technological competition, at least in innovative economies, and typically do not constitute significant barriers to entry.

There are no definitive principles for deciding the appropriate mix of public funding for research and proprietary rights to research outcomes. The conventional solution in most societies has been to draw a fuzzy line between investments in basic scientific knowledge and applied research in specific processes and products with commercial applicability. The former might be considered true public goods in that they are both non-rival and offer general knowledge that can support multiple uses. Thus, private markets would fail to invest sufficient resources in their generation, requiring a public solution. The latter are more properly construed as quasi-private goods because of their specificity and relative ease of technical or legal excludability. Regulatory instruments typically should be sufficient to ensure their market provision.

This distinction is central both to conventional research policy and IPRs. The bulk of U.S. government-performed research findings and data, save that reserved for security purposes, traditionally has been placed into the public domain. University scientists operate in a vigorous "open source" mode in which their findings are thoroughly vetted and published for wider use. The gains to successful scientists in this environment stem from building reputation assets rather than property rights, though reputations may be manifested in higher incomes. Thus, by tradition, basic scientific results have found their way readily into the public domain for all of competence to use.

Conventional conceptions of IPRs firmly embody this distinction between basic knowledge and commercial applications. Most prominently, in most countries outside the United States it is impermissible to patent basic discoveries (as opposed to inventions), mathematical algorithms, and genetic research tools. Further, for a patent to be awarded, the invention must meet a utility standard under which the technology must be reduced to some industrially or commercially useful form. A rigorous utility standard is a central means by which public authorities deny proprietary rights to basic scientific knowledge. For its part the United States has greatly weakened its utility standard and thus permitted property rights on basic enabling technologies.

Also relevant is the inventiveness standard under which a new technology must display real creativity in order to achieve exclusive rights. However, the EU's Database Directive vitiates this requirement by awarding patent-like protection to simple compilations of data and information on the theory that "sweat of the brow" effort is sufficiently creative to deserve reward. This diminution of an essential standard also threatens to remove important research results from the public domain.

Taken together, the U.S. and EU expansive treatments of public research findings as commodities on which property rights may be asserted has significantly eroded the practical distinction between basic knowledge and applied R&D. The political rhetoric justifying the mixing of public funding with private rights is couched in competitiveness terms. Specifically it is argued that the mixture will procure greater economic rents from the public investment in research and provide more incentives for product development, to the ultimate benefit of society.

It remains to be seen whether this salutary outcome emerges in the context of technologically advanced countries. It is by no means a settled issue among scientists, economists, and legal scholars, and there are several who have expressed significant concerns (Mowery and Sampat, 2004; Nelson, 2004). A careful look at licensing regimes emerging from patents on basic tools in biomedical research suggests that transactions costs in organizing scientific inquiry have been increased markedly (Heller and Eisenberg, 1998).[7] Ideas are inherently heterogeneous and the costs of merging them into useful technologies likely are raised by IPRs on scientific knowledge. The essential point is that providing patents to basic technologies can be harmful to dynamic competition by raising licensing costs and extending reach-through proprietary rights to all potential new innovations using those technologies (Aoki and Sadaoka, 2003).

[7]However, see Walsh, Arora and Cohen (2003) for a survey-based skeptical view of the proposition that these costs deter scientific research. Their data, however, indicate that there were sometimes costs and delays, associated with, for example, obtaining licenses, conducting research offshore, running the risk of infringement, or modifying research strategies.

This concern is buttressed by the fact that open-access basic science and commercial R&D traditionally have enjoyed a complementary relationship, with the former having little direct utility in terms of marketable goods but supporting a range of innovative products and follow-on technologies. Put differently, open science tends to raise the expected returns to private investment in proprietary R&D (David, 2003). One reason is that access to existing knowledge provides developers with a roadmap of promising areas in which to invest and likely dead ends, increasing the efficiency of capital allocation. Another is that public funding for university research and training generates high-quality technical personnel that often move into industrial employment, a key element of technology transfer in the United States. These spillovers serve essentially as a general subsidy to applied R&D, but as proprietary rights are extended on public research results, the scope for such spillovers is likely to be diminished.

Ultimately the mix of intellectual property rights and public provision of research needs to strike a balance between resolving problems of appropriability in order to induce investment and commercialization and ensuring that basic knowledge is accessible and widely distributed. The diffusion and use of knowledge itself serves as a platform for both further basic research and development of applied technologies.

We have argued that the increasing application of proprietary rights to publicly funded and basic research results may be problematic even within the United States and the European Union. Research universities and large firms in those countries may be able to engage in sufficient patent pooling and cross-licensing that their research programs are not greatly inhibited (Walsh, Arora and Cohen, 2003). However, it has been observed that startup firms and small enterprises in developed countries have been placed at a distinct disadvantage in this context (Reichman and Uhlir, 2003).

This problem is compounded when one considers research processes in most developing countries. Public research institutes, university science and education, and the development and diffusion of applied technologies all are dependent on access to basic knowledge, which is overwhelmingly generated in the rich nations

(Evenson, 2004). Increasing privatization of basic data by entities in the developed countries threatens to retard the diffusion of such knowledge into science and competition in developing countries. Few of the latter are in a position to mount significant public funding for basic research in their own universities and institutes. Thus, we remain concerned that one significant outcome of IPRs policy in the United States and the European Union will be higher costs for, and diminished access to, the fundamental scientific results that have been a foundation for technical change.

Put another way, public research traditionally has generated large spillover benefits across borders in the forms of improved education, research, and competition. Technological change is the main engine of growth, but learning from such change, and contributing to it, requires basic educational, scientific and technological capabilities. Thus, access to knowledge is central to prospects for growth and transformation, especially for developing countries. The more such knowledge is protected by exclusive rights, the lower these spillovers are likely to be.

3.2. *Why is a Multilateral Agreement Needed?*

We have argued that knowledge is, in fact, a global public good because its non-rivalry clearly transfers across borders. Thus, widespread international access to basic knowledge generates multilateral gains in terms of science, education, technological change and dynamic competition, even as it reduces the economic rents that might be available to those entities that would control its diffusion. Seen in this light, it is unsurprising that those technologically advanced countries generating the major share of global knowledge increasingly view it as a source of competitive advantage if it can be protected on a preferential basis.

However, if we conceive of basic science and technology as an essential component of the public and global commons then we must recognize that national governments cannot be expected to provide such investments at the optimal international level. At one level, governments may fail to provide sufficient resources to support nationally optimal levels of basic science and technology research, even for their own economies. There are several reasons for this situation, ranging from greater priorities for limited budgets to political-economic structures disfavoring the provision of national public goods. It also reflects, in some part, the non-rivalry problems discussed above, which manifest themselves in too-little domestic science and a heavy reliance on access to international science. Certainly, most developing countries may be characterized as having inadequate education and research infrastructures, along with little effort to promote scientific research. Without adequate access to general technologies, their attempts to develop effective "national innovation systems" are unlikely to bear much fruit.

A more fundamental problem for our purposes, however, is that national governments, acting on their own to establish support policies for basic science and technology, will jointly under-invest in that area from a global standpoint. Again, the essential difficulty is the non-rivalrous nature of basic science and technology. In setting their own subsidies to the development of basic knowledge, individual countries would not take into account the spillover benefits to other countries (Drahos, 2004). That is, each country would rationally fund research up to the point at which the marginal cost of developing knowledge equals the sum of the marginal valuations of all domestic users. From a global view, however, the domestic marginal cost would be less than the sum of international valuations in any nationally determined equilibrium. As a result, too little investment in basic knowledge would be undertaken, even in the absence of free riding. That is, even if international valuations were known, individual countries would be loath to fund additional knowledge to meet international needs. If we add free riding (an unwillingness of importing countries to reveal their social valuations for knowledge) to this equation, the situation is made yet worse.

In asserting private ownership rights to basic knowledge, the United States and the European Union essentially are trying to solve the international free riding problem, but in an inefficient manner. While extracting some international surplus for the fruits of its investment, the U.S. policy does so on behalf of private interests rather than the public purse. More importantly for welfare, the provision of private exclusive rights can be expected to reduce net international access to information and, if transactions costs are raised sufficiently, to inhibit future investments in technology. Put in the language of economics, strong IPRs endowed upon basic research results are the wrong instrument for meeting the dual targets of expanding investments in knowledge as a global public good and making it widely available.

In principle there are two appropriate solutions to this set of problems. The first would be for a centralized global authority to fund the development of basic technologies that cost less than the sum of global valuations. However, no such central authority exists and this solution cannot be considered feasible.

The second would be a multilateral agreement on access to basic science and technology, which is the proposal we advance in this paper. A bargaining approach could permit countries to exchange concessions about access to their own basic technologies, recognizing that their own educational and technological enterprises would benefit from reciprocal access. The reciprocity associated with bargaining can mitigate some of the spillover problems deterring investments. There may be some decrease in political incentives to invest in research, because the competitiveness motivations would be less intense, but this decrease would be minimized by the reciprocity requirement, and any impact that it may have should be counterbalanced by the increased productivity of the global research enterprise. Moreover, to the extent that bargaining permits rationalization of basic research, in the sense of avoiding wasteful duplication of similar programs, bargaining can mitigate the underlying provision problem as well. Research costs in some forms of basic science may be so large that private rights could not generate sufficient revenue to cover them. In some cases shared financing across governments may be required.

In the following section we set out the framework of an ABST and discuss why the WTO is the appropriate forum for achieving it. We also consider its feasibility in terms of the political economy of market access exchanges.

4. An Agreement on Access to Basic Science and Technology

We have argued that, as things stand today, too little basic science is developed, too little of that flows to developing countries, and the intellectual property system needs to be complemented with a mechanism for resolving these failures. We therefore offer a proposal for an international agreement aimed at meeting these needs.

In our view an ABST would be best fashioned within the WTO, for several reasons. First, without a multilateral agreement to discipline free riding, any bilateral or plurilateral agreement is liable not to be sustainable. Second, the WTO already has responsibility for major agreements governing intellectual property, subsidies, standards, and trade in services, all of which would be interrelated strongly with transfer of scientific results. Third, the WTO offers a recognized format for arbitrating and settling disputes arising between governments, which would be primary players in this treaty. Fourth, the organization has a dynamic negotiating process that permits tradeoffs in concessions across sectors and functional agreements. Fifth, the organization has become a focal point for the strengthening of national constituencies seeking the benefits of multilateral agreements. Finally, many of the essential WTO principles may be applied to ABST, as we discuss next.

4.1. *Format of an Agreement*

Several provisions would need to be addressed in the treaty. The first, and most basic, would be its scope in terms of subject matter and processes. We have used the term "basic" science and technology but it is not easy to determine the dividing line between basic and applied research. In principle, one would describe basic knowledge as that which is truly non-rival and, by itself, has limited commercial utility. Examples are numerical formulas, algorithms, discoveries, surgical methods, research tools, and genetic sequences. Note that such forms of knowledge are not patentable under most legal jurisdictions. Another class of basic technologies would be those supporting the provision of global public goods, such as environmentally sound processes and health care.

However, there is no clear practical sense in which these characteristics might be defined. One way to manage the distinction would be to include research processes and results and data that are largely publicly funded, whether through direct research in government laboratories or grants to universities, NGOs, or other institutions. Observe that this distinction between technological characteristics and funding may not be critical, for presumably most basic and public-goods technologies require public financing in any event. Thus, focusing on publicly funded research and data may be sufficient.

A second aspect of scope relates to the forms in which access is to be granted, or the nature of liberalization. In principle, three levels of commitment could be entertained. First, "input liberalization" would permit researchers from other countries to participate in, or compete with, local research teams for grants and subsidies. This could be combined with increased opportunities for temporary migration of scientific personnel and additional student visas. Thus, particular provisions could prohibit preferences for national firms and institutions as regards publicly funded research consortia and access to research-based tax advantages. Commitments could be reached narrowing visa restrictions that inhibit the ability of students to study at universities in another nation or restrict the scope for scientists and engineers to participate in conferences or training programs. However, under this alternative, governments could choose to reserve their research results for preferential use by local firms and the registration of intellectual property rights. While this approach could expand research efficiency and transfer more skills abroad, its scope for raising access to new information would be limited.

Second, "output liberalization" would entail offering researchers in other countries access to nationally generated science and data, without increasing their ability to use underlying funding or research facilities. This approach would usefully expand the public commons and increase knowledge transfers but would not directly expand efficiency or transfer research skills. A key provision here would promote access to scientific databases and would ensure that intellectual property regulations not limit access to basic scientific knowledge.

Finally, "full liberalization" would combine these regimes, both expanding international flows of research contracts and personnel and increasing global access to outcomes. It is evident that we favor full liberalization to the extent it is politically feasible. In getting there, however, it may be necessary to adopt something like a GATS approach, permitting governments to reserve sensitive areas of technology and to designate different levels of commitment to open access.

Consistent with other WTO agreements, national treatment would be a key legal provision of the treaty, requiring that, in as many ways as possible, foreign scientists and firms be treated the same way as national ones with respect to access to a country's scientific and technical support programs and outcomes. Like TRIPS, ABST could also build on an MFN commitment, unless there were compelling reasons for regional preferences.

It is evident that a treaty of this kind would need to be balanced by safeguard clauses. One issue involved in international scientific and technological collaboration relates to the equitable and efficient distribution and management of intellectual property that could emerge from subsequent applied innovation. Another is that concerns regarding national security and technology proliferation would need to be addressed. For example, the United States has moved to establish new security classifications for biological data and restrict some foreign students from studying particular areas of biotechnology. Such restraints need to be balanced with the advantages of promoting the scientific and technological commons, a balance that could be set out in an international agreement.

In recognition of the need for encouraging a "sound and viable technological base", it would be possible to build in certain preferential advantages for the developing economies. For example, to the extent that data and research results are to be made available at some cost, differential pricing schemes for governments and institutions in poor countries could be encouraged. Efforts to encourage research participation by scientists and engineers from developing countries could be written into proposal solicitations. Marginal visa allocations could be aimed at students and researchers from poor countries. More generally, developed countries could commit themselves to help developing nations build capacity for improving educational and scientific processes, including their ability to benefit from available international information. Assistance in development and use of electronic resources, especially the internet, could be particularly valuable.

While perhaps not fitting within the parameters of a treaty, additional commitments would be beneficial. Donor nations might consider establishing special trust funds to encourage research and science in developing countries and to help them benefit from ABST and other international commitments (Roffe, 2002). Of particular importance would be expanded commitments to support programs for providing public scientific and technological goods to developing countries. Examples include the Consultative Group on International Agricultural Research and the new public–private partnerships for research on HIV, TB, and malaria. Such programs are greatly under-funded (Commission on Macroeconomics and Health, 2001). Financial commitments could be supplemented by agreements to help manage potential problems with intellectual property rights, including gaining access to patented technologies.

Two other issues arise for construction of an ABST agreement. First, careful consideration is needed of how its provisions relate to other WTO agreements and even such non-WTO accords as the Convention on Biodiversity. Within the WTO, efforts to harmonize ABST and TRIPS would be required. In effect, ABST would be an attempt to re-balance the strong privatization of rights in information implicit in TRIPS. Similarly, specification of ABST could usefully sort out the meaning of pre-competitive research subsidies and how they might be provided internationally. However, some forms of research support could be problematic in terms of the WTO Agreement on Subsidies and Countervailing Measures.[8] The GATS Agreement ("Mode 4") might need to accommodate more liberal treatment of temporary research and educational visas.

Second, there would need to be provisions for regular meetings, for a small secretariat or council to evaluate the extent of scientific and technological cooperation and its benefits, and for ongoing

[8]Hoekman and Kostecki (2001) provide a comprehensive discussion.

negotiations. These negotiations could provide a forum for scientific and technical communities to pursue further expansion of the global information commons.

4.2. *Arguments for Concluding an ABST Treaty*

We see several arguments in favor of developing a global treaty on science and technology. First, we reiterate our concern that the United States, the EU, and other developed economies are rapidly moving toward establishing private property rights in basic knowledge and data, even when developed from public resources. This tendency threatens to reduce excessively the amount and scope of information in the public domain, with deleterious impacts on the provision of public goods and on prospects for dynamic competition. This "new enclosure movement" may be detrimental even for the countries that practice it (Boyle, 2003). However, it raises significant difficulties for the diffusion of information to developing countries and the prospects for building a capability in science and technology. In short, we offer the proposal as a means of recapturing publicly funded and basic research for the public domain and for encouraging knowledge transfers to developing economies.

Second, as noted above, many countries remain disappointed with the apparent inability of TRIPS and the other WTO Agreements to induce additional and large transfers of technology. Those provisions are largely aimed at transfer of commercial technologies and an ABST agreement would focus on transfer of basic (and somewhat less commercial) technologies. In that context, it would provide a re-balancing of benefits under TRIPS in favor of technology users. We emphasize that ABST would serve as a complementary and supporting mechanism for market-mediated technology transfers, rather than substitute for them.

Third, TRIPS itself recognizes the need for building "... a sound and viable technological base" in poor countries without providing any means for achieving it. Surely one important reason that there evidently has not been much expansion of technology flows to developing countries since the implementation of TRIPS is that this technological base often is not sufficiently viable to reduce the transactions costs of transfers. No doubt much capacity needs to be built, in terms of governance, transparency, infrastructure, and other environmental characteristics, before much scope for private technology transfers to some countries could emerge. However, a central component must be the construction of a national science and technology capability for purposes of adapting technologies to local needs and for performing local research. If the ultimate outcome of TRIPS and stronger protection in the rich nations is to restrict access to basic research and data from abroad, the prospects for building such a capability could be damaged irrevocably. Our proposal aims at sustaining and enhancing such access.

Fourth, TRIPS Article 66.2 makes reference solely to the least developed countries as potential beneficiaries of government efforts to encourage private technology transfers. There are 30 WTO Members that are designated "least developed", a roster that hardly exhausts the list of countries that could benefit from greater access to scientific opportunities and results. Thus, Article 66.2 can do virtually nothing to provide access to basic technology on behalf of the broad scope of developing countries. In contrast, ABST would be a global agreement.

Finally, we believe that such a treaty could be structured in a way that, while restraining the movement toward privatizing rights in basic information, would respect exclusive rights in commercial information and applied technologies. Inevitably there would be difficult negotiations over international benefit-sharing and licensing terms at the interface between basic science and technology and its commercial applications. However, such negotiations should be more efficient when undertaken against a backdrop of an international legal agreement regarding access and rights allocation.

4.3. *Political Economy Issues*

Our proposal aims at resolving the problems described earlier but is keyed on the free riding problems that have pushed science providers into an increasingly protectionist treatment of information. That treatment is "protectionist" in that it limits foreign access to basic knowledge, in a manner analogous to trade restrictions that limit market access. As such, a WTO approach that offers reciprocal and non-discriminatory access to knowledge is appropriate for restoring the public domain and promoting international competition.

The nature of the treaty we have in mind would increase global access to the fruits of public research funding. An obvious difficulty is that research decisions are endogenous and funding might decline in authorities in the major countries perceive that an ABST would dilute the ultimate economic benefits from such investments without reciprocal benefits from abroad.

Thus, analysis of national economic interests in an ABST treaty is relevant for considering its construction and feasibility. Mutual trade liberalization in the WTO has been achieved through a mercantilist agenda in which countries were willing to offer greater market access to foreign firms in return for reciprocal access abroad. A similar reciprocity in which access of foreign researchers to grants and research results is provided in return for related opportunities abroad could appeal to competitiveness concerns. A broader scope of opportunities and research competition presumably would expand the efficiency with which public science and technology are generated, resulting in mutual gains from trade. And the opportunity to negotiate liberalization will focus the attention of those in the scientific and technological communities to press politically for the benefits of liberalization. Moreover, with a wider base of basic technologies from which to work, and with these technologies largely in the public domain, the scale of product innovation built on such information should increase.

At the same time, countries are highly asymmetric in terms of their abilities to finance and develop basic science and technology. The United States, the EU and Japan may see some complementarities in mutually integrating access to these resources. Some large developing countries such as Brazil, India and China could be attractive as well. However, small and developing countries with limited research resources offer little in the way of "export" interest to researchers in the main technology-developing nations. In this environment, governments of the developed countries might effectively restrict access by encouraging exclusive rights in research outcomes and data, as they are doing. Yet, even the United States, by far the preeminent technological power, is now seeing the need to depend on scientists from other nations (National Science Foundation, 2004). Thus, it is not yet clear whether there is a balance of mercantilist interests to support a multilateral ABST treaty.

In consequence, a WTO treaty to ensure access to basic science might require technology importers to offer other, perhaps complementary, concessions in such areas as services, investment and product-market access. In addition, the case could be made that firms in the poorest countries pose no competitive threat in the medium term and that permitting them to join on a preferential basis could help develop their research and innovation capabilities, in line with other development assistance.

There is another reason to think that an agreement may be supported by powerful economic interests. Unlike the situation 30 years ago, multinational enterprises now often transfer technology in order to build export products in developing countries. The costs of doing so would diminish when local researchers have access to basic technologies and can effectively deploy them. Thus, multinational enterprises might be expected to lobby for such an agreement, particularly to the extent that it can be accompanied by appropriate policy responses in recipient countries regarding governance and infrastructure. Further, the treaty would provide legal certainty about the scope of the public and private domain across countries, which would benefit global enterprises.

4.4. *A Smaller Substitute*

While we are optimistic about the prospects for such an agreement to work, we recognize that the problems noted above are serious and raise obstacles to the negotiation of a maximal treaty. In that context, we provide brief comments on a less ambitious approach that could achieve some of the objectives we have identified for greater international access to basic science and technology. This approach essentially would work through the intellectual property system itself.

First, from the standpoint of distinguishing between basic and applied research, it would be beneficial for countries to reassert the importance of a rigorous utility standard in patenting. For those jurisdictions, such as the United States, that offer patents to basic research tools and software, careful consideration to limiting the breadth of reach-through claims is in order. For their part, developing countries should avoid adopting weak eligibility standards for patents and broad claims of this kind, while sustaining research and education exemptions in the use of patented knowledge. Negotiations at WIPO regarding the Substantive Patent Law Treaty offer scope for rebalancing patent rules on behalf of restoring the public domain in this way.

Next, if governments insist on permitting the registration of private rights on the results of publicly funded basic research, they could limit the risk that such rights would support monopoly restrictions on access by establishing careful guidelines on licensing as suggested in Nelson (2004) or a wide domain for automatic licensing on receipt of a modest fee. This form of "liability rule" would permit those who wish to use fundamental scientific results to do so upon payment of a well understood licensing fee or royalty (Lewis and Reichman, 2004). In establishing such a system it should be feasible to organize a fee schedule that discriminates in favor of users in developing countries and educational institutions. This kind of price differentiation is an especially attractive form of gaining some returns on investment in research while encouraging wider dissemination.

5. Concluding Remarks

We have outlined a proposal for negotiating a multilateral agreement on open access to publicly funded research results and data. In our view this agreement could have at least five globally beneficial impacts. First, it would help resolve the free rider problems that reduce investments in science and technology relative to a global optimum. Second, it could restrain the tendency of governments to restrict access and to encourage privatization of basic knowledge. This rebalancing of technology development norms in favor of expanding the public domain could help vitalize scientific research in many countries, while promoting applied innovation.

Third, the treaty could provide an important plank for the construction of modern technological capabilities in poor countries, while sustaining access to information for educational purposes. Fourth, it should not unduly restrict the rights of firms to exploit intellectual property in applied technologies and products. Finally, it could help restore confidence on the part of developing countries that TRIPS and the WTO are institutions that facilitate, rather than hinder, technology transfer.

This essay sets out the positive case for such an agreement. Undoubtedly numerous practical difficulties would arise in working out specific provisions and defining terms. Even the notion of what is "publicly provided" or "basic science and technology" is not easily determined in each case. Nevertheless, we think the principles set out here can offer a platform for moving forward.

References

Aoki, Reiko and Sadao Nagaoka. 2003. The Utility Standard and the Patentability of Intermediate Technology. Hitotsubashi University, manuscript.

Arrow, Kenneth J. 1962. Economic Welfare and the Allocation of Resources for Inventions. In Richard R. Nelson, ed., *The Rate and Direction of Inventive Activity: Economic and Social Factors* (Princeton: Princeton University Press).

Barton, John H. 2003a. Patents and the Transfer of Technology to Developing Countries. Prepared for OECD Conference on IPR, Innovation and Economic Performance, Paris, 28–29 August 2003.

Barton, John H. 2003b. Preserving the Global Scientific and Technological Commons. Stanford University, manuscript.

Barton, John H. 2004. Issues Posed by a World Patent System. *Journal of International Economic Law* 7: 341–358.

Becker, Gunnela. 2002. Transfer of Technology and the TRIPS Agreement: A Study of Article 66.2 in the WTO/TRIPS Agreement. Report for Swedish International Development Agency.

Boyle, James. 2003. The Second Enclosure Movement and the Construction of the Public Domain. *Law and Contemporary Problems* 66: 33–74.

Commission on Macroeconomics and Health. 2001. *Macroeconomics and Health: Investing in Health for Economic Development* (Geneva: World Health Organization).

David, Paul A. 1993. Intellectual Property Institutions and the Panda's Thumb: Patents, Copyrights and Trade Secrets in Economic Theory and History. In M. Wallerstein, M. Mogee, and R. Schoen, eds., *Global Dimensions of Intellectual Property Rights in Science and Technology* (Washington DC: National Academy Press).

David, Paul A. 2003. The Economic Logic of 'Open Science' and the Balance between Private Property Rights and the Public Domain in Scientific Data and Information: A Primer. Stanford Institute for Economic Policy Research, Discussion Paper No. 02–30.

David, Paul A. 2004. *Koyaanisqatsi* in Cyberspace. In Keith E. Maskus and Jerome H. Reichman, eds., *International Public Goods and Transfer of Technology under a Globalized Intellectual Property Regime* (Cambridge: Cambridge University Press), forthcoming.

Drahos, Peter. 2004. The Regulation of Public Goods. *Journal of International Economic Law* 7: 321–339.

Evenson, Robert E. 2004. Agricultural Development and Intellectual Property Rights. In Keith E. Maskus and Jerome H. Reichman, eds., *International Public Goods and Transfer of Technology under a Globalized Intellectual Property Regime* (Cambridge: Cambridge University Press), forthcoming.

Heller, Michael A. and Rebecca S. Eisenberg. 1998. Can Patents Deter Innovation? The Anticommons in Biomedical Research. *Science* 280: 698–701.

Hoekman, Bernard M. and Michel M. Kostecki. 2001. *The Political Economy of the World Trading System: 2d Edition* (Oxford: Oxford University Press).

Hoekman, Bernard M., Keith E. Maskus, and Kamal Saggi. 2004. Transfer of Technology to Developing Countries: Unilateral and Multilateral Policy Options. World Bank, manuscript.

Lewis, Tracy and Jerome H. Reichman. 2004. Using Liability Rules to Stimulate Local Innovation in Developing Countries: A Law and Economics Primer. In Keith E. Maskus and Jerome H. Reichman, eds., *International Public Goods and Transfer of Technology under a Globalized Intellectual Property Regime* (Cambridge: Cambridge University Press), forthcoming.

Maskus, Keith E. 2000. *Intellectual Property Rights in the Global Economy* (Washington DC: Institute for International Economics).

Maskus, Keith E. and Jerome H. Reichman. 2004. The Globalization of Private Knowledge Goods and the Privatization of Global Public Goods. *Journal of International Economic Law* 7: 279–320.

Mowery, David C. and Bhaven Sampat. 2004. The Bayh–Dole Act of 1980 and University-Industry Technology Transfer: A Model for Other OECD Governments? In David C. Mowery, Richard R. Nelson, Bhaven N. Sampat and Arvids A. Ziedonis, eds., *Ivory Tower and Industrial Innovation: University-Industry Technology Transfer Before and After the Bayh–Dole Act* (Stanford: Stanford University Press).

National Science Foundation (National Science Board). 2004. *Science and Engineering Indicators 2004* (Arlington, VA: National Science Foundation).

Nelson, Richard R. 2004. The Market Economy, and the Scientific Commons. *Research Policy* 33: 455–471.

Reichman, Jerome H. and Paul Uhlir. 2003. A Contractually Reconstructed Research Commons for Scientific Data in a Highly Protectionist Intellectual Property Environment. *Law and Contemporary Problems* 66: 315–462.

Roffe, Pedro. 2002. Preliminary Note on the WTO Working Group on Trade and Transfer of Technology. Geneva: United Nations Conference on Trade and Development, manuscript.

Smith, Pamela J. 2001. How Do Foreign Patent Rights Affect U.S. Exports, Affiliate Sales, and Licenses? *Journal of International Economics* 55: 411–439.

Walsh, John P., Ashish Arora, and Wesley M. Cohen. 2003. Research Tool Patenting and Licensing and Biomedical Innovation. In *The Operation and Effects of the Patent System*, Report of the STEP Board of the National Academy of Sciences, National Research Council (Washington DC: National Academy Press).

Chapter 17

Developing and Distributing Essential Medicines to Poor Countries: The DEFEND Proposal*

Mattias Ganslandt, Keith E. Maskus and Eina V. Wong

1. Introduction

Perhaps the most critical task currently facing the global economy is to devise mechanisms that both encourage research aimed at finding treatments for diseases that are common in impoverished nations and that achieve widespread international distribution of these treatments at sufficiently low costs to be effective and affordable. This issue has achieved prominence by virtue of the severe epidemic of the HIV virus, which inevitably leads to the onset of AIDS, in Sub-Saharan Africa and, increasingly, in South Asia and Southeast Asia.

HIV/AIDS is not the only disease that plagues poor nations, where malaria, tuberculosis, and other maladies are equally lethal and debilitating. Indeed, HIV/AIDS is unusual in that strong incentives for pharmaceutical companies to develop treatments for sufferers in high-income economies have resulted in medicines that effectively permit patients to function well for many years before onset of the disease. In that regard, the current debate is about how best to transfer these medicines to poor countries. In contrast, there is virtually no R&D aimed at producing new treatments for malaria or tuberculosis. This situation arises largely because those

*Reprinted from *The World Economy*, Vol. 24, No. 6, pp. 779–795. Copyright 2001, with permission from John Wiley and Sons.

who suffer are overwhelmingly poor and could not afford medicines in sufficient quantities to cover R&D costs. The problem is accentuated by weak patent protection in potential markets, further reducing the willingness of pharmaceutical enterprises to develop new drugs and vaccines.

To put it in economic terms, under the current system the incentives to achieve efficient dynamic and static provision of medicines are grossly inadequate in the face of massive poverty. To deal with this problem essentially two programs have been advanced in recent years, which are considerably at odds with each other. On the one hand, the Agreement on Trade-Related Aspects of Intellectual Property Rights (TRIPS) within the World Trade Organization requires member countries to grant and enforce patents for new pharmaceutical products (Maskus, 2000a; Gorlin, 1999). More precisely, developers of new drugs have enjoyed exclusive marketing rights (EMRs) in all WTO members since January 1, 1995. While product patents are not required until the year 2005 in the least-developed countries, EMRs provide similar protection. Various economic studies suggest that this new regime could raise prices of new drugs markedly in developing countries (Watal, 1999; Fink, 2000; Lanjouw, 1998; Subramanian, 1995), though substantial uncertainty remains on this point.[1] Thus, some possibility exists that patents will raise incentives for R&D in these neglected diseases (Lanjouw, 1998). However, this policy shift does nothing directly to increase incomes of sufferers, who would, if anything, become less able to afford new medicines.[2]

Thus, on the other hand, considerable pressure has mounted on pharmaceutical companies to provide their drugs at marginal production cost (or less) to poor countries. Several firms have responded, such as Merck & Co., Bristol-Myers Squibb Co., GlaxoSmithKline PLC, and Abbott Laboratories. For example, Merck & Co. recently announced it would reduce the prices of two AIDS-controlling drugs in Africa by 40% to 55%, adding to sharp price cuts announced in

[1]See Rozek and Berkowitz (1998) for a dissenting view.

[2]See Abbott (2000) for a legal analysis of the pharmaceutical aspects of TRIPS, claiming that the agreement raises difficult contradictions between the trading system and needs for protecting public health.

2000.[3] Abbott announced that it would sell its two AIDS drugs, Norvir and Kaletra, at a price that would earn the company no profit.[4] In some degree these actions are a competitive response to offers by two Indian producers of generic AIDS drugs, Cipla Ltd. and Hetero Drugs Ltd., to provide medicines at even lower prices. As we note in the next section, however, even at these prices the drugs may be beyond the reach of most patients.

The research-intensive pharmaceutical firms that invented these drugs have three concerns about low-cost distribution programs. First, provision at marginal cost or lower adds nothing to their ability to cover the fixed costs of R&D. Second, while they may be willing to circulate their medicines cheaply, the firms are anxious to retain the exclusive distribution rights inherent in patents and EMRs. Indeed, this preference to forestall generic competition is the root of the ongoing lawsuit raised by 39 drug makers in South Africa aimed at striking down that country's 1997 Medicines and Related Substances Control Act.[5]

Third, and perhaps most significantly, original drug developers worry that the availability of far-cheaper treatments in poor countries could erode their price-setting power in rich countries. This erosion could happen directly through unauthorized parallel trade in drugs or indirectly through political pressure mounted by patients and insurance companies on health authorities to require significant price reductions. Because the vast bulk of returns to R&D are realized in the United States, the European Union, and other industrialized nations, pharmaceutical companies argue that such price spillovers would significantly hamper their incentives to develop new treatments.[6]

Control over patent rights in AIDS treatments is now before the WTO in a dispute raised by the United States against Brazil

---

[3] *Wall Street Journal*, "Price War Breaks Out Over AIDS Drugs in Africa as Generics Present Challenge," 7 March 2001.
[4] *Wall Street Journal*, "Abbott to Cut Prices on AIDS Drugs Distributed in Sub-Saharan Africa," 27 March 2001.
[5] *Wall Street Journal*, "Big Drug Firms Defend Right to Patents on AIDS Drugs in South African Court," 6 March 2001.
[6] *New York Times*, "Group Says Discount AIDS Drugs Endanger Research," 13 February 2001.

in February 2001. Under Article 71 of Brazil's 1997 Patent Act, foreign firms must manufacture patented drugs within Brazil before three years have elapsed from patent grant. Failure to meet these "working requirements" could result in an order by the Brazilian Health Ministry to local firms to manufacture generic substitutes, a threat that currently faces makers of the AIDS drugs Efavirenz (Merck & Co.) and Nelfinavir (Roche).[7] The TRIPS Agreement would seem to restrict considerably Brazil's ability to enforce working requirements. Thus, this case could set an important precedent concerning the ability of countries to limit private rights to exploit patents.

Putting these elements together, drug development and distribution involve tradeoffs that implicate important principles underlying protection of intellectual property rights (IPRs). To begin, there is a strong global public interest in providing sufficient incentives for the continual development of new medical treatments for diseases afflicting the poor. Within the intellectual property system these incentives stem largely from exclusive production and distribution rights provided by EMRs and patents. However, such rights may be inadequate for meeting the needs of extremely poor patients that do not have enough income to purchase them even at low prices.

Further, such rights are national or territorial in scope, meaning that governments may choose their own regimes concerning whether rights holders can prevent parallel trade.[8] Indeed, the TRIPS Agreement affirms that countries have the authority to decide whether exclusive rights are exhausted at national borders. The threat that products may be shipped from lower-priced countries to higher-priced countries reduces the enthusiasm of rights holders to supply them at low cost.

The current system generates numerous undesirable outcomes. First, there are not enough incentives to develop new treatments for endemic diseases in impoverished markets. The resulting high rates of infection and contagion impose external costs on others both within

---

[7] *New York Times*, "Brazil May Defy U.S. and Make More AIDS Drugs," 14 February 2001.

[8] Maskus (2000b) provides an overview of the economics of parallel trade.

and across borders, in part because of lower productivity. Surely the industrialized economies suffer some costs from slower growth in the afflicted countries. In this sense, development and provision of effective drugs is a global public good.

Second, demands that drugs be provided at marginal cost in some countries force patients in higher-price countries to accept a disproportionate share of the burden of financing R&D cost recovery. Put another way, patients in lower-cost nations effectively free ride on the pricing systems of the United States and other industrialized nations. In fact, the free riding has at least two dimensions. In addition to the low prices in poor countries, price controls in Canada, Europe, and elsewhere mean that patients in those nations provide limited contributions to recovering fixed R&D costs.[9] In that context, American patients and insurance companies bear the brunt of paying for R&D and any losses associated with distribution programs abroad. Thus, neither pharmaceutical companies nor their patients may be expected to embrace the costs of distribution and development.

Third, pharmaceutical firms chronically under-supply the medicinal needs of poor countries, partly because of limited exclusivity in rights, including the need to restrain parallel trade.

These problems point squarely at the need for further public involvement in encouraging new drugs and in procuring and distributing medicines. In this paper we set out a proposal for addressing the fundamental problems in a manner that is least disruptive to the international system of IPRs. It involves, first, increases in public assistance or public health budgets in the rich countries in order to fund purchases by a body such as the World Health Organization (WHO) of exclusive licenses to distribute selected medicines in poor countries. The license fees should be sufficient to cover all or a substantial portion of fixed R&D costs, thereby establishing a strong incentive for pharmaceutical and vaccine firms to produce new treatments. In terms of distributing these products in poor markets,

[9] The general nature of this problem is reflected in recent legislative proposals in the United States partially to deregulate restrictions on parallel imports of prescription pharmaceuticals in order to permit U.S. patients to gain access to cheaper foreign sources of supply.

the WHO would be free to do so at a per-unit price below its marginal private costs in recognition of the external benefits from improved health status. Finally, each country or region that avails itself of this program would be required to assert strong controls on parallel exports in order to safeguard prices in markets in high-income economies.

The procurement portion of our proposal is similar to the idea for a vaccine-purchase fund put forward by Sachs *et al.* (1999). However, their proposal involves a guaranteed price per dosage without contemplating difficulties in effecting distribution or in segmenting markets. It also bears similarity to current proposals for ensuring "tiered pricing" of existing HIV/AIDS drugs (Barton, 2001; Subramanian, 2001) but these programs make no provisions for managing dynamic R&D incentives. Thus, we offer our proposal as complementary to both of these ideas.

In the next section we provide basic evidence on the extent of the R&D, distribution, and pricing problems in the current system. In Section 3 we discuss the economics of optimal provision in recognition of the significant externalities involved. In Section 4 we set out the proposal explicitly and discuss ideas for its implementation. We conclude in Section 5.

2. Scope of the Problem

The incidence and costs of endemic diseases in poor countries are staggering. These maladies not only afflict high rates of mortality but also significantly reduce the health status and productivity of the affected population. Table 1 provides estimates by the World

Table 1. Deaths and DALYs caused by HIV/AIDS, tuberculosis, and malaria, 1999 (000s).

	World		Africa		Americas		SE Asia	
Disease	Deaths	DALYs	Deaths	DALYs	Deaths	DALYs	Deaths	DALYs
HIV/AIDS	2673	89819	2154	74449	81	2815	360	8866
TB	1669	33287	357	8721	59	1114	723	14101
Malaria	1086	44998	953	36838	2	76	69	3071

Source: World Health Organization, *World Health Report 2000* (Geneva, 2000).

Health Organization of deaths and productive time lost (measured in disability-adjusted life years, or DALYs) to three major diseases in 1999 for Africa, the Americas, and Southeast Asia. HIV/AIDS is thought to have killed 2.7 million people globally in 1999, with 2.2 million of these in Africa. It claimed 81,000 victims in the Americas and 360,000 victims in Southeast Asia, where the problem is rising rapidly. The disease was also responsible for 89.8 million adjusted life-years lost to morbidity and mortality. Again, this loss was concentrated in Africa, where 74.4 million life-years were foregone.

The victims of tuberculosis (TB) are spread more evenly through the developing world. It killed 1.7 million people in 1999, with 357,000 in Africa, 59,000 in the Americas, and 723,000 in Southeast Asia. Importantly, TB is frequently contracted by HIV/AIDS sufferers and surveys suggest that up to 70 percent of tuberculosis patients are infected with HIV.[10] Such joint cases are concentrated in Sub-Saharan Africa. Malaria is also concentrated in Africa, killing perhaps 953,000 people in 1999 and sacrificing 36.8 million life-years. According to the WHO, the direct and indirect costs of malaria in Sub-Saharan Africa exceed $2 billion per year.[11] Malaria is not at this time a large problem in the Americas.

Additional figures illustrate the scope of HIV/AIDS in Africa. There are now 25.3 million Africans living with HIV or AIDS.[12] In eight countries at least 15 percent of adults are infected. Infection rates in African women in their early 20s are three times higher than in men of the same age group. In Botswana, 36 percent of adults are now infected with HIV, while in South Africa the figure is 20 percent. South Africa has 4.2 million infected people, the largest number in the world. These figures are rising at alarming

[10]UNAIDS Press Release: "World TB Day 2001: Access to TB Cure a Human Rights Imperative," at www.unaids.org/whatsnew/press/eng/pressarc01/TB_220301.html.
[11]WHO, "Fact Sheet: Malaria," at www.who.int/inf-fs/en/fact094.html.
[12]These figures are from "Fact Sheet: HIV/AIDS in Africa," at www.unaids.org/fact_sheet/files/FS_Africa.htm. There are also 5.8 million living with HIV/AIDS in South Asia and Southeast Asia, see "Regional HIV/AIDS Statistics and Features, End of 2000," www.unaids.org/wac/2000/wad00/files/WAD_epidemic_report/css/WAD_epidemic_report_5.htm.

rates.[13] Among the 1.4 million children under the age of 15 living with HIV/AIDS at the end of 2000, 1.1 million reside in Sub-Saharan Africa. Perhaps 12.1 million children have been orphaned by the disease in that region.

Economic studies suggest that South African GDP will be perhaps 17 percent lower in 2010 than it would be without AIDS, removing $22 billion in output from the economy. In Botswana, there could be a 13–15 percent reduction in the income of the poorest households. The fiscal cost of the disease is also debilitating. It has been estimated that in seven of 16 African countries surveyed, public health spending for AIDS alone exceeded two percent of GDP in 1997, against total spending for health care of three to five percent of GDP.

These three diseases display different characteristics in terms of treatment costs and R&D incentives. Tuberculosis is curable with a single drug treatment that costs as little as $10–$15 per patient.[14] Unfortunately, TB is an airborne virus and in crowded environments with large numbers of sufferers, it is difficult and expensive to achieve eradication. The effective approach to TB is procurement programs to purchase and distribute these treatments widely in order to eradicate its presence, a task that lies beyond the economic reach of many health ministries in poor countries. Note also that there is little research into new treatments for TB. The World Health Organization (1996) estimated that of the $56 billion spent globally on medical R&D in 1994, less than 0.2 percent was spent on TB, diarrheal maladies, and pneumonia. Virtually all of the latter research was performed by public agencies and military authorities.

Malaria can be partially prevented through sanitation programs and prophylaxis, while it can be treated with available drugs. Again, these drugs may be out of the reach of poor patients. Moreover, because the disease tends to build resistance to drugs over time there is a continuous need for research into new medicines. The most effective long-term solution, in addition to vector control strategies,

[13]At the same time, successful prevention programs in a few African countries, such as Uganda, have reduced national infection rates.
[14]See UNAIDS, note 9 supra.

is the development of malaria vaccines, which could be administered to children.[15] However, there is insufficient R&D in anti-malarial vaccines or drugs. Sachs *et al.* (1999) cite a Wellcome Trust study that found that public and non-profit malaria research amounted to $84 million in 1993, with vaccine research amounting to a small portion of that spending. Private sector spending was lower still. We should note that more research into vaccines and anti-malarial drugs is underway under the auspices of the Multilateral Initiative on Malaria, involving the UNDP, the World Bank, and the WHO, and the Medicines for Malaria Venture, a public–private sector cooperative initiative. However, funding for the former comes to perhaps $3 million per year and the latter group is soliciting support from foundations in the hopes of achieving $30 million per year. These amounts seem inadequate for the job, given the underlying costs of developing and testing new drugs, and also fail to exploit private incentives within the intellectual property system.

As a final observation on the current problem, note that even though many pharmaceutical firms have slashed their prices for HIV/AIDS treatments in poor African countries, the prices on offer still do not reduce per-patient cost burdens relative to those in rich nations. In Panel A of Table 2 we show the current average prices in U.S. dollars for six AIDS drugs of an annual treatment for a single patient in the United States, Sweden, and South Africa. For South Africa we show both the prices offered by pharmaceutical companies that own patents on these drugs in the United States and prices offered by Indian generic producers. For example, the drug Viramune costs $3,508 in the United States, $2,565 in Sweden, and is now offered at $483 (original version) and $340 (generic version) in South Africa. In that context the prices are far lower in South Africa than in the United States.[16]

However, as shown in Panel B, when these prices are divided by the U.S.-dollar value of per-capita GDP in 1998, the burden of these

[15]Sachs *et al.* (1999).

[16]Note that Bristol-Myers Squibb Co. quite recently offered Zerit to South Africa for $54 per patient per year, making that price much less than indicated. See *Wall Street Journal*, "Bristol-Myers Squibb Offers to Sell AIDS Drugs in Africa at Below Cost," 15 March 2001.

Table 2. International price comparison for a selection of HIV/AIDS drugs, 2001.

Panel A: Prices in the USA, Sweden and South Africa, March 2001 (in USD)

Product (U.S. brand)	Sweden	USA	South Africa	
	Original mnf.	Original mnf.	Original mnf.	Generic subs.
3TC	1709	3271	232	98–190
Zerit	3078	3589	252	47–70
Viramune	2565	3508	483	202–340
Stocrin	3231	4730	500	1179
Combivir	4535	7093	730	293–635
Crixivan	3339	6016	600	2300

Note: Prices are for yearly treatment of a single adult patient with regular dosage.
Source: *Wall Street Journal*, 3/7/2001; LINFO, http://www.linfo.se/fass/.

Panel B: Prices as Share of GDP Per Capita (in percent)

Product (U.S. brand)	Sweden	USA	South Africa	
	Original mnf.	Original mnf.	Original mnf.	Generic subs.
3TC	6.3	10.1	7.2	3.0–5.9
Zerit	11.4	11.1	7.8	1.5–2.2
Viramune	9.5	10.8	15.0	6.3–10.6
Stocrin	11.9	14.6	15.5	36.6
Combivir	16.8	21.9	22.6	9.1–19.7
Crixivan	12.3	18.6	18.6	71.4

Note: GDP per capita, 1998. The exchange rates in 1998 were on average 1 USD = 5.54 ZAR and 1 USD = 7.95 SEK and in 2001 (until 3/25/2001) on average 1 USD = 7.81 ZAR and 1 USD = 9.69 SEK.
Source: The World Bank, *World Development Indicators 2000*; OECD, *Main Economic Indicators May 2000.*

drugs in income units is essentially the same in all three markets. Indeed, the price as a proportion of per-capita GDP is lower in South Africa than in the United States only in two drugs and is higher in three.[17] The range of prices of available generic substitutes generally

[17]We use 1998 GDP per capita for this purpose because it is the latest year available. Note that the South African Rand depreciated by 29 percent (and the Swedish Krona by 18 percent) relative to the dollar from 1998 to 2001, making the Rand-denominated burdens yet higher to the extent that nominal depreciation reflects GDP changes.

lies below the original manufacturer's price in South Africa, but not in all cases.

3. The Economics of Developing and Distributing Drugs

A key health-policy objective of most countries is to give patients access to existing pharmaceutical drugs at a reasonable cost. From a welfare point of view, effective medicines have a value both to the individual and to society as a whole. First and foremost, pharmaceutical drugs have value to the individual, in some cases as a treatment of symptoms, in other cases as a cure. But they also have additional value to society as a method to limit the risk for healthy individuals to be harmed by infectious diseases. Total welfare is maximized in the short-run if existing drugs are provided at a price equal to, or in some cases below, the marginal cost of production.

The problem, however, is that developing new drugs typically involves substantial investments in research and development. The average cost to develop a new pharmaceutical drug is approximately \$300 million and in some cases substantially higher.[18] These costs are mainly fixed and sunk once the drug is developed.

If prices were set equal to, or even below, marginal cost of production the pharmaceutical companies would not be able to recoup their investments and the economic incentives for research and development would disappear. The result of marginal-cost-pricing is, therefore, that too little investment in research and development takes place and too few drugs are developed in the long run. To correct for this market imperfection, patents exist to reduce competition and allow pharmaceutical companies to exercise some market power in order to recover their investments in R&D.

The welfare optimization problem in a closed economy, thus, involves a trade-off between giving patients access to existing drugs at reasonable costs versus profits for pharmaceutical companies, which are incentives for researching and developing new drugs in

[18] Sachs *et al.* (1999) estimate the average cost for a new drug to be \$300 million and predict that developing vaccines for HIV, tuberculosis and malaria would "potentially cost several times as much given the scientific challenges involved.", p. 8.

the future. Unfortunately, monopoly pricing of existing drugs causes static problems of insufficient market access for patients. Such problems can be solved, at least in theory, if the short-run and long-run objectives are separated. The first-best solution from a welfare perspective is to reward new innovations with a fixed lump-sum transfer to the innovating firm and to distribute existing drugs at competitive, or even below competitive prices.

While a policy to separate fixed and variable costs of pharmaceutical drug production might be unpractical or even impossible to implement in most cases, it can be useful in particular situations. More precisely, cost-based pricing and lump-sum payments for innovations could be the only way to achieve both the current and future health objectives in the poorest countries of the world.

So far we have discussed the problem of static distortions and dynamic efficiency in general terms. It is, however, important to recognize the international dimension of this issue. First of all, the trade-off between different objectives is not identical in all countries and, consequently, the optimal policy differs across nations. Moreover, in a global economy with trade in pharmaceutical products, health-care policy in one country has important implications for policy in other countries.

Starting with the issue of different objectives in industrialized and developing countries, it is crucial to note that the weights put on short-run and long-run objectives depend on several factors and the optimum is likely to vary across countries with different levels of income. Countries with high average income are likely to put more weight on new and improved drugs relative to countries with medium or low average income. As long as future drugs are normal goods, rich countries can be expected to have a higher willingness to pay for research and development. Lower rates of time preference in developed countries could also affect the trade-off in the same direction. Governments in industrialized countries are therefore more willing to accept high profits in the pharmaceutical industry to promote future innovations and improved drugs, while governments in developing countries to a larger degree prefer to give patients access to existing drugs at low costs.

Restricting our attention to the pricing problem of pharmaceutical companies, the optimal prices in local markets typically depend on the price elasticity of demand as well as the potential for arbitrage between markets. If the average income differs across two segmented markets, optimal prices for a monopolist are likely to be different in the two locations. Giving rebates to consumers with low income is often profitable for the monopolist as long as the rebated price is above the marginal cost of production and the scope for re-sale to high-income consumers is limited. When discounts for a homogenous good are the same for all consumers within a specific market but vary across different markets, the pricing strategy corresponds to third-degree price discrimination.

Arbitrage between markets — often referred to as parallel imports — limits the scope for third-degree price discrimination. If both markets are served by the monopolist the price in the low-income country is likely to rise as a result of parallel trade, while the price in the high-income country is likely to fall. The pharmaceutical company receives less revenue from both the low-income as well as the high-income market when parallel imports result in equalized prices. With large differences in average income across markets, as is the case with developing and industrialized countries, it is quite possible that parallel trade makes it unprofitable to serve low-income markets. Under such circumstances, it is beneficial for all parties — more precisely a Pareto improvement — to restrict parallel imports and increase the degree of price discrimination.[19]

The trade regime affects not only the scope for monopoly price discrimination but also, and more generally, the range of differences in health policies in different countries. More precisely, parallel trade undermines the independence of health authorities in both industrialized and developing countries. In practice, most industrialized countries maintain a policy that allows the pharmaceutical companies to recover their investment in research and development through

[19]As noted in Varian (1988), this result is quite robust. If price discrimination results in a new market being opened up, then it is typically a Pareto-improving welfare enhancement. Hausman and MacKie-Mason (1988) study this problem in the context of new patents.

monopoly mark-ups on existing drugs primarily in the U.S., Japanese and European markets. In this context, it is clear that marginal-cost-based pricing in developing countries could have serious effects on the incentives to introduce drugs in the poorest countries unless the price spillover to industrialized countries is limited.

A necessary, but not sufficient, condition for an effective solution to the access-development problem for pharmaceutical products in developing countries is, therefore, to limit parallel exports from the developing countries as well as parallel imports into the industrialized countries.

4. A Proposal for a Developing Economies' Fund for Essential New Drugs

In this section we set out a new proposal that would help resolve the incentive problems plaguing development and dissemination of drugs under the current system. We term our initiative the DEFEND Proposal, for "Developing Economies' Fund for Essential New Drugs".

4.1. *Criteria*

The magnitude of the problem with HIV/AIDS, TB and malaria in the least developed countries, particularly in Sub-Saharan Africa, suggests that any proposal to solve the problem must meet several criteria.

First and foremost, giving the poorest countries access to existing therapies and drugs would require prices equal to, or in most cases below, marginal cost. The magnitude of the epidemic and the low level of income in the poorest countries make low prices a necessity. This point can easily be illustrated with a hypothetical experiment. Assuming that all HIV-positive individuals in sub-Saharan Africa were treated with a typical AIDS cocktail therapy (Crixivan, AZT and 3TC) bought at U.S. prices, the total expenditure for these drugs would be more than total GDP in the Sub-Saharan countries put together.[20]

[20]The total GDP in sub-Saharan Africa is approximately $285 billion according to the most recent figures from the World Bank (World Economic Indicators 2000). A therapy

Moreover, for countries with very low median income it can be expected that even a small or moderate monopoly mark-up would generate a substantial allocative inefficiency and dead-weight loss. This is a fundamental reason to separate the incentives for development of new drugs from the distribution of existing drugs. The distribution of existing drugs in the poorest countries should, therefore, be founded on cost-based pricing while the incentives for development of new drugs have to be effected by other means. We will turn to this latter problem next.

The second criterion for good policy is that it has to include incentives to encourage innovation and development of new therapies and drugs. The problem is not that it is too profitable to innovate for poor countries, but rather that it is too unprofitable. For the world's three most deadly infectious diseases — AIDS, tuberculosis and malaria — effective vaccines still have to be invented. Moreover, most of the existing treatments for HIV/AIDS have serious and sometimes lethal side effects. In other words, more research on new drugs as well as improvements of existing drugs for the poorest countries is needed in the future.

The typical incentive for research and development of new pharmaceutical products is the prospect of future profits. But we have previously argued that it would be inexpedient and unrealistic to generate sufficient incentives for R&D through monopoly mark-ups in the world's poorest countries. There are three reasons why reliance on future monopoly profits is not a desirable incentive scheme: the monopoly mark-up is distortionary, the potential rents are too small and the political risks involved are too large for the pharmaceutical companies (e.g., the risk of compulsory licencing or generic substitution). The solution to these problems is to design a scheme with fixed lump-sum payments for new innovations, partly subsidized by the industrialized countries with a long-term guarantee to the pharmaceutical companies that they will receive some reasonable return on their investment in new and effective drugs.

with Crixivan, AZT and 3TC is $11,800 per patient per year at U.S. prices and the total for the sub-Saharan countries would be $299 billion per year, if all 25.4 million HIV-infected individuals were treated.

The third criterion any realistic proposal must meet is that it has to be developed within the limits of international law and treaties and must be supported by established international organizations. The most important examples are the rules of the WTO and the offices of the WHO. In particular, the TRIPS agreement requires all parties to give patent protection to new innovations, including pharmaceutical products. But it also leaves the question of the legality of parallel imports to national governments.

As we have previously stressed, the problem of access to existing and new drugs in the least developed countries is not only a question of trade, patents and pricing but also requires financial aid from industrialized countries. This latter task is best carried out as a coordinated program by the World Health Organization. The main functions of the WHO are to give worldwide health guidance, set global health standards, cooperate with governments to strengthen national health programs and, finally, to develop and transfer appropriate health technology, information and standards.

The fourth and final criterion is to limit coverage of inexpensive distribution to well-defined and restricted geographical areas. The health policies of most developed countries have to be taken as given and must be isolated from the strategy for access to pharmaceutical drugs in the least-developed countries. In order to avoid spillovers to the high-income, high-price OECD markets the policy should include official restrictions on parallel imports of the program drugs into the industrialized nations. Moreover, the least-developed countries need to impose restrictions on parallel exports from their own markets in order to deter slippage into countries that are not designated as recipients. Put briefly, we envision a regime of regional exhaustion within the WHO-designated program areas but tight controls to prevent the low-cost drugs from escaping those areas.

4.2. *Outline of the Proposal*

In the previous section we stressed that a successful strategy to give people in developing countries access to effective medicines has to involve four components. First, the cost of giving patients access to existing drugs has to be separated from the incentives for pharmaceutical companies to improve and develop new drugs. Second, the financial incentives to invent new drugs for the world's least developed countries must be subsidized by the industrialized countries. Third, a coordinated strategy should be jointly financed by the developed countries and implemented by an established international organization within the limits of international treaties. Fourth, the strategy should be focused on the least-developed countries and price spillovers should be limited by restrictions on parallel exports. A fund for essential new drugs could potentially help to solve this problem.

The principal structure of the strategy would be an international fund managed by UNAIDS or WHO. With contribution from the developed countries — and possibly some middle-income developing countries — the fund would buy licenses to produce and sell patented essential drugs in those least-developed nations that choose to be part of the program. Contributions to the fund should be in the form of cash to finance current expenditure. Equally important would be binding commitments to pay for future drugs, in particular vaccines for HIV, TB and malaria.

The program should be open to the least developed countries and all countries in Sub-Saharan Africa.[21] Any government, international organization or non-governmental organization should be allowed to use the license in the participating countries under three conditions: the original patent is respected in non-participating markets, the distribution is restricted to patients in the participating countries and parallel trade to other markets is prohibited.[22] The portfolio of licenses managed by the international fund should be limited to the

[21]The World Bank defines low-income economies as countries with a 1999 GNP per capita of $755 or less. In the most recent classification there were 64 countries in this category.

[22]Production and distribution under these licenses should not be allowed for companies that produce generic substitutes competing with the patented product in non-participating markets. The main reason for this restraint is to avoid strategic spillovers due to cost efficiencies in the production of the licensed product. If, for example, a firm in country A were certified to produce a drug under a publicly procured license, with sales intended for designated recipient countries, its expanded output could provide it with a competitive advantage in non-participating countries by virtue of increasing returns to scale.

most essential drugs. A board representing donors would regularly review the portfolio of current and future licenses.

Payments to patent holders should be in the form of a fixed, yearly, lump-sum transfer that would feature three characteristics. First, it should guarantee successful drug and vaccine developers a net present value over the life of the program that should equal expected R&D costs. Second, it should be positively related to the social value (associated with reduced mortality, morbidity, and spillovers) of the drug in the licensed areas in order to tie R&D incentives to underlying needs. Third, given that there may be broader markets for the new drugs and vaccines, it should be positively related to the global share of patients in the licensed areas.

In addition to paying patent holders for licenses, the Fund could provide subsidies to purchase and distribute essential drugs in countries where a large fraction of the population is infected or production cost of the drug is too high in relation to the average income. For available life-extending treatments — such as the existing AIDS therapies — a possible policy would be to subsidize purchases so that a specific treatment does not cost more than a pre-defined share (e.g. 40 percent) of the average GNP per capita in a particular country (the remainder would have to be financed by the local government, NGOs, donors or the patients as a form of co-payment). For vaccines these purchases could be subsidized to a larger degree (up to 100 percent) as widespread access to vaccines has positive externalities both in the local and global community.

4.3. *Implementation of the Proposal*

The implementation of the proposal could be gradual. Starting with HIV/AIDS treatment, the Fund could buy a portfolio of five or six licenses for the most important AIDS/HIV drugs.[23] For Sub-Saharan Africa a reasonable payment for these licenses could be in the range of $500 million to $1 billion per year.[24] Adding a subsidy for distribution

[23]Examples of drugs for an initial portfolio include 3TC, Zerit, Viramune, Stocrin, Combivir (AZT plus 3TC) and Crixivan.

[24]The lower bound of these license payments would be $442 million per year and is based on the assumption of a portfolio with ten patents, an average R&D cost of $360 million

of the drugs, which would guarantee that the treatment does not cost more than 40 percent of GDP per capita in a specific country, would require additional funds. Based on prices of generic substitutes, a cocktail of three AIDS/HIV drugs may be expected to cost between $400 and $600 per patient per year. Thus, the subsidy from the Fund would sum up to a maximum total cost of $4.7 billion to $8.1 billion per year for all HIV-infected individuals in Sub-Saharan Africa. This funding, however, would not be a substitute for the $3 billion WHO and UNAIDS estimate to be needed for basic care and prevention efforts. The total cost for an international strategy is, therefore, in the range of $8.2 billion to $12.1 billion annually. According to the OECD, total levels of official development assistance from bilateral donors and multilateral agencies amounted to $84.9 billion in 1999, two percent of which was devoted to basic health needs.[25] Thus, this commitment would represent a substantial portion of the current aid funding. However, it would correspond only to 0.03–0.05 percent of total GDP in the OECD countries in 1998. To put this in further perspective, if this amount were fully paid by the United States, the European Union, and Japan, it would come to only $13.50 per person per year. In another view, $12.1 billion may be compared with the anticipated loss in South African GDP of $22 billion in the year 2010.

5. Concluding Remarks

The poorest nations of the world suffer from extreme disease burdens, which go largely untreated because weak incomes and the prevailing system of intellectual property rights fail to provide sufficient incentives to develop new treatments and distribute them at low cost. Recent price reductions for HIV/AIDS drugs are encouraging but offer only a limited solution.

per drug (Danzon, 1997), a patent length of 20 years with approval coming eight years after the patent was filed, a five percent discount rate and a 75 percent contribution to development costs from the Fund (with the remainder being covered by profits from high-income markets). The upper bound would be $1.165 billion per year and is based on the same assumptions with the discount rate changed to ten percent and the Fund's contribution to development costs raised to 100 percent.

[25]See www.oecd.org/dac/htm/dacstats.htm.

In this paper, we analyzed the economic tradeoffs involved in supporting drug and vaccine research through exclusive rights and distributing the fruits of that research to poor countries. Such research is expensive and would not be undertaken by private firms without some prospect for recovering expected R&D costs. However, even if they were developed, private property rights to the distribution of these drugs, in the form of patents and EMRs, could support inefficiently high prices and generate large deadweight welfare losses compared to the social optimum in poor countries. This system fails to account for the strong external benefits of providing additional treatments and vaccines in poor countries. These benefits accrue also to the rich countries, both for reasons of humanity and because lower economic activity in developing countries is costly in trade terms.

We offer a proposal to overcome these incentive problems. Our DEFEND ("Developing Economies' Fund for Essential New Drugs") proposal would work within the existing international legal structure but significantly would raise the returns to R&D in critical medicines and expand distribution programs. A public international organization would purchase the license rights for designated areas and distribute the drugs at low cost with a required co-payment from local governments. Furthermore, governments would restrict parallel trade to support desirable price discrimination. Costs would be funded largely by increased foreign assistance from the developed nations, but these costs would be low in relation to current aid budgets. We believe a strong program could be mounted for $8 billion to $12 billion per year and would be an extremely effective use of foreign aid.

References

Abbott, Frederick M. 2000. The TRIPS-Legality of Measures Taken to Address Public Health Crises: Responding to USTR-State-Industry Positions that Undermine the WTO. Florida State University, manuscript.

Barton, John R. 2001. Draft on Tiered Pricing for WG-4. Stanford University Law School, manuscript.

Danzon, Patricia M. 1997. *Pharmaceutical Price Regulation: National Policies versus Global Interests* (Washington DC: The American Enterprise Institute).

Fink, Carsten. 2000. Patent Protection, Transnational Corporations, and Market Structure: A Simulation Study of the Indian Pharmaceutical Industry. The World Bank, manuscript.

Gorlin, Jacques J. 1999. *An Analysis of the Pharmaceutical-Related Provisions of the WTO TRIPS (Intellectual Property) Agreement* (London: Intellectual Property Institute).

Hausman, Jerry and Jeff MacKie-Mason. 1988. Price Discrimination and Patent Policy. *Rand Journal of Economics* 19, 253–265.

Lanjouw, Jean O. 1998. The Introduction of Pharmaceutical Patents in India: Heartless Exploitation of the Poor and Suffering? Boston: National Bureau of Economic Research working paper 6366.

Maskus, Keith E. 2000a. *Intellectual Property Rights in the Global Economy* (Washington DC: Institute for International Economics).

Maskus, Keith E. 2000b. Parallel Imports. *The World Economy: Global Trade Policy 2000* 23, 1269–1284.

Rozek, Richard P. and Ruth Berkowitz. 1998. The Effects of Patent Protection on the Prices of Pharmaceutical Products: Is Intellectual Property Protection Raising the Drug Bill in Developing Countries? *The Journal of World Intellectual Property* 1, 179–243.

Sachs, Jeffrey, Michael Kremer and Amar Hamoudi. 1999. The Case for a Vaccine Fund. Center for International Development, Harvard University, at www.cid.harvard.edu/cidsocialpolicy/policypapers.htm.

Subramanian, Arvind. 1995. Putting Some Numbers on the TRIPS Pharmaceutical Debate. *International Journal of Technology Management* 10, 252–268.

Subramanian, Arvind. 2001. The AIDS Crisis, Differential Pricing of Drugs, and the TRIPS Agreement: Some Notes. International Monetary Fund, manuscript.

Varian, Hal. 1988. Price Discrimination. In R. Schmalensee and R. Willig, eds., *Handbook of Industrial Organization* (Amsterdam: North-Holland).

Watal, Jayashree. 1999. Introducing Product Patents in the Indian Pharmaceutical Sector: Implications for Prices and Welfare. *World Competition* 20, 5–21.

World Health Organization. 1996. *Investing in Health Research and Development: Report of the Ad Hoc Committee on Health Research Relating to Future Intervention Options* (Geneva: World Health Organization).

About the Author

Keith E. Maskus is Professor of Economics and former Associate Dean for Social Sciences at the University of Colorado, Boulder, USA. He has been a Lead Economist in the Development Research Group at the World Bank. He is also a Research Fellow at the Peterson Institute for International Economics, a Fellow at the Kiel Institute for World Economics, and an Adjunct Professor at the University of Adelaide. He has been a visiting professor at the University of Bocconi and a visiting scholar at the CES-Ifo Institute at the University of Munich and the China Center for Economic Research at Peking University. He serves also as a consultant for the World Bank and the World Intellectual Property Organization and is currently chairing a panel of the National Academy of Sciences on intellectual property management in standards-setting organizations.

Maskus received his Ph.D. in economics from the University of Michigan in 1981 and has written extensively about various aspects of international trade. His current research focuses on the international economic aspects of protecting intellectual property rights. He is the author of *Intellectual Property Rights in the Global Economy*, published by the Institute for International Economics, and co-editor of *International Public Goods and the Transfer of Technology under a Globalized Intellectual Property Regime*, published by Cambridge University Press. A new volume, *Private Rights and Public Problems: The Global Economics of Intellectual Property in the 21st Century*, was published in 2012 by the Peterson Institute for International Economics.